ROTORDYNAMICS OF TURBOMACHINERY

John M. Vance

Department of Mechanical Engineering
Texas A & M University
College Station, Texas

WILEY

A Wiley-Interscience Publication

JOHN WILEY & SONS

New York / Chichester / Brisbane / Toronto / Singapore

Copyright © 1988 by John Wiley & Sons, Inc.

Library of Congress Cataloging in Publication Data:

Vance, John M.
  Rotordynamics of turbomachinery.

  "A Wiley-Interscience publication."
  Includes bibliographies and index.
  1. Turbomachines—Dynamics. 2. Rotors—Dynamics.

I. Title.
TJ267.V26 1987   621.406   87-34055
ISBN 0-471-80258-1

Printed in the United States of America

10 9 8 7 6 5 4 3 2 1

Preface

I first became involved in turbomachinery applications of rotordynamics in the summer of 1969, at Pratt & Whitney Aircraft (P&W) in West Palm Beach, Florida. This work began as a natural and interesting evolvement of my previous work in the aerospace industry on the dynamics of spin-stabilized space vehicles, including projects on spin balancing of multipiece missile flight assemblies. I found turbomachinery rotordynamics to be so fascinating and rewarding that I have worked almost exclusively in this field ever since. In fact, the initial summer at P&W proved to be the beginning of a long and still continuing series of summer appointments, engineering consulting jobs, and research contracts dealing with the subject. These activities included three more summers at P&W, five summers as Battelle Scientific Advisor to the U.S. Army helicopter propulsion branch of USARTL in Ft. Eustis, Virginia, two summers working with the industrial turbomachinery consulting group at Southwest Research Institute (SWRI) in San Antonio, Texas, and two summers at the Shell Westhollow Research Center in Houston, Texas, developing improved computer programs for predicting critical speeds of turbomachines used in petrochemical plants.

While working for the Army, I was privileged to visit every manufacturer of turboshaft engines in the United States, mostly in connection with rotordynamics problems encountered with helicopter engines and auxiliary power plants. At SWRI, my work was focused more toward petrochemical and gas pipeline applications, a diversification which has been accentuated since I joined the Turbomachinery Laboratories at Texas A&M University in 1978.

In educating myself along the way, I noted the absence of any comprehensive explanatory reference or textbook on rotordynamics and relied heavily on the literature of published technical papers [especially the *Transactions of the ASME*

(American Society of Mechanical Engineers)], unpublished company reports, and the oral accounts of engineers working in the field. In the late 1960s and early 1970s, the best single published source was NASA SP-113 by Dr. E. J. Gunter, Professor of Mechanical Engineering at the University of Virginia. Rotordynamics has evolved as a union of classical dynamics with bearing technology, and Dr. Gunter is primarily responsible for arranging the marriage and tolling the bells, in my opinion. Although many others have made valuable contributions, Dr. Gunter's work had the advantage of being readable, practice oriented, and interesting. More recently, the greatest advances in computational methods have been made and inspired by the work of Dr. J. W. Lund at the Technical University of Denmark.

After I had published some papers in the field, I began to receive requests and suggestions from working engineers to write a book. The nucleus for the book itself had its beginning in 1981 as notes for a short course at Texas A&M University. Teaching this course, mostly to engineers from industry, has reinforced my conviction that what is really needed is a bridge from research which uses advanced mathematical simulations and analyses over to the practice-oriented world of the working engineer in turbomachinery-related industries. The bridge should have two lanes, since research should be guided and motivated by the experience and needs of engineering practice.

Most engineers in industry and government laboratories have terminated their formal education at the Bachelor's level, and most have never taken a formal course in rotordynamics. A lack of familiarity with the requisite mathematical tools makes it very difficult for them to read the contemporary technical literature on rotordynamics. Part of the blame lies with the authors of the technical papers, who tend to write for a self-imaging audience of college professors and graduate students. But the problem remains, and a unified understanding of rotordynamic phenomena cannot be obtained without some comprehension of mathematical modeling and simulation, which I hope this book will provide.

I have attempted to write an informative book for both ends of the bridge. After a brief section on motivation and objectives, Chapter I begins immediately to describe some of the simplest mathematical models that give insight into the principal rotordynamic phenomena which can be experimentally observed on a rotating machine. Thereafter, each chapter and section begins with a purely verbal account of what is known about the topic and then begins to develop a foundation for the more advanced analytical models and mathematical methods which have been found useful and powerful. The last part of Chapter VII gives a detailed description of the most recently developed methods for rotordynamic stability analysis, using a digital computer. This latter section can be skipped by unprepared readers, but the preceding material in the chapter will, I hope, provoke some into digging out and studying the references supplied so as to develop the required competence. Chapter VIII describes recently developed electronic measurement tools and measurement techniques, which not only can be used for troubleshooting but are also invaluable for research where predictions of the theory can now be verified or modified as required.

I believe this book will be helpful to the design engineer or troubleshooter who needs to develop a working knowledge of turbomachinery with an insight into rotordynamics. It may also be used as a reference by engineering students or even as a textbook for graduate courses in rotordynamics or turbomachinery (although homework problems would have to be provided by the instructor). Most of the analytical methods presented for rotordynamics are applicable to other types of rotating machinery, as well as to turbomachinery.

It is my hope that the book will make the way both easier and even more interesting for those who follow.

JOHN M. VANCE

College Station, Texas
October 1987

Acknowledgments

A number of sections of this book, notably in Chapters VI, VII, and VIII, are adapted from ASME* papers written (or coauthored) by the author. Some of the ASME paper numbers are 78-PET-26, 80-GT-149, 76-DE-29, 85-GT-191, 74-GT-54, 84-GT-140, 84-DET-55, 86-WA/APM-23, 85-DET-145, and 85-DET-146. Some were also published in the *Transactions of the ASME*.

The polynomial method for computing damped eigenvalues of rotor-bearing systems, described in Chapter VII, was first published in the *Proceedings of the 10th Annual Turbomachinery Symposium, Texas A&M University, December 1981*, and later in the *Transactions of the ASME*.

The sections in Chapter II on Scaling and Turboshaft Engines with Front Drive are adapted from Report TR-74-66 written by the author for the U.S. Army Air Mobility Research and Development Laboratory (now USARTL) at Ft. Eustis, Virginia.

The experimental measurements of rotor natural frequencies and critical speeds described in Chapters IV and VIII were made at the Shell Westhollow Laboratories in Houston, Texas, under the supervision of Mr. Harley Tripp. The results of that research were first published in the *Proceedings of the 13th Turbomachinery Symposium*, Texas A&M University, November 1984, and later in the *Transactions of the ASME*.

J.M.V.

*American Society of Mechanical Engineers, 345 E. 47th St., New York, NY 10017.

Contents

Chapter I

Introduction to Rotordynamics of Turbomachinery

Modern turbomachines produce or absorb an amazing amount of power in a relatively small package. Undoubtedly the most impressive example is NASA's Space Shuttle main engine turbopumps, which produce 70,000 hp in two turbine stages about the size of a frisbee. In more common applications, turbojet engines provide propulsion for supersonic airplanes, turbine-compressor trains accomplish astounding rates of process in petrochemical industries, and steam turbines produce megawatts of electrical power for utilities.

The property of turbomachinery which allows these high energy densities and flow rates to be accomplished is high shaft speed, relative to other types of machines of the same physical size. Along with high speeds come high inertial loads, and potential problems with shaft whirl, vibration, and rotordynamic instability, the subjects to which this book is addressed.

The engineering design challenge presented by aerodynamic and hydrodynamic flows, design of blading, and the like sometimes causes the rotordynamic requirements of turbomachinery design to be overlooked. Too often it has happened that expensive turbomachines have been built and found to be incapable of producing their rated performance, or even of running at all, because of an ignorant assumption that making the rotor run smoothly and reliably at the design speed is a trivial problem.

Even when the rotordynamics problems are considered, the empirical and intuitive methods often employed in the fluid dynamics and heat transfer aspects of the design will get the rotordynamics engineer into trouble. Two characteristics of rotordynamics analysis are that its predictions are quite accurate when compared against experimental measurements (as long as accurate values for machine parameters are used in the mathematical model) and that its predictions are also quite often contrary to human intuition.

One example of the latter characteristic is the prediction (verified by experiment) that the unbalanced mass of a rotor will *not* "fly to the outside" of the shaft whirl orbit at high speed. (It will come around to the inside and stay there.) Another example is the verified prediction that *damping* in the rotor[1] of a turbomachine can produce a violently unstable whirling motion at high speed.

Given the potential accuracy of rotordynamics analysis and the fallibility of human intuition to replace it, the reader should recognize the usefulness of learning some of its mathematical predictions and familiarizing him/herself with the principle results of the experimental investigations that have been made.

OBJECTIVES OF ROTORDYNAMICS ANALYSIS

In designing, operating, and troubleshooting turbomachinery, rotordynamics analysis can help accomplish the following objectives:

1. *Predict Critical Speeds.* Speeds at which vibration due to rotor imbalance is a maximum can be calculated from design data, so as to avoid them in normal operation of the machine.

2. *Determine Design Modifications to Change Critical Speeds.* Whenever design engineers fail to accurately accomplish objective 1 above, or it becomes necessary to change the operating speed range of a turbomachinery, design modifications may be required to change the critical speeds.

3. *Predict Natural Frequencies of Torsional Vibration.* This objective usually applies to the entire drive train *system* in which the turbomachine is employed. For example, a centrifugal compressor rotor driven by a synchronous electric motor through a gearbox may participate in a mode of torsional vibration excited by pulsations of the motor during start-up. In such a case, it might be desirable to change the natural frequency to a value which has the least possible excitation (in magnitude and/or time duration).

4. *Calculate Balance Correction Masses and Locations from Measured Vibration Data.* This capability allows "in-place" rotor balancing to be accomplished, thereby reducing the amplitude of synchronous vibration.

5. *Predict Amplitudes of Synchronous Vibration Caused by Rotor Imbalance.* This is one of the most difficult objectives to accomplish accurately since the amplitude of rotor whirling depends on two factors which are both very difficult to measure: (a) the distribution of imbalance along the rotor, and (b) the rotor–bearing system damping. What can be done, however, is to predict the relative effects of rotor imbalance and system damping at specific locations.

---

[1]In this book the word "rotor" will be used to designate the assembly of rotating parts in a turbomachine, including the shaft, turbine wheels, compressor disks, or pump impellers.

6. *Predict Threshold Speeds and Vibration Frequencies for Dynamic Instability.* This objective is another challenging one at present, since a number of the destabilizing forces are still not understood well enough for accurate mathematical modeling. However, the instability caused by journal bearings, known as "oil whip," can be predicted quite accurately.

7. *Determine Design Modifications to Supress Dynamic Instabilities.* This objective can be met more readily than objective 6 above, since computer simulations can predict the relative stabilizing effect of various hardware modifications, even if the models for destabilizing force are only approximations.

THE SPRING–MASS MODEL

The simplest possible model for vibration analysis is a rigid mass mounted on a linear spring, with only one degree of freedom (see Fig. 1.1c). The first critical speed of some rotor–bearing systems can be approximated by the natural frequency of this model converted to revolutions per minute:

$$N_1 = \frac{60}{2\pi} \sqrt{\frac{k}{m}} \text{ rpm,} \qquad (1\text{-}1)$$

where k is the effective stiffness for the first mode of whirling and m is the effective mass.

For a rotor which is relatively rigid compared to the bearing supports, the effective mass is the total mass of the rotor and the effective stiffness is the stiffness of all the bearing supports taken in parallel (see Fig. 1.1a).

For a rotor which has a relatively flexible shaft, compared to the bearing support stiffness, Fig. 1.1b shows that the effective stiffness is determined by the bending stiffness of the shaft. In this case, only a portion of the shaft mass contributes to the effective mass of the single-degree-of-freedom model, since the shaft mass near the bearing supports does not fully participate in the vibratory motion.

Such a simple model has a number of serious limitations for rotordynamics analysis.

First, the single-degree-of-freedom spring–mass model can execute a translational motion in only one direction, whereas the rotor–bearing system can execute whirl *oribts*, which may have complex shapes and patterns. This shortcoming can be partially removed by considering a spring–mass system with two degrees of freedom, allowing it to vibrate simultaneously in two directions, say X and Y.

The combination of vibrations in two orthogonal directions can produce several different types of motion of the mass. The type of motion produced depends on the relative amplitudes and phase relationship of the X and Y motions (see Fig. 1.2a). If the vibration has a single frequency, the motions produced are circular

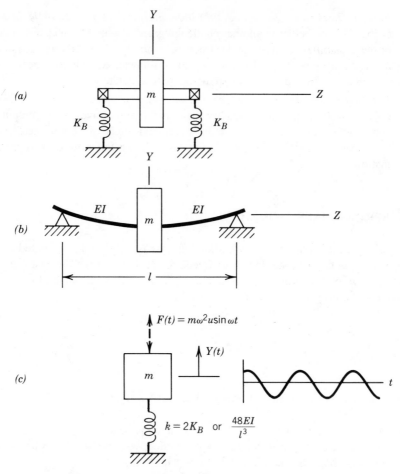

Figure 1.1. Rigid and flexible rotors viewed as a spring–mass system.

orbits (Fig. 1.2b), elliptical orbits (Fig. 1.2c), and straight line motions at any angle to the X axis (Fig. 1.2d).

This two-degrees-of-freedom model was used by Rankine in 1869 for the first published analysis of machinery rotordynamics [1] in an attempt to explain the "critical speed" behavior of rotor–bearing systems. The system model consisted of a rigid mass whirling in a circular orbit, with an elastic spring acting in the radial direction (see Fig. 1.3). Rankine used Newton's second law incorrectly in a rotating coordinate system, and predicted that rotating machines would never be able to exceed their first critical speed.

Although the two-degrees-of-freedom spring–mass model can execute the orbital motions of a rotor–bearing system, it does not contain a realistic representation for the rotating imbalance in the rotor. Since a perfectly balanced rotor never occurs in real machines, and since it is the rotating imbalance which excites the most

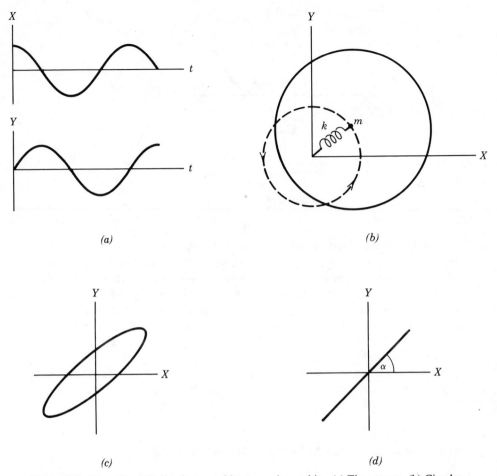

Figure 1.2. How X and Y vibrations combine to produce orbits. (a) Time traces. (b) Circular orbit. (c) Elliptical orbit. (d) Translational vibration.

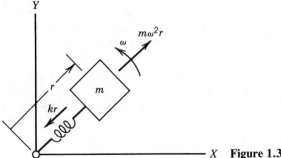

Figure 1.3. Rankine's model.

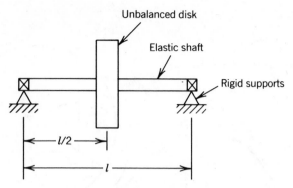

Figure 1.4. The Jeffcott rotor.

commonly observed type of vibration (synchronous) in turbomachines, it follows that the rotating imbalance is an essential ingredient of one of the most useful models for rotordynamic analysis. This model is called the *Jeffcott rotor* (see Fig. 1.4), named after the English dynamicist who first used the model in 1919 to analyze the response of high speed rotating machines to rotor imbalance [2]. It consists of a massive unbalanced disk mounted midway between the bearing supports on a flexible shaft of negligible mass. The bearings are rigidly supported, and viscous damping acts to oppose absolute motions of the disk.[2]

Jeffcott's analysis explained how the rotor whirl amplitude becomes a maximum value at the critical speed but diminishes as the critical speed is exceeded due to the "critical speed inversion" of the imbalance.

SYNCHRONOUS AND NONSYNCHRONOUS WHIRL

The frequencies present in the measured vibration signal constitute some of the most useful information obtainable for diagnosing rotordynamics problems in turbomachinery. For example, a common source of shaft whirling is rotor imbalance, and imbalance always produces whirling which is synchronous with shaft speed. Hence, large amplitudes of synchronous vibration usually indicate a rotor imbalance problem.

Synchronous whirl excited by imbalance was the problem addressed by Jeffcott's analysis, described above, and presented in detail in the next section.

But not all shaft whirling is synchronous; in fact the more destructive rotordynamic problems involve nonsynchronous whirl.

Figure 1.5 shows an end view of a whirling rotor and describes the essential difference between the two types of motion. The shaded element represents an unbalanced mass. In Fig. 1.5a the time rate of change of the angle $\phi(\dot\phi)$ is the

[2]In the Jeffcott model, the only source of this type of damping is air drag on the disk. The viscous representation is a useful approximation.

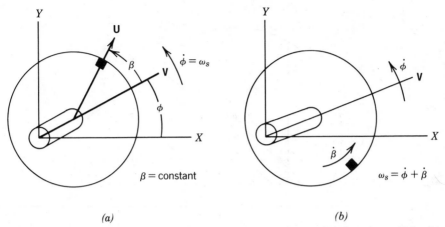

Figure 1.5. Synchronous versus nonsynchronous whirl. (a) Synchronous whirl. (b) Nonsynchronous whirl.

whirl speed. The angle β remains constant, so the whirl speed and the shaft speed are the same (synchronous whirl). Thus the rotor imbalance **U** leads the rotor whirl vector **V** by a constant angle β.

In Fig. 1.5b the time rate of change of the angle $\beta\,(\dot{\beta})$ is the spin velocity of the rotor, *relative to the rotating whirl vector* **V**, so the shaft speed is the sum of $\dot{\beta}$ and $\dot{\phi}$. In this case the whirl speed and shaft speed are not the same (nonsynchronous whirl).

The distinction betwen these two types of rotor motions provides the most basic classifying factor for frequency spectrum analysis. A later chapter in this book (IV) on critical speeds and imbalance response deals only with synchronous whirl. Chapter VII on rotordynamic instability deals almost exclusively with nonsynchronous whirl.

ANALYSIS OF THE JEFFCOTT ROTOR

Figure 1.6 shows an end view of the whirling Jeffcott rotor, with coordinates that describe its motion.

The center of mass of the unbalanced disk is at M. The point C locates the geometric center of the disk. Thus the amount of static imbalance is denoted by $u = \overline{CM}$ (inches), and the shaft bending deflection due to dynamic loads is $\overline{OC}$. Gravity loads are neglected in this analysis. They are insignificant compared to the dynamic (inertial) loads in many turbomachines.[3]

---

[3]Large multistage steam turbines in electric utility plants constitute the most notable exception, in which gravity effects can cause a peak response to occur at a speed of about one-half the first critical speed.

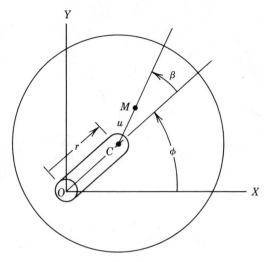

Figure 1.6. End view of the Jeffcott rotor, with coordinates.

The shaft has a bending stiffness of k lb/in, the disk has a mass of m lb-sec^2/in, and air drag on the whirling disk and shaft is approximated by a viscous damping coefficient of c lb-sec/in.

The dynamic system has three degrees of freedom; an assumption of constant speed reduces them to two.

The polar coordinates r, ϕ, β have an advantage of giving the synchronous whirl solution in terms of constants which are readily interpreted, but the equations of motion are nonlinear and hence are not well suited to an analysis of rotordynamic instability (introduced in a later section of this chapter).

The cartesian coordinates X, Y (of the shaft center) along with the angle β produce linear differential equations. Furthermore, the solution in terms of X and Y as functions of time is what vibration probes usually measure in a turbomachinery installation.

The differential equations and their solutions in both sets of coordinates are presented here, for the case of a constant shaft speed ω.

Polar Coordinates

The differential equations of motion are [3]

$$\ddot{r} + \frac{c}{m}\dot{r} + \left(\frac{k}{m} - \dot{\phi}^2\right)r = \omega^2 u \cos(\omega t - \phi), \tag{1-2}$$

$$r\ddot{\phi} + \left(2\dot{r} + \frac{c}{m}r\right)\dot{\phi} = \omega^2 u \sin(\omega t - \phi). \tag{1-3}$$

The solution for synchronous whirling is

$$r_s = \frac{\omega^2 u}{\sqrt{(k/m - \omega^2)^2 + (c\omega/m)^2}},$$ (1-4)

$$\omega_s t - \phi_s = \beta_s = \tan^{-1}\left[\frac{c\omega}{m(k/m - \omega^2)}\right].$$ (1-5)

The constant whirling amplitude r_s and phase angle β_s satisfy the differential equations (1-2) and (1-3) for any constant shaft speed ω. Typical plots of r_s and β_s versus ω are shown on Fig. 1.7 for two different values of damping.

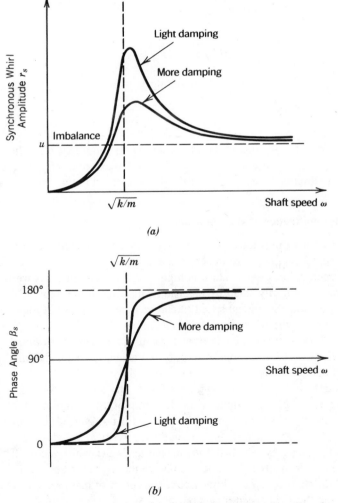

(a)

(b)

Figure 1.7. Imbalance response of a Jeffcott rotor.

Cartesian Coordinates

The differential equations of motion are [4]

$$m\ddot{X} + c\dot{X} + kX = m\omega^2 u \cos \omega t \qquad (1\text{-}6)$$

$$m\ddot{Y} + c\dot{Y} + kY = m\omega^2 u \sin \omega t. \qquad (1\text{-}7)$$

The solution for synchronous whirling is

$$X = \frac{\omega^2 u}{\sqrt{(k/m - \omega^2)^2 + (c\omega/m)^2}} \cos(\omega t - \beta_s), \qquad (1\text{-}8)$$

$$Y = \frac{\omega^2 u}{\sqrt{(k/m - \omega^2)^2 + (c\omega/m)^2}} \sin(\omega t - \beta_s), \qquad (1\text{-}9)$$

$$\beta_s = \tan^{-1}\left[\frac{c\omega}{m(k/m - \omega^2)}\right]. \qquad (1\text{-}10)$$

From the geometry of Fig. 1.6 it can be seen that the shaft deflection is

$$r = \sqrt{X^2 + Y^2}. \qquad (1\text{-}11)$$

Inspection of equations (1-4), (1-8), and (1-9) show that the solutions satisfy equation (1-11). Hence Fig. 1.7a is also a typical plot of the amplitudes of horizontal vibration x [equation (1-8)] or of vertical vibration y [equation (1-9)].

Physical Significance of the Solutions

Figure 1.7a shows how the amplitude of synchronous whirl increases as the critical speed is approached, and then decreases and approaches the value of static imbalance at supercritical speeds. Thus, at high speed, the synchronous whirl amplitude can be made arbitrarily small by precision balancing of the rotor.

At speeds near the critical speed, it can be seen that the most important parameter for reducing the whirl amplitude is damping.

Figure 1.7a also provides the most useful definition of a critical speed: the "speed at which synchronous response to imbalance is maximum." Note that increased damping raises the critical speed slightly.

Inspection of Fig. 1.6 in conjunction with the solution for the synchronous phase angle β_s (Fig. 1.7b) yields an explanation for the asymptotic approach of the whirl amplitude toward u. As the critical speed is transversed, the angle β_s passes through 90° and approaches 180° at highly supercritical speeds. Figure 1.6 shows that β is the angle by which the imbalance leads the whirl vector. Thus at high speed the center of mass M comes around to the *inside* of the whirl orbit, stands still, and the shaft center C whirls around the center of mass M. This phenomenon is called the "critical speed inversion."

Note that the center of mass stays to the outside of the whirl orbit only at low speed, and that the imbalance leads the whirl vector by exactly $90°$ when the shaft speed equals the *undamped* critical speed ($\sqrt{k/m}$). This latter fact is the basis of a method for accurate measurement of the undamped critical speed (which would have an unbounded whirl amplitude, difficult to measure).

Three Ways to Reduce Synchronous Whirl Amplitudes

A review of the Jeffcott rotor analysis yields three approaches to the problem of minimizing amplitudes of synchronous whirl: (1) balance the rotor, (2) change the speed (away from the critical speed), and (3) add damping to the rotor–bearing system. Although the Jeffcott rotor is a simple model, the same three approaches are effective for controlling synchronous whirl in more complex machines. The proper approach to use depends on the practical constraints of the problem at hand.

Balancing the rotor is the most direct approach, since it attacks the problem at its source. Methods for balancing are described in Chapter V. It should be pointed out here, however, that in practice a rotor cannot be balanced perfectly and that the best achievable state of balance tends to degrade during operation of a turbo-machine.

The second approach, moving machine operation farther away from the critical speed, can be achieved either by changing the operational shaft speed or by changing the critical speed itself. In practice the latter is usually accomplished by modifying rotor support stiffness. This parameter is not included in the Jeffcott model, but it has the same effect on critical speeds as the shaft stiffness k. A detailed analysis of the effect of flexible bearing supports is presented in a section to follow. In general, changing the critical speed is most useful for constant speed machines or for machines with a narrow range of operational speeds.

If a critical speed must be traversed slowly or repeatedly, or if machine operation near a critical speed cannot be avoided, then the most effective way to reduce the amplitude of synchronous whirl is to add damping. This would be difficult in the Jeffcott rotor, since the only source of damping is aerodynamic drag, but fortunately most turbomachines have flexible bearing supports in which damping can be added or oil-film bearings in which damping is inherent and can be changed by design modifications. Note that internal damping, or hysteresis, in the rotor shaft does *not* provide the type of damping modeled by equations (1-2), (1-3), (1-6), and (1-7), since it acts only on motions relative to the whirl vector $\overline{OC}$. In fact, it wll be shown later how internal friction in rotating parts is a source of self-excited nonsynchronous whirling (rotordynamic instability).

SOME DAMPING DEFINITIONS

Since damping forces are difficult to measure directly and since a rotor-bearing system can have several different sources of damping, it has become common practice to quantify the total damping in terms of a percentage of "critical damping.'' The critical damping coefficient c_{cr} is the value required to completely suppress any free vibration of the system. Thus the "damping ratio" ξ is c/c_{cr} and the "percent damping" is $100\ \xi$. For the Jeffcott rotor, the critical damping coefficient c_{cr} has the value $2\sqrt{km}$ and is assumed to be concentrated at the central disk. Using these definitions, equation (1-4) gives the whirl amplitude at $\omega = \sqrt{k/m}$ as $u/2\xi$ and the critical speed as $\sqrt{k/m(1\text{-}2\xi^2)}$. The imbalance multiplier $1/2\xi$ is sometimes referred to as the "magnification factor," or "Q factor," of the rotor–bearing system. For small (<10 percent) damping, it gives a good approximation to the maximum whirl amplitude at the critical speed.

Equation (1-4) can be put into dimensionless form by using the damping ratio ξ and the natural frequency $\omega_n = \sqrt{k/m}$ as parameters. The result is

$$\frac{r_s}{u} = \frac{(\omega/\omega_n)^2}{\sqrt{\left[1 - (\omega/\omega_n)^2\right]^2 + (2\xi\ \omega/\omega_n)^2}}. \tag{1-12}$$

EFFECT OF FLEXIBLE SUPPORTS

The bearing supports[4] of any real turbomachine are necessarily flexible, since every engineering material has elasticity. Furthermore, it is desirable from a rotordynamics standpoint to have the supports more flexible than the rotor. The two major reasons are as follows:

1. Low support stiffness reduces the dynamic loads transmitted through the bearings to the nonrotating structure, thus prolonging bearing life and minimizing structural vibration.
2. Low support stiffness allows the damping in hydrodynamic bearings or dampers to operate more effectively, thus attenuating rotor whirl amplitude at the critical speed.

The first reason can be explained by an analysis of a short rigid rotor[5] on symmetric flexible supports with damping in the supports (see Fig. 1.8). The differential equations of motion and their solution for this model are exactly the same as for the Jeffcott rotor [equations (1-2) through (1-11)]. Only the definitions of stiffness k and damping c are changed: $k = 2K_B$ and $c = 2C_B$ (the sum of the stiffness and damping of both bearing supports).

Figure 1.6 still defines the coordinates which describe the motion, but now $r = \overline{OC}$ is the deflection of the bearing supports rather than of the rotor shaft.

The damping in bearing supports is produced by the oil film in hydrodynamic

---

[4]The term "bearing support" here will denote the bearing together with its supporting structure or housing.
[5]From the standpoint of transmitted bearing loads, the rigid rotor is the worst case.

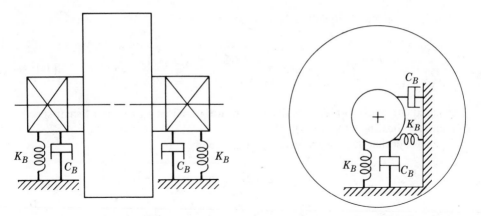

Figure 1.8. Short rigid rotor on damped flexible bearing supports.

bearings or squeeze film dampers, by specially designed elastomeric dampers, and/ or by internal friction in the bearing and its housing structure assembly.

Consider the force F_B transmitted through each bearing to the machine structure during a synchronous whirl motion in which point C (Figs. 1.6, 1.8, and 1.9) traces out a circular orbit of radius r_s. Force F_B is the vector sum of the radial stiffness force F_k and the tangential damping force F_c. Figure 1.9 shows these two force components

$$F_k = K_B r_S \quad \text{and} \quad F_c = C_B \omega r_s, \tag{1-13}$$

which have a resultant of

$$F_B = \sqrt{F_k^2 + F_c^2} = r_s \sqrt{K_B^2 + (C_B \omega)^2}, \tag{1-14}$$

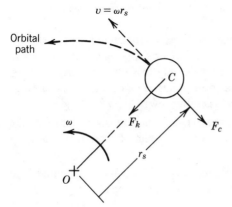

Figure 1.9. Forces on a bearing journal executing synchronous whirl.

where r_s is given by equation (1-4). Thus

$$F_B = \frac{1}{2} m\omega^2 u \sqrt{\frac{(2K_B)^2 + (2C_B\omega)^2}{(2K_B - m\omega^2)^2 + (2C_B\omega)^2}} \tag{1-15}$$

is an equation which gives the dynamic bearing load as a function of shaft speed and rotor–bearing system parameters.

Consider that the bearing load F_B would be given by

$$F_\infty = \tfrac{1}{2} m\omega^2 u \tag{1-16}$$

if the supports were rigid. A few sample numerical calculations of this rigid support bearing force F_∞ for typical high speed turbomachines will convince the reader that it is intolerably high. With the proper selection of rotor–bearing parameters, the flexible support bearing force F_B can be made considerably smaller than the rigid support force F_∞. The ratio of these forces can be obtained by dividing equation (1-15) by equation (1-16). The result, expressed in terms of dimensionless ratios, is

$$\frac{F_B}{F_\infty} = \sqrt{\frac{1 + (2\xi\omega/\omega_n)^2}{[1 - (\omega/\omega_n)^2]^2 + (2\xi\omega/\omega_n)^2}}, \tag{1-17}$$

where $\omega_n = \sqrt{2K_B/m}$ is the undamped critical speed and $\xi = \sqrt{C_B/m\omega_n}$ is the damping ratio.

Equation (1-17) gives the transmissibility of the imbalance force to the bearing support structure. Figure 1.10 is a plot of equation (1-17) as a function of speed ratio for two values of damping. Note that the two curves intersect at a speed ratio of $\sqrt{2}$. This is because equation (1-17) gives the same transmissibility value (1.0) for *any* value of damping at the shaft speed $\omega^* = \sqrt{2}\,\omega_n$.

The following are some observations of practical interest from Fig. 1.10:

1. Bearing support flexibility can greatly reduce the dynamic load transmitted through the bearings, provided that the supports are made soft enough to keep the undamped critical speed considerably less than 70 percent of the operating speed. In Fig. 1.10, this corresponds to an operating speed range of $\omega/\omega_n \gg \sqrt{2}$ to keep the transmissibility low.

2. Bearing support damping increases the dynamic load transmitted through the bearings at high speeds ($\omega > \omega^*$) where the effect of support flexibility is favorable.

3. Bearing support damping may be necessary to keep the transmitted load within acceptable limits while traversing the critical speed.

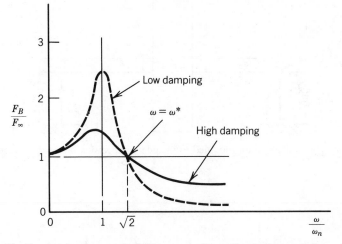

Figure 1.10. Bearing force transmissibility vs. shaft speed ratio for two values of damping.

4. Low support stiffness is not an unconditional panacea; improperly chosen support parameters can produce dynamic bearing loads in excess of the rigid support values.

Note also that the effects of damping on transmitted force is different from its effect on rotor whirl amplitude. Remember that Fig. 1.7 gives the whirl amplitude for the rotor–bearing model of Fig. 1.8, and shows that the effect of damping on whirl amplitude is favorable over the entire speed range.

If shaft flexibility is incorporated into the model (e.g., a Jeffcott rotor modified to include damped flexible supports, as shown in Fig. 1.11 the analysis becomes more complicated, even when internal damping in the rotor shaft is neglected.[6] Reference [5] gives a solution for optimum support damping from the standpoint of minimizing whirl amplitude, as a function of the ratio of support stiffness to rotor shaft stiffness. This solution is represented here by Figs. 1.12 and 1.13.

Figure 1.12 shows how the optimum amount of support damping varies with stiffness ratio. Note that the optimum support damping coefficient is less than the critical value[7] when the support stiffness is less than the shaft stiffness. Too much damping (more than the optimum value) "locks up" the supports, allowing the shaft, which has no damping, to act as the dominant spring in the system.

Figure 1.13 shows how the synchronous whirl amplitude r_s increases with support stiffness, even with the optimum amount of damping. This is because

---

[6]Internal friction in a shaft is usually insignificant compared to bearing support damping. If not, its major effect is to produce a subsynchronous whirl instability.

[7]The "critical value" of damping is defined here as $2\sqrt{km}$, the amount required to completely suppress free vibration of the spring–mass system of Fig. 1.11 with rigid supports (i.e., the Jeffcott rotor with damping on the central disk). The total bearing damping is $2 C_B$.

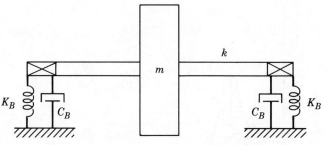

Figure 1.11. Jeffcott rotor modified to include damped flexible bearing supports.

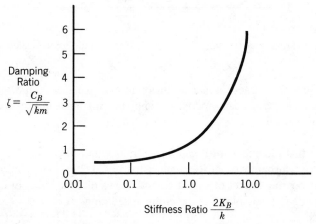

Figure 1.12. Optimum support damping for the rotor–bearing system of Fig. 1.11. From Ref. [5].

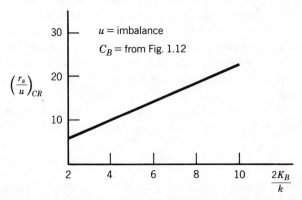

Figure 1.13. Whirl amplitude at the critical speed, with optimum damping, vs. stiffness ratio. From Ref. [5].

increasing support stiffness inhibits the motion of the support and therefore does not allow the damping to operate effectively.

ROTORDYNAMIC INSTABILITY

The great majority of rotordynamic problems encountered with turbomachinery involve synchronous whirl, i.e., response to imbalance. The approaches to these problems are straightforward, as described earlier in this chapter, even though the typical turbomachine has a rotor–bearing system that is more complex than the models which have been analyzed here.

The remaining minority of problems involving nonsynchronous whirl or nonsynchronous vibration can be subdivided into three classifications:

1. Supersynchronous vibrations due to shaft misalignment (the frequency is often twice shaft speed).
2. Subsynchronous and supersynchronous vibrations due to cyclic variations of parameters, mainly caused by loose bearing housings or shaft rubs.
3. Nonsynchronous (usually subsynchronous) rotor whirling that becomes unstable, typically when a certain speed called the threshold speed of instability is reached.

Problems of the first and second classifications have solutions which are obvious: align the shafts, tighten the bearing housings, or eliminate the rub.

Problems of the third classification, although relatively uncommon, have a history of causing some very expensive[8] turbomachinery failures, with elusive causes and cures. This is the main classification of problems referred to as rotordynamic instability.[9]

At this point, the reader may wonder why rotordynamic instability occurs in turbomachinery and not in other types of rotating machines—for example, in the reciprocating internal combustion engines used in automobiles. The answer lies in the high rotational speeds and the flexible, relatively unsupported shafts common to turbomachinery.

"High speed" is a relative term here. It could be only a few hundred revolutions per minute, provided it is significantly higher than a natural whirling frequency (eigenvalue) of the rotor–bearing system. Rotordynamic instability is manifested by shaft whirling, and the shaft will tend to whirl at its natural frequency (as modified by gyroscopic moments). Since it has already been said that instability frequencies are subsynchronous, it follows that they almost always occur when shaft speeds are higher than the natural whirling frequency.

Historically, the occurrence of serious instability problems has been sufficiently

---

[8]In terms of both machinery damage and lost production due to machine downtime. See Chapter VII.
[9]Chapter VII, which treats rotordynamic instability in more detail, also includes some material on the second classification.

infrequent to preclude its consideration as a primary factor in turbomachinery design by engineers and manufacturers. In the case of aircraft turbojet engines, this is generally still true, but the U.S. Army recently funded a research program to quantify destabilizing forces from shaft splines in turboshaft engines for helicopters [6].

There is now a well-documented history of rotordynamic instability problems in centrifugal compressors used by the process industries (see Chapter VII).

During the period of development of turbopumps for the cryogenic fuels used in roc'.et engines, approximately 50 percent of these machines suffered rotordynamic instability problems at some stage during their development [7].

Rotordynamic instability is a special case of the more general theory of dynamic instability, or instability of dynamic systems. Another and simpler special case is found in classical vibration theory. Both the theory and experiments show that the amplitude of free unidirectional vibration grows with time if the damping is negative. In fact, it is well known that damping is stabilizing unless it becomes negative. Figure 1.14 shows a simple apparatus (from Den Hartog [8]) in which negative damping can be generated. The flow of air around the beam of semicircular cross section produces a pressure distribution which pushes the beam in the same direction as its instantaneous velocity (for motion in a vertical plane), as shown in Fig. 1.15. The differential equation for the vertical translational displacement Y of vibration is

$$m\ddot{Y} - c\dot{Y} + kY = 0, \tag{1-18}$$

where m is the mass of the beam, c is the (negative) damping coefficient, and k is the total effective stiffness of all the springs. The solution of equation (1-18) is

$$Y(T) = Ae^{st}, \tag{1-19}$$

where A is a constant, and the values of s which satisfy equation (1-18) are the complex conjugate eigenvalues. They are

$$s = \frac{c}{2m} \pm i\sqrt{\frac{k}{m} - \left(\frac{c}{2m}\right)^2}. \tag{1-20}$$

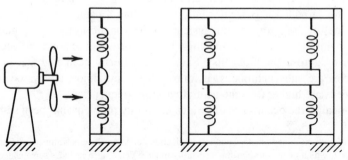

Figure 1.14. Apparatus to demonstrate unstable vibration. From Den Hartog [8].

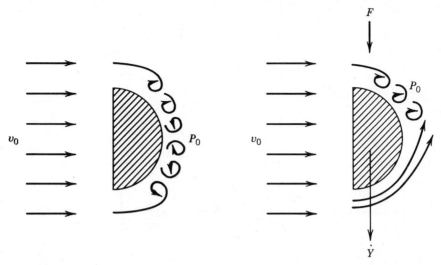

Figure 1.15. Negative damping produced by aerodynamic flow separation.

The positive real part of the eigenvalue indicates that the vibratory motion of the beam will be unstable, with an amplitude which grows exponentially with time. In practice, the motion will become bounded at some finite amplitude large enough to render the linear equation (1-18) no longer valid.

Note that every term in equation (1-18) contains the coordinate Y or one of its derivatives. This makes the equation "homogeneous," which is a general property of the type of equations used to predict instabilities.

In turbomachines, rotordynamic instability is usually produced by forces which are tangential to the rotor whirl orbit, acting in the same direction as the instantaneous motion. They are sometimes called follower forces, or destabilizing forces. If the magnitude of the follower force is proportional to the instantaneous whirl velocity, it is classified as a negative damping force, just as in classical vibration theory. If the magnitude of the force is proportional to the rotor displacement (instantaneous orbit radius), it is classified as a cross-coupled stiffness force. The "cross-coupled" terminology comes from the form of the force expressions in a nonrotating X–Y coordinate system. A rotor displacement in the X direction produces a force in the Y direction, and vice versa.

Figure 1.16 shows a tangential follower force F_ϕ resolved into F_X and F_Y components produced by cross-coupled stiffness coefficients K_{XY} and K_{YX}.

The representation of destabilizing forces by stiffness and damping coefficients implies that the force is linearly proportional to the rotor displacement or velocity. The usefulness of this representation for analysis, along with its limitations, are discussed in Chapter VII.

Most of the known destabilizing forces in turbomachinery are represented by cross-coupled stiffness coefficients. A surprising example is the follower force

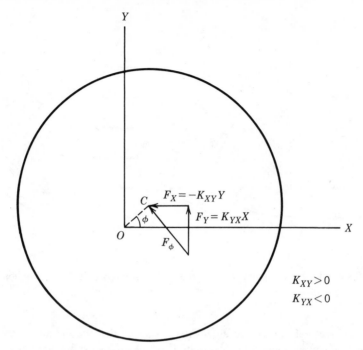

Figure 1.16. Cross-coupled stiffness representation of destabilizing force on a deflected rotor disk.

produced by internal friction in a rotor shaft, when the internal friction forces are represented by a viscous damping model![10]

Figure 1.17 illustrates the distribution of tensile stress over a cross section of the shaft in a whirling Jeffcott rotor. The compressive stress, which occurs over the unshaded half of the cross section, is a negative mirror image of the tensile stress. Figure 1.17b shows the stress distribution produced by bending strain. The higher stress areas are shaded darker, with the highest tensile stress at point T. This stress distribution produces a radial restoring moment which opposes the centrifugal force $m\dot{\phi}^2 r$.

Figure 1.17c shows the stress distribution produced by the time variation of strain for the case in which the whirling speed is less than the shaft speed ($\dot{\phi} < \omega$). In this case the material in compression at point P (Fig. 1.17b) moves around counterclockwise toward point T, thus changing its rate of stress from compression to tension. The maximum rate of change, and thus the maximum tensile damping stress, occurs at point T'. This distribution of stress produces a tangential moment which opposes the external damping force $cr\dot{\phi}$ (aerodynamic drag on the disk).

---

[10]Experimental measurements have shown that the damping in steel is not viscous, but the model is a useful approximation for stability analysis over a narrow range of frequencies.

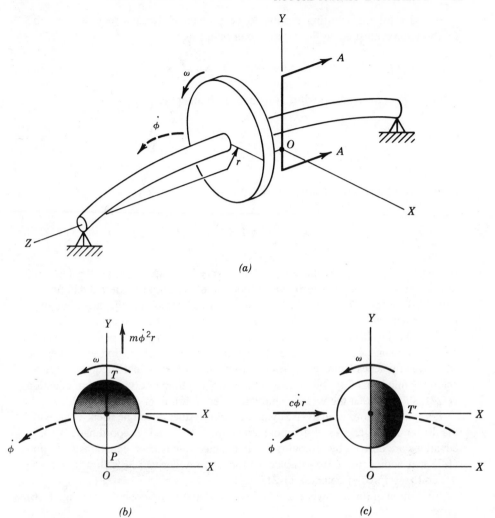

Figure 1.17. Internal stresses induced by nonsynchronous whirl. (a) Whirling Jeffcott rotor. (b) Distribution of tensile stress over section A–A due to shaft bending strain. (c) Distribution of tensile stress over section A–A due to rate of strain.

A detailed mathematical analysis of these stresses, presented in Appendix A of Ref. [3], shows that the internal moment produced by strain rate effects is equivalent to a tangential follower force given by

$$F_\phi = c_i(\omega - \dot{\phi})r, \qquad (1\text{-}21)$$

where c_i is the coefficient of internal damping and r is the whirl radius.

In order for the whirling motion to be a state of dynamic equilibrium, the follower force must equal the external damping force:

$$c_i(\omega - \dot{\phi})r = c\dot{\phi}r. \tag{1-22}$$

Since the rotor tends to whirl at its natural frequency,

$$\dot{\phi} = \sqrt{\frac{k}{m}} = \omega_n \tag{1-23}$$

can be substituted into equation (1-22) to give

$$\frac{\omega}{\omega_n} = 1 + \frac{c}{c_i}. \tag{1-24}$$

Equation (1-24) gives the ratio of shaft speed to natural frequency that will produce subsynchronous whirling as a function of the ratio of external damping to internal damping. Thus, for example, if the two damping coefficients are equal, the rotor will whirl when the shaft speed reaches twice the natural frequency.

What has just been shown is that subsynchronous whirling is a possible dynamic equilibrium state for a balanced Jeffcott rotor with internal friction whenever the shaft speed reaches "threshold speed." Such an analysis is not adequate to prove rotordynamic instability, but note that the whirl amplitude r divides out of equation (1-22), showing that the whirl is in dynamic equilibrium at *any* amplitude.

For a more rigorous analysis, dynamic stability theory can be used, employing the linear homogeneous differential equations in X and Y coordinates. These equations are derived by removing the imbalance forces from equations (1-6) and (1-7) and adding the cross-coupled stiffness terms derived as follows from the destabilizing force of equation (1-21).

Resolution of the follower force (1-21) into X and Y components (see Fig. 1.16) gives

$$F_X = -F_\phi \sin \phi = -c_i \omega Y \tag{1-25}$$

and

$$F_Y = F_\phi \cos \phi = c_i \omega X, \tag{1-26}$$

since $\dot{\phi} = 0$ for the equilibrium state to be investigated.

These force components can also be expressed in terms of stiffness and damping coefficients, for which the general form is

$$F_X = -K_{XX}X - K_{XY}Y - C_{XX}\dot{X} - C_{XY}\dot{Y}, \tag{1-27}$$

$$F_Y = -K_{YX}X - K_{YY}Y - C_{YX}\dot{X} - C_{YY}\dot{Y}. \tag{1-28}$$

Comparison of equations (1-25) and (1-26) with equations (1-27) and (1-28) shows that

$$K_{XY} = c_i \omega \qquad (1\text{-}29)$$

and

$$K_{YX} = -c_i \omega, \qquad (1\text{-}30)$$

with the remaining coefficients equal to zero. Equations (1-29) and (1-30) give the cross-coupled stiffness coefficients to represent the destabilizing force produced by internal damping. The resolution of the X and Y components into a tangential follower force is illustrated in Fig. 1.16.

Now the differential equations of motion for the balanced Jeffcott rotor with internal friction can be written as

$$m\ddot{X} + c\dot{X} + kX + K_{XY}Y = 0, \qquad (1\text{-}31)$$

$$m\ddot{Y} + c\dot{Y} + kY + K_{YX}X = 0, \qquad (1\text{-}32)$$

where c is now the sum of the external aerodynamic and internal rotor damping coefficients.

These equations are linear, homogeneous, coupled, and have constant coefficients. Their general solution is [9]

$$X = A_1 e^{st} \qquad (1\text{-}33)$$

$$Y = A_2 e^{st}, \qquad (1\text{-}34)$$

where s is the eigenvalue and A_1, A_2 are determined by the initial amplitude of a perturbation.[11]

Substitution of equations (1-33) and (1-34) into equations (1-31) and (1-32) converts the differential equations into algebraic equations in the unknowns A_1 and A_2. Expressed in matrix form, these equations are

$$\begin{bmatrix} (ms^2 + cs + k) & K_{XY} \\ K_{YX} & (ms^2 + cs + k) \end{bmatrix} \begin{Bmatrix} A_1 \\ A_2 \end{Bmatrix} = \begin{Bmatrix} 0 \\ 0 \end{Bmatrix}. \qquad (1\text{-}35)$$

Since the equations are homogeneous, nonzero solutions for the ratio A_1/A_2 can exist only if the determinant of the matrix is zero. Equating the determinant to zero gives the "characteristic polynomial" in the eigenvalue s:

$$(ms^2 + cs + k)^2 - K_{XY}K_{YX} = 0. \qquad (1\text{-}36)$$

---

[11]The equilibrium state to be perturbed is the zero whirl solution $x = y = 0$, which satisfies the differential equations. The perturbed motion must also satisfy the equations.

The eigenvalues of the rotordynamic system are the roots of the characteristic polynomial. They are generally complex numbers. That is, each root will have the form

$$s = \lambda + i\omega_d, \tag{1-37}$$

where λ is the damping exponent, and ω_d is the damped natural frequency (i.e., the whirling frequency) due to the form of the solutions (1-33) and (1-34). For example, equation (1-33) becomes

$$X(t) = A_1 e^{\lambda t}(\cos \omega_d t + i \sin \omega_d t). \tag{1-38}$$

If $\lambda > 0$, the perturbed motion grows exponentially with time and is therefore said to be unstable.

The algebraic sign of λ depends on the relative magnitude of the cross-coupled stiffness (i.e., the destabilizing force) in equation (1-36).

As was shown for internal rotor damping, most of the destabilizing sources satisfy the condition

$$K_{XY} = -K_{YX}. \tag{1-39}$$

Using this condition, and representing the magnitude of K_{XY} and K_{YX} by $\mathcal{K}$,[12] the real and imaginary parts of the roots (1-37) are found to be

$$\lambda = -\frac{c}{2m} \pm \sqrt{\left(\frac{c}{2m}\right)^2 + \left(\omega_d^2 - \frac{k}{m}\right)} \tag{1-40}$$

$$\omega_d^2 = \frac{k}{2m} - \frac{c^2 \pm \sqrt{(c^2 - 4km)^2 + 16\mathcal{K}^2 m^2}}{8m^2}. \tag{1-41}$$

Figures 1.18 and 1.19 show how the whirl frequency ω_d and the damping exponent λ vary with the strength of the cross-coupled stiffness $\mathcal{K}$ for three different values of damping ratio ξ. The latter is defined by

$$\xi = \frac{c_d + c_i}{2\sqrt{km}}, \tag{1-42}$$

where c_d is the external damping coefficient and c_i is the internal (rotor) damping coefficient.

Figure 1.18 helps to explain why the measured whirling frequency in violently

---

[12]Equations (1-29) and (1-30) show $\mathcal{K} = c_i \omega$ for internal rotor damping, but there are also a number of other sources of cross-coupled stiffness which produce different magnitudes of $\mathcal{K}$.

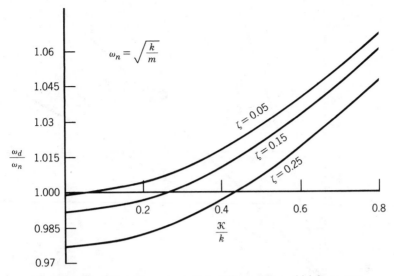

Figure 1.18. Effect of cross-coupled stiffness $\mathcal{K}$ on whirl frequency.

unstable machines is usually higher than the associated critical speed, since the whirling frequency increases with the magnitude of the destabilizing force.[13]

Figure 1.19 shows that a rotor–bearing system with 5 percent damping (a typical value) can become unstable with a cross-coupled stiffness of only 10 percent of the effective shaft stiffness k. (Both stiffness values must be measured at the same point on the rotor).

Chapter VII discusses various sources of instability. In most, the cross-coupled stiffness increases with shaft speed. An example is internal rotor friction, where $\mathcal{K}$ is speed dependent according to equations (1-29) and (1-30). Reformulated in terms of the undamped natural frequency ω_n, this relationship can also be written as the ratio

$$\frac{\mathcal{K}}{k} = 2\xi_i \frac{\omega}{\omega_n}, \tag{1-43}$$

so that the abscissa of Figure 1.19 can be rescaled in terms of the speed ratio ω/ω_n, for any particular value of the *internal* damping ratio

$$\xi_i = \frac{c_i}{2\sqrt{km}}. \tag{1-44}$$

Choosing $\xi_i = 0.025$, for example, Fig. 1.20 shows that the rotor–bearing

---

[13]Gyroscopic moments, not present in the Jeffcott model, are a much stronger influence to raise the whirl frequency at high shaft speeds.

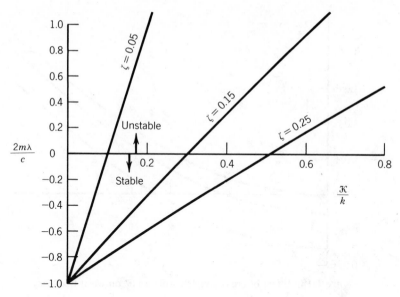

Figure 1.19. Effect of cross-coupled stiffness $\mathcal{K}$ on whirl stability.

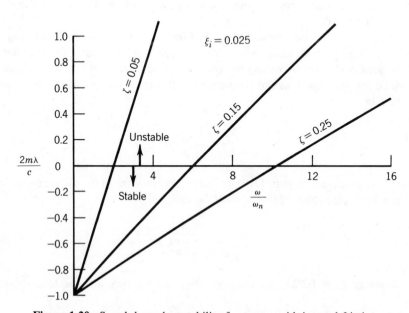

Figure 1.20. Speed-dependent stability for a rotor with internal friction.

system with 5 percent total damping becomes unstable at a shaft speed of about twice the undamped natural frequency. This speed is called the threshold speed of instability. The other two curves in Fig. 1.20 show that increasing the external damping raises the threshold speed.

THE "GRAVITY CRITICAL"

Large massive rotors with a long bearing span sometimes exhibit an apparent critical speed at a shaft speed of one half the true first critical speed. The vibration is characterized by both synchronous and supersynchronous (twice synchronous) frequencies. It is a direct result of the fluctuating moment caused by imbalance, a moment that was ignored in the Jeffcott rotor analysis where constant speed was assumed in order to reduce the number of degrees of freedom to 2.

The gravity critical can be analyzed using a Jeffcott rotor model, but incorporating the moment equation, i.e., writing a differential equation for the dynamic equilibrium of torsional moments. The following analytical development was adapted from Ref. [10] by Mr. Jim Williams, a graduate student in the Turbomachinery Laboratories at Texas A&M University.

Figure 1.21 shows the model and variables required. Basically, the model consists of an undamped Jeffcott rotor on rigid supports. Damping is not required

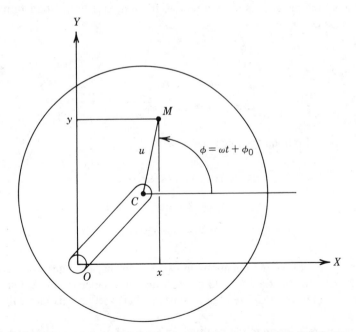

Figure 1.21. Coordinate definitions for the gravity critical analysis.

for a description of the gravity critical phenomenon and is, therefore, excluded from the analysis. It should be noted that x and y denote the coordinates of the center of mass of the rotor and not the geometric center of the rotor as earlier in this chapter. In this case, the acceleration of the mass center is just $\ddot{x}$ and $\ddot{y}$, and the only forces on the rotor are from the stiffness of the shaft and the rotor weight. From Newton's second law, the equations of translatory motion can be written

$$m\ddot{x} = -k(x - u \cos \phi), \tag{1-45}$$

$$m\ddot{y} = -k(y - u \sin \phi), \tag{146}$$

or,

$$m\ddot{x} + kx = ku \cos \phi, \tag{1-47}$$

$$m\ddot{y} + ky = ku \sin \phi. \tag{1-48}$$

Summing moments about the support axes gives the last equation of motion for the current three degree of freedom system.

$$\bar{I}\ddot{\phi} + m(x\ddot{y} - y\ddot{x}) = -mgu \cos \phi \tag{1-49}$$

It should be seen that equation (1-49) includes only the gravity moment and neglects any torsional moments caused by a motor, friction, etc. For a rotor turning at a constant speed,

$$\ddot{\phi} = 0, \tag{1-50}$$

$$\dot{\phi} = \omega, \tag{1-51}$$

$$\phi = \omega t + \phi_0. \tag{1-52}$$

The equations of motion then become

$$m\ddot{x} + kx = ku \cos(\omega t + \phi_0), \tag{1-53}$$

$$m\ddot{y} + ky = ku \sin(\omega t + \phi_0), \tag{1-54}$$

$$m(x\ddot{y} - y\ddot{x}) = -mgu \cos(\omega t + \phi_0). \tag{1-55}$$

Since this is a set of three equations in two unknowns, the integration constants from the first two equations should be such that they satisfy the last. Only if this is true can the above model hold at a constant shaft speed. This fact will now be used.

The complete solution to the first two equations of motion consists of a complementary and a particular solution. Assuming the following solution for the complementary contribution

$$x = Ae^{st},$$
$$y = Ae^{st}, \tag{1-56}$$

the characteristic equation yields

$$ms^2 + k = 0, \tag{1-57}$$

or

$$s = i\sqrt{k/m} = i\omega_n. \tag{1-58}$$

Therefore,

$$x_c = A\cos\omega_n t + B\sin\omega_n t. \tag{1-59}$$

The two trigonometric terms are combined into one with a phase angle β, defined as follows to simplify the problem later:

$$\beta = \gamma + \phi_0 = \tan^{-1}(B/A). \tag{1-60}$$

Thus,

$$x_c = C\cos(\omega_m t + \gamma + \phi_0) = \sqrt{A^2 + B^2}\cos(\omega_n t + \gamma + \phi_0). \tag{1-61}$$

A similar result is obtained for the y component since its equation is identical in form to the x equation; thus,

$$y_c = C\sin(\omega_n t + \gamma + \phi_0). \tag{1-62}$$

The particular solution is found by assuming the following solution similar to the forcing functions.

$$x = X_p\cos(\omega t + \phi_0). \tag{1-63}$$

Substituting gives

$$X_p = \frac{u\omega_n^2}{\omega_n^2 - \omega^2}\cos(\omega t + \phi_0). \tag{1-64}$$

Likewise, for the y component,

$$y = Y_p \sin(\omega t + \phi_0), \tag{1-65}$$

$$Y_p = \frac{u\omega_n^2}{\omega_n^2 - \omega^2} \sin(\omega t + \phi_0). \tag{1-66}$$

The complete solution, then, is

$$x = C \cos(\omega_n t + \gamma + \phi_0) + \frac{u\omega_n^2}{\omega_n^2 - \omega^2} \cos(\omega t + \phi_0), \tag{1-67}$$

$$y = C \sin(\omega_n t + \gamma + \phi_0) + \frac{u\omega_n^2}{\omega_n^2 - \omega^2} \sin(\omega t + \phi_0). \tag{1-68}$$

Normally, the constants are found from initial conditions, but, if constant speed and no external moment are to be enforced, the torque equation (1-55) must be satisfied. Therefore, we substitute

$$m(x\ddot{y} - y\ddot{x}) = m\left[\left(Cc_1 + \frac{u\omega_n^2}{\omega_n^2 - \omega^2} c_2 \right) \left(-C\omega_n^2 s_1 - \frac{u\omega_n^2\omega^2}{\omega_n^2 - \omega^2} s_2 \right) \right.$$
$$\left. - \left(Cs_1 + \frac{u\omega_n^2}{\omega_n^2 - \omega^2} s_2 \right) \left(-C\omega_n^2 c_1 - \frac{u\omega_n^2\omega^2}{\omega_n^2 - \omega^2} c_2 \right) \right], \tag{1-69}$$

where c_1, c_2, s_1, and s_2 are cosines and sines of their respective arguments in equation (1-67) and (1-68). This yields

$$m(x\ddot{y} - y\ddot{x}) = m\left[Cu\omega_n^2 (c_1 s_2 - s_1 c_2) \right]. \tag{1-70}$$

To simplify, C is defined as follows

$$C = M/ku, \tag{1-71}$$

where M is another arbitrary constant. Then,

$$m(x\ddot{y} - y\ddot{x}) = m\left[\frac{Mu\omega_n^2}{ku} (c_1 s_2 - s_1 c_2) \right]$$
$$= M(c_1 s_2 - s_1 c_2)$$
$$= M\left[\cos(\omega_n t + \gamma + \phi_0) \sin(\omega t + \phi_0) - \sin(\omega_n t + \gamma + \phi_0) \right.$$
$$\left. \cdot \cos(\omega t + \phi_0) \right]. \tag{1-72}$$

Using a trigonometric identity gives

$$m(x\ddot{y} - y\ddot{x}) = M \sin[(\omega - \omega_n)t - \gamma].$$ (1-73)

Using equation (1-55) and assuming at $t = 0$, $\phi = \phi_0 = 0$, gives

$$m(x\ddot{y} - y\ddot{x}) = -mgu \cos(\omega t + \phi_0) = -mgu \cos \omega t$$

$$= mgu \sin(\omega t - \pi/2).$$ (1-74)

Therefore,

$$mgu \sin(\omega t - \pi/2) = M \sin[(\omega - \omega_n)t - \gamma].$$ (1-75)

The above can be true only if,

$$M = mgu, \qquad C = mgu/ku = g/\omega_n^2,$$ (1-76)

$$\gamma = \pi/2,$$ (1-77)

$$\omega t = (\omega - \omega_n)t,$$

or,

$$\omega = \tfrac{1}{2}\omega_n.$$ (1-78)

Thus, the only way for the rotor to run at a constant speed with the moment equation satisfied is to have

$$x = \frac{g}{\omega_n^2} \cos\left(\omega_n t + \frac{\pi}{2}\right) + \frac{4}{3} u \cos\left(\frac{1}{2}\omega_n t\right),$$ (1-79)

$$y = \frac{g}{\omega_n^2} \sin\left(\omega_n t + \frac{\pi}{2}\right) + \frac{4}{3} u \sin\left(\frac{1}{2}\omega_n t\right).$$ (1-80)

The fluctuating moment caused by the rotor weight is seen to produce vibrations in the shaft as shown above. This can only occur at shaft speed equal to one half of the first critical speed. The amplitude is determined by the natural frequency and the imbalance. The amplitude of the complementary term can be expressed using simple beam theory as follows.

$$\frac{g}{\omega_n^2} = \frac{mg}{k} = \frac{mg}{48EI/l^3} = \frac{mgl^3}{48EI}$$ (1-81)

where l is the bearing span and EI is the bending stiffness of the shaft.

Equation (1-81) shows why the gravity critical is significant only in heavy rotors with a high ratio of bearing span to bending stiffness.

ADDED COMPLEXITIES

Although analyses of symmetric single-disk rotor models, as heretofore presented, provide valuable insight into many of the fundamental questions of rotordynamics, there are additional characteristics of real machines which are not included in these models. Some of them are the following:

1. Gyroscopic effects, which modify the critical speeds.
2. More than one disk and a significant distributed mass of the shaft, which produce a multiplicity of critical speeds and which may require multiplane balancing.
3. Speed-dependent rotor support stiffness and damping, produced by oil film bearings and fluid seals, which modify the critical speeds and the response to imbalance.
4. Asymmetric mass and stiffness properties of the rotor and of the bearing supports, which produce parametrically excited whirl or backward whirl at various frequencies.

Although all of these characteristics (and more) can be put into a computer simulation of rotordynamics, it is difficult to gain an understanding of their individual effects from a model which is too complex. A more enlightening approach is to analyze the simplest possible model that includes the factor or characteristic of special interest. The understanding thus obtained can be used to interpret the output from a computer simulation of the machine with all factors included, or vibration measurements from the machine itself.

It is often helpful to analyze a model in which the factor of interest has a magnified effect. For example, gyroscopic effects are most pronounced in models with an overhung (cantilevered) disk. It is significant that one of the most common models analyzed in the rotordynamics literature is the single overhung disk on a cantilevered shaft. This model actually simulates a certain class of real machines fairly well, such as single stage compressors and pumps with inlets unrestricted by bearing housings.

Gyroscopic effects are also significant in a long rigid rotor on flexible supports. In Chapter IV, a long rigid rotor model will be used to illustrate gyroscopic effects, conical whirl modes, backward whirl, and the distinction between critical speeds and eigenvalues. Chapter III will give a detailed description of eigenvalue analysis using torsional vibration as an introductory application.

But, first, the next chapter will describe how rotordynamic considerations influence the design of turbomachines.

Chapter V will define rotor imbalance and will describe several methods for balancing turbomachinery rotors in place.

Chapter VI will address the effects of bearings and seals, including the phenomenon known as "oil whip," and Chapter VII will present a comprehensive treatment of rotordynamic instability.

Finally, Chapter VIII will describe some of the instrumentation and measurement techniques that have made it possible to verify and improve the predictions of rotordynamics analysis and that now make turbomachinery troubleshooting more of a science than an art.

REFERENCES

1. Rankine, W. A., "On the Centrifugal Force of Rotating Shafts," *Engineer* (*London*), **27,** 249 (1869).
2. Jeffcott, H. H., "The Lateral Vibration of Loaded Shafts in the Neighborhood of a Whirling Speed: The Effect of Want of Balance," *Philosophical Magazine*, Ser. 6, **37,** 304 (1919).
3. Gunter, E. J., Jr., *Dynamic Stability of Rotor–Bearing Systems*, NASA SP-113, 29 (1966).
4. Rieger, N. F., *Vibrations of Rotating Machinery*, Pt. I: *Rotor–Bearing Dynamics*, The Vibration Institute, Clarendon Hills, IL, 1982, Sec. 1.6.
5. Barrett, L. E., Gunter, E. J., and Allaire, P. E., "Optimum Bearing and Support Damping for Unbalance Response and Stability of Turbomachinery," *Journal of Engineering for Power*, pp. 89–94 (January 1978).
6. Marmol, R. A., *Engine Rotor Dynamics: Synchronous and Nonsynchronous Whirl Control*, USARTL-TR-79-2, U.S. Army Research and Technology Laboratories, Ft. Eustis, VA, pp. 69–125 (February 1979).
7. Ek, M. C., "Dynamics Problems in High Performance Rocket Engine Turbomachinery," Workshop on Rotordynamics Technology for Advanced Turbopumps, NASA–Lewis Research Center, Cleveland (February 23, 1981).
8. Den Hartog, J. P., *Mechanical Vibration*, 4th ed., McGraw-Hill, New York, 1956, p. 301.
9. Beachley, N. H., and Harrison, H. L., *Introduction to Dynamic System Analysis*, Harper & Row, New York, 1978, pp. 43–48.
10. Timoshenko, S., *Vibration Problems in Engineering*, Van Nostrand, New York, 1928, pp. 184–188.

Chapter II

Rotordynamic Considerations in Turbomachinery Design

In the design of turbomachinery, performance considerations and rotordynamic considerations are often in conflict. The performance engineer usually wants to maximize the volume flow rate and pressure, and minimize the fluid energy losses, through a machine of limited size and weight. This generally suggests high shaft speeds, multiple stages, highly loaded rotating components, large spacing between stages, and unrestricted inlets to the stages. All of these features tend to create rotordynamic problems, as will be seen. The design of a successful and reliable turbomachine requires cooperation and compromises between the two disciplines.

From a rotordynamics standpoint, the successful design of a turbomachine involves:

1. Avoiding critical speeds, if possible

2. Minimizing dynamic response at resonance, if critical speeds must be traversed

3. Minimizing vibration and dynamic loads transmitted to the machine structure, throughout the operating speed range

4. Avoiding turbine or compressor blade tip or seal rubs, while keeping tip clearances and seals as tight as possible to increase efficiency

5. Avoiding rotordynamic instability

6. Avoiding torsional vibration resonance or torsional instability of the drive train system

The mth critical speed of a rotor–bearing system may be expressed as

$$\omega_m = \sqrt{k_m/m_m},\qquad(2\text{-}1)$$

where k_m and m_m are the modal stiffness and modal mass, respectively, of the mth whirling mode. These quantities may be simply thought of as the effective stiffness and mass for the critical speed of interest. The effective stiffness is that of the shaft and supports in series ("stacked"), and the effective mass is generally some fraction of the total rotor mass.

Thus, the design of a rotor–bearing system to meet criteria 1–5, above, involves considerations of both the rotor and the bearings and supports.

ROTOR DESIGN

Figure 2.1 shows a typical multistage compressor rotor. Increasing the axial spacing between the stages (wheels) of a compressor, pump, or turbine (to improve fluid for flow efficiency) implies a long rotor and a long bearing span. Since the lateral bending stiffness of a shaft is inversely proportional to the cube of the span of its supports, and the mass of the rotor increases with its length and with its number of wheels, it can be seen that the natural frequencies and critical speeds of the rotor will be lowered. Thus the machine will have to operate through more critical speeds and into a supercritical region where rotordynamic instability is sometimes encountered. Reducing the shaft diameter between stages on a rotor to reduce the restriction on fluid flow has the same deleterious effect on the critical speeds.

One way to get unrestricted flow into the inlet of a stage is to mount the stage wheel on the end of the rotor, outside of the closest bearing (see Fig. 2.2). This general type of "overhung," or cantilevered, design is a rotordynamic challenge and has therefore been the subject of numerous technical papers describing rotor-dynamics analysis. The lack of support on one end of the wheel lowers the critical speeds and produces a conical whirling mode of the wheel that is susceptible to instability from internal friction, torquewhirl, seal forces, and fluid forces on the impeller (see Chapter VII). Nevertheless, a number of successful machines of this type have been built by several manufacturers (including the one shown) and are in service [1].

For manufacturing and assembly reasons, turbomachine rotors are usually built up from shafts, disks, blades, couplings, etc., rather than machined as one piece. The interfaces between components are a source of added internal friction or damping in the rotor assembly, which tends to induce rotordynamic instability as shown in Chapter I. From a rotordynamics standpoint it is advisable to keep the number of such interfaces to a minimum and to keep shrink fits as tight as possible within stress limitations. In his pioneering work on this subject, Kimball [2] suggests undercutting the bore of shrunk-on wheels to reduce internal friction,

Figure 2.1. Five-stage centrifugal compressor rotor. Courtesy of Transamerica Delaval.

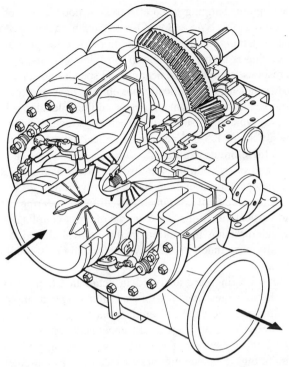

Figure 2.2. Single-stage centrifugal compressor with an overhung wheel. Cutaway courtesy of The Single Stage Products Division of Ingersoll-Rand's Air Compressor Group.

leaving a tight fit only at the ends of the bore. Modern practice is to make one end tighter than the other.

BEARING SELECTION AND SUPPORT DESIGN

The two bearing support properties of major interest are stiffness and damping. From the discussion above, it might be thought that the bearings and related structure should be made as stiff as possible to raise the critical speeds above the operating range. However, when this is done the system damping is greatly reduced, since nearly all of the useful damping in most rotor–bearing systems is in the supports, and a high stiffness will not allow sufficient motion for this damping to act. Thus, many modern turbomachines are designed with soft supports which place the first two "rigid-rotor"[1] critical speeds below the operating speed range

---

[1]See Chapter IV.

and allow the support damping to be effective. A notable exception is cryogenic applications, such as the NASA Space Shuttle main engine (SSME) turbopump, in which support damping is negligible due to the low temperature.

Support damping is highly desirable in most applications, as it can reduce the synchronous response to imbalance near critical speeds and suppress rotordynamic instability. Although it is theoretically possible to have too much damping in a rotor–bearing system, the optimum amount is a difficult level to achieve in practice. For example, a typical all-steel turbomachine structure with rolling-element bearings has only about 1.5–2.5 percent of critical damping, some of which is in the rotor and is therefore destabilizing.

Asymmetry and cross-coupling are also important aspects of bearing support properties, as they have profound effects on rotordynamics, especially with regard to stability. An example of asymmetry is different bearing stiffnesses in the horizontal and vertical directions. An example of cross-coupling is force generated in the horizontal direction by motion in the vertical direction. Hydrodynamic bearings generally have much greater asymmetry and cross-coupling than rolling-element bearings.

Hydrodynamic bearings are analyzed in detail in Chapter VI. They provide shaft support through pressure in a thin oil film between the shaft journal and bearing. The fluid may be either a liquid, usually oil, or a gas, usually air. The pressure in hydrodynamic bearings is self-generated from rotation of the journal. Metal-to-metal contact between the journal and bearing does not occur during normal operation; thus these bearings are characterized by extremely long life when properly designed.

Because of the oil film, damping is much easier to obtain in hydrodynamic bearings than in rolling-element bearings. In fact, the most common way of incorporating damping into machines with rolling-element bearings is to mount the (nonrotating) outer race in a thin oil film called a "squeeze film damper."

An inherent disadvantage of oil film journal bearings (but not squeeze film dampers) is that they produce a rotordynamic instability, known as "oil whirl," at high speeds. The oil whirl problem can be solved by the application of newer bearing designs, such as the "tilt-pad bearing," at the expense of some system damping and at a higher manufacturing cost.

For aircraft engine design, rolling-element bearings are used almost exclusively. The author is aware of only one case in which a hydrodynamic bearing was approved for flight. The principal reasons are that (1) a rolling-element bearing usually fails in a gradual way, which gives warning time before aircraft power is lost, and (2) rolling-element bearings reject less heat to the lubricant, thus allowing a smaller heat exchanger for cooling.

The latter comparison does not apply to gas bearings. This, together with the small radial space requirement and the ready availability of air as a lubricant, has provided considerable incentive for research and development of intershaft gas bearings for aircraft turboshaft applications. Gas bearings have neglible damping, however, and the rotordynamics become quite complex. Until considerable progress is made in this area, ball and roller bearings will continue to be the rule for aircraft engines.

It is often desirable to reduce the overall support stiffness of rolling-element bearings by mounting them in a housing or cage, which is specially designed to have low and predictable stiffness properties. The damping of steel supports is, of course, very low, so a squeeze film damper becomes attractive.

When squeeze films are used, they are mounted either in series or in parallel with the soft mechanical spring supports. The latter are usually designed to provide the minimum stiffness practical while maintaining the required strength and reliability. Most fall roughly into one of three categories:

1. The "squirrel cage," so named because of its appearance, made as a cylinder of thin ribs (see Fig. 2.3).

2. Welded rod support. In this design the bearing mount is cantilevered on several metal rods parallel to the shaft centerline.

Figure 2.3. "Squirrel-cage" bearing support in an aircraft turbine engine. Photograph courtesy of General Electric Co., Aircraft Engine Business Group.

3. Curved beam segments. Variations of this idea can be constructed to fit snugly around the outer bearing race and provide a mechanical cushion through deflection of small segments or protruding elements. The difficult trick is to produce a predictable and reproducible stiffness [3].

The stiffness properties of all these designs are determined by analysis, testing, and iterative redesign as required. The capability to predict these properties analytically in preliminary design stages needs to be improved. One way of doing this would be to standardize some designs throughout industry, although the practicality of accomplishing this in a competitive environment is questionable.

Another type of mechanical bearing support used in noncritical applications to provide both low stiffness and some degree of damping is the common "O-ring." Grooves in the bearing housing bore are made to hold the O-rings so that the cross section is compressed into an elliptical shape when the roller bearing outer race is inserted. The O-rings are made of elastomeric materials with fatigue properties and thermal degradation properties still to be determined. Researchers at MTI [4] have published information on the stiffness and damping properties of some elastomers and have experimented with a multiple support pad configuration in place of the O-ring.

The most discouraging characteristic of elastomeric bearing supports is a propensity to lose their damping properties at high temperature, a phenomenon which has been quantified by experimental research [4] and which the author has observed in a rotordynamics test apparatus at Texas A&M University.

ROTORDYNAMIC DESIGN EVALUATION

Several dimensionless numbers can be used for comparison of turbomachinery rotor–bearing designs from a rotordynamic standpoint.

One such number is

$$R_K = \frac{k_B}{k_R}, \tag{2-2}$$

the ratio of bearing support stiffness to rotor stiffness. Barrett et al. [5] have shown that response to imbalance and the tendency toward instability can both be reduced by making R_K as small as practical. The necessity to avoid blade rubs under shock loading or vehicle maneuvering conditions may impose lower limits on the bearing support stiffness.

A number which is closely related to R_K is

$$R_\omega = \frac{\omega_s}{\omega_1}, \tag{2-3}$$

the ratio of operating shaft speed to the lowest critical speed that involves a significant amount of rotor bending or flexure. Both experience and analysis have shown that R_ω should be kept smaller than 2.0, and as small as practical, especially from the standpoint of rotordynamic stability.

Another relative measure is the percentage of total rotor–bearing strain energy represented by rotor deflection, for each whirling mode in the operating speed range [6]. For a rotor–bearing system with N bearings the strain energy in the supports is

$$U_B = \tfrac{1}{2} \sum_{n=1}^{N} k_{B_n} Y_n^2 \qquad (2\text{-}4)$$

where Y_n is the whirling amplitude at the nth bearing support.

The strain energy in the rotor with modal deflection curve $y(z)$ is

$$U_R = \int_0^L \frac{EI}{2} \left(\frac{d^2 y}{dz^2}\right)^2 dz, \qquad (2\text{-}5)$$

where the integral over the length L of the rotor is discontinuous due to changes in the area moment of inertia I along the rotor.

Thus the number of interest is

$$R_U = \frac{U_R}{U_R + U_B}, \qquad (2\text{-}6)$$

which should be minimized for good rotordynamic design. Here R_U may be a more useful number than R_K for machines with complex rotor assemblies in which k_R may be difficult to define.

For bearing life considerations or to reduce vibration transmitted to the nonrotating structure, the transmissibility

$$\tau = \frac{F_B}{F_\infty}, \qquad (2\text{-}7)$$

defined in Chapter I, can be a useful measure to minimize.

SCALING OF EXISTING DESIGNS[2]

When a particular design is successful, it is tempting to scale it larger or smaller in size as required to meet new performance or weight requirements. Purely geometric scaling, however, can have disastrous consequences to the rotordynamics of the machine.

---

[2]This and the following section are adapted from Ref. [7].

An appreciation for the effect of geometric scaling on rotordynamics can be gained through dimensional analysis of a rotor–bearing system. As an example, consider a turbine or compressor wheel assembly (disks, spacers, and blades) centrally located on a flexible shaft, as shown in Fig. 2.4, and assume that it is desired to scale the system down in physical size without reducing airflow rates. This can be done by increasing shaft speed.

It has been the practice of turbomachinery engineers, with only a few recent exceptions, to maintain operating shaft speeds at least 20 percent below the first critical speed in shaft bending (usually the third critical speed). If the bearings of this example are rigidly supported, the bending critical speed is approximated by

$$\omega_{cr} = 1.8 \frac{d}{l^2} \sqrt{\frac{E}{(0.5 + a_1^2 a_2)\rho}}, \tag{2-8}$$

where

$E =$ Young's modulus
$\rho =$ mass per unit length
$a_1 = d/D$
$a_2 = l/L$

and d, D, l, and L are defined by Fig. 2.4.

For a given material and geometric configuration, equation (2-8) can be rewritten as

$$\omega_{cr} = \overline{C} \frac{d}{l^2}, \tag{2-9}$$

where $\overline{C}$ is a constant.

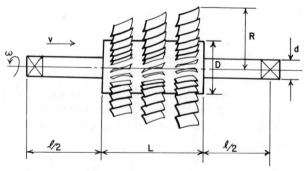

Figure 2.4. Significant rotor dimensions. From Ref. [7].

In addition to the variables defined above, other variables which are pertinent to a scaling analysis of the rotor–bearing system are

ω = shaft speed
R = maximum blade radius
V = average axial air velocity
Q = air volume flow rate

The most important dimensionless groups are

$$\pi_1 = a_i \left(\frac{d}{D}, \frac{l}{L}, \text{etc.} \right),$$

$$\pi_2 = \frac{\omega}{\omega_{cr}} \approx \frac{\omega l^2}{\overline{C}d},$$

$$\pi_3 = \frac{\omega R^3}{Q} \approx \frac{\omega R}{V}.$$

To preserve similitude [8] for performance prediction, all of the π groups must be held constant when size is reduced. The first π group (π_1) determines the "scale factor" n. That is,

$$\frac{d_1}{D_1} = \frac{d_2}{D_2}$$

requires

$$\frac{D_1}{D_2} = \frac{d_1}{d_2} = n, \tag{2-10}$$

where the subscript 1 refers to the large machine, and the subscript 2 refers to the small machine.

The second π group (π_2) preserves the same critical speed margin (say, 20 percent) in the small machine as in the large machine:

$$\frac{\omega_1 l_1^2}{\overline{C}d_1} = \frac{\omega_2 l_2^2}{\overline{C}d_2}$$

requires

$$\omega_2 = n\omega_1. \tag{2-11}$$

The third π group (π_3) requires the same velocity profiles in the small machine as in the large machine. If, in addition, the same air volume flow rate is required (assuming equivalent gas temperatures), the thrid π group requires

$$\frac{\omega_1 R_1^3}{Q} = \frac{\omega_2 R_2^3}{Q},$$

or

$$\omega_2 = n^3 \omega_1. \tag{2-12}$$

Clearly, equations (2-11) and (2-12) are incompatible.

If aerodynamic performance requirements [equation (2-12)] are given priority over rotordynamics requirements [equation (2-11)], a very slight reduction in machine size can eliminate the critical speed margin.

In a similar way, scaling a machine up in size while keeping shaft speed constant to increase airflow rates can cause critical speed margins to be lost.

The model of Fig. 2.4 is oversimplified, but it illustrates the limitations and potential danger of simply scaling existing designs in size.

EXAMPLE: TURBOSHAFT ENGINES WITH FRONT DRIVE

There are a number of special requirements connected with rotordynamics that are peculiar to rotor–bearing systems in the class of small turboshaft engines for aircraft applications requiring front drive (e.g., helicopters):

1. *Increasingly Higher Shaft Speeds.* It is desired to increase airflow and power output without increasing physical size.

2. *Light Weight and Small Physical Size.* The latter is especially important in frontal area and wheel diameter.

3. *Front Drive.* It is usually desired to locate the power takeoff shaft out through the front of the compressor section.

4. *Maintainability.* Individual components making up the rotor–bearing assembly should be easily replaceable.

5. *Long Life.* It is desired to increase the engine operating life. Frequency of overhauls should be reduced.

Taken on an individual basis, these requirements would generate significant problems for the rotor–bearing design engineer. Taken simultaneously, they are a severe challenge. For example, the combination of requirements 1 and 2 results in a blade tip clearance problem in small engines. The very short blades require extremely small tip-to-housing clearances to maintain high aerodynamic efficiency and high power output. This in turn requires very small rotor shaft excursions to avoid blade–housing interferences, a design condition that is incompatible with low dynamic bearing loads at high speeds.[3]

Another example of design conditions that are difficult to meet simultaneously in small high speed engines is the combination of the front-drive requirement and the maintainability requirement. Referring again to Fig. 2.4, the critical speed equation (2-8) is derivable from the more basic equation (2-1):

$$\omega_{\mathrm{cr}} = \sqrt{\frac{k_m}{m_m}},$$

in which k_m is the shaft stiffness effective at the disk (wheel assembly), and m_m is the effective mass at the disk. It is seen that maintaining a high critical speed depends on maintaining high shaft stiffness. The stiffness of the shaft is given by

$$k_m \propto \frac{EI}{l^3}, \tag{2-13}$$

where I is the cross-sectional area moment of inertia; the other symbols were previously defined. It is clear that the stiffness, and consequently the critical speed, is a very strong inverse function of the span l between bearings.

In a front-drive turboshaft engine, either the distance between the bearings supporting the power turbine shaft cannot be shorter than the length of the compressor spool or an intershaft bearing must be employed, since the power turbine shaft must pass through the inside of the compressor spool to reach the front of the engine (see Fig. 2.5 and 2.6). Intershaft bearings, supporting relative rotation between the compressor spool and the power turbine shaft, have been noted to have unique problems of rotordynamics and maintainability [9, 10].

As a result of the above considerations and constraints, some front-drive turboshaft engines have power turbine shafts operating through and above a critical speed in shaft bending. In rotordynamics, this is called "supercritical operation." Safe operation of aircraft rotor–bearing systems at supercritical speeds can be obtained by the following means:

1. Accurate high speed balancing of rotor shaft assemblies. Passage through bending criticals may require multiplane balancing (more than two planes).
2. Provision for significant amounts of external damping.

Satisfaction of condition 1 allows passage through critical speeds without

---

[3]Low dynamic bearing loads are obtained through low bearing support stiffness, which allows larger rotor excursions.

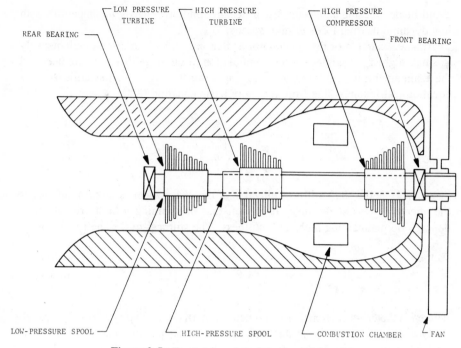

Figure 2.5. Front drive aircraft engine schematic.

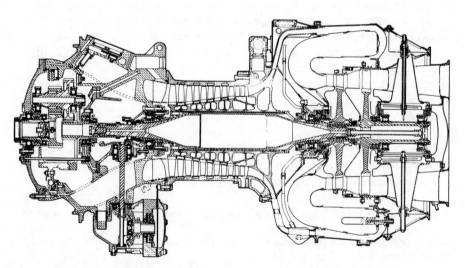

Figure 2.6. Front drive turboshaft engine. Section drawing. From Ref. [7].

destructive whirl amplitudes. Satisfaction of condition 2 also reduces synchronous whirl amplitudes, but additionally it suppresses rotordynamic instability.

The alternatives to a supercritical power turbine shaft in small high speed front-drive engines are as follows:

1. Intershaft bearings.
2. Gas generator redesign to allow a large-diameter power turbine shaft.
3. Use of a power turbine shaft material with a significantly higher stiffness/density ratio than steel.

These alternatives will now be discussed sequentially.

Intershaft bearings have been researched since the 1970s for application to large multispool turbojet engines and have been considered for the small turboshaft application. However, a design problem of radial space availability between the inner and outer shafts is encountered. The space problem suggests an oil-film bearing, but lubrication problems, high temperature, and reliability questions have precluded a successful application to small engines. It is sometimes desirable from an aerodynamic design standpoint to have the gas generator shaft and the power turbine shaft rotating in opposite directions, but this may be another problem for the intershaft bearing, especially a fluid-film bearing. A fluid-film bearing with journal and sleeve counterrotating at equal speeds has zero load capacity (see Fig. 2.7 and Chapter VI).

Rolling-element bearings also present problems for intershaft service. In addition to taking up more radial space, they are subject to shaft bowing from thermal effects, inacessibility for lubrication, and skidding of the rolling elements.

Intershaft bearings of all types also are a source of nonsynchronous vibration,

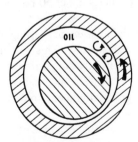

JOURNAL AND BEARING
COUNTERROTATING;
TURBULENT EDDIES INDUCED
IN OIL FILM; NO SUPPORT
PRESSURE GENERATED.

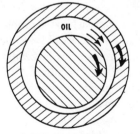

JOURNAL AND BEARING
COROTATING; OIL DRAWN
INTO CONVERGING WEDGE
TO GENERATE PRESSURE.

Figure 2.7. Dependence of intershaft bearing load capacity on rotation direction of both shafts. From Ref. [7].

since they can transmit dynamic loads from one shaft to the other at any of the predominant frequencies.

The second alternative to supercritical shaft design involves either an increase of gas generator bearing diameter (with a consequent increase in bearing *DN* values[4]), to accommodate a larger-diameter power turbine shaft, or a bias toward a purely centrifugal compressor design to allow a shorter shaft. Either of these options raises the power turbine shaft critical speed, as desired, but they both tend to increase the cross-sectional size of the engine. It is interesting to note that a large-diameter compressor spool bearing is compatible with some of the unique requirements of gas-film bearings. For example, gas bearings require large bearing areas and high journal velocities to generate significant load capacity.

The third and final alternative was the subject of some preliminary studies at Williams Research Corporation about 15 years ago. Equation (2-8) shows that the shaft critical speed is proportional to $\sqrt{E/\rho}$. Most engineering metals in common use (e.g., steel, aluminum) have almost identical E/ρ values, thus offering little selectivity with respect to critical speed properties. Engineers at Williams pointed out that beryllium is an exception to this rule, with an E/ρ value of about three times that of steel or aluminum. In designing a beryllium power turbine shaft, problems to overcome are notch sensitivity and the brittleness of the material. If these problems were overcome through proper design or through modification of material properties, a power turbine shaft of this material could operate at significantly higher speeds (perhaps 20 percent) without passing through resonance in bending.

More recently, the U.S. Army (Aviation R&D, Ft. Eustis, Va.) has sponsored research at General Electric to develop a metal matrix composite shaft. At the time of this writing, the most promising combination appears to be titanium (Ti-6-4) with silicon carbide fibers, oriented at about 25° from the shaft axis to improve the torsional strength. The E/ρ ratio of the composite is about 50 percent higher than that for steel.

Another approach to the problem is to develop higher strength shafting materials so the wall thickness can be reduced for a given load capacity. This will also raise the natural frequencies in bending. The Army is currently supporting a research program at Pratt & Whitney Aircraft with this objective. The material is a Ni-Mo-Al textured grain superalloy, with a preferred grain alignment, produced with powdered metallurgy technology.

This discussion of small aircraft engine design is just one example of how rotordynamics considerations can impact turbomachinery design. A similar discussion could be given of the problems encountered in designing large, multistage, ultra-high-pressure industrial compressors, or a high performance cryogenic fuel pump for rocket engines, or any one of several other turbomachinery design challenges.

Another class of problems is encountered when turbomachines are connected

---

[4]*DN* is the product of bearing diameter and shaft speed. It is the most important parameter affecting the life of high speed rolling-element bearings.

together in a drive train. Such a system is susceptible to torsional vibration, or even to torsional instability.

TORSIONAL VIBRATION CONSIDERATIONS

Individual turbomachine rotors are generally stiff enough in torsion to put their natural frequencies of torsional vibration above the range of most torsional excitations. An exception is the torsional excitation in some steam turbines that fatigues blades.

When turbomachines are connected together by shaft couplings, however, each of the individual rotors can act as a single massive inertia. The torsional stiffnesses of the couplings and connecting shafts are then often low enough to bring natural frequencies in torsion down into the range of excitation frequencies. This may also occur when a turbomachine is connected to some other type of rotating inertia (e.g., an electric motor or a reciprocating engine). Such drive trains generally include gearboxes as well. In fact, it is the non-turbomachinery components in such drive trains that usually provide the torsional excitation. The excitation frequencies are generally quite low, often below 20 Hz.

An example is illustrated in Fig. 2.8, which shows a synchronous electric motor driving a centrifugal or axial flow compressor through a gearbox. The components in such trains are often quite large. The motor is typically 5000 hp or larger.

As already mentioned, the rotor of the individual turbomachine generally participates in the torsional vibration as a single rigid inertia, swinging against the other inertias in the drive train. The single most important component which the designer has control of is usually the shaft coupling.

For heavy industrial machinery, a Holset [11] or Geislinger [12] coupling is sometimes used to provide very low torsional stiffness and thus drop the natural frequencies far below the excitation frequencies. In vibration terminology, such a device is called a "torsional isolator." It is the torsional equivalent of the soft bearing supports described earlier in this chapter. A disadvantage of these couplings is that they are so massive that they sometimes lower the lateral critical speeds, which is usually undesirable (as also described earlier).

It is theoretically possible to design light centrifugal link couplings having zero torsional stiffness. This seeming anomaly has been explained [13], as has an application of the couplings to helicopter drive trains [14]. Although zero stiffness may be undesirable, (due to a "hunting" characteristic), a small enough positive stiffness can practically eliminate the two-node resonance in a three-inertia drive train [14]. The main problem yet to be overcome is the short life of the centrifugal link bearings, especially at the high speeds encountered in the turbomachinery applications.

Since it is difficult to build a reasonably light coupling with low torsional stiffness and high torque capacity, damping is incorporated into some designs as the mechanism for vibration suppression. A torsional damper dissipates the kinetic

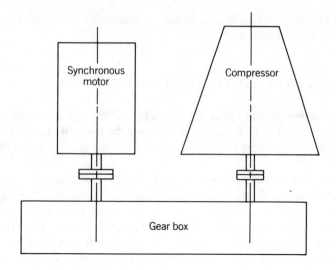

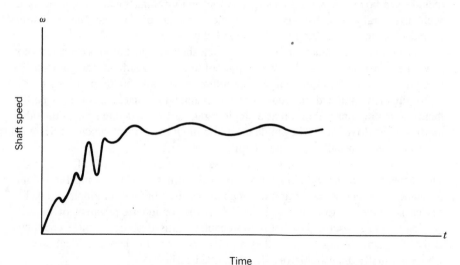

Figure 2.8. Torsional vibration excited by synchronous motor start-up.

energy of torsional vibration into heat. From vibrational theory, a damper works best at resonance and therefore is the device of choice if excitation at a natural frequency cannot be avoided.

Torsional dampers can be subdivided into two types: devices in which the damping element transmits none of the useful torque, and devices in which the element transmitting the torque also contains the damping. The second type always utilizes a specified and controlled stiffness in addition to the damping.

A unique variation of the first type is the Houdaille damper [15], which utilizes an internal free flywheel. Friction, or viscous shear, is the only mechanism by which torque can be transmitted between the flywheel and its coupling housing. Since the inertia of the flywheel prevents it from following the torsional oscillations, vibration energy is dissipated.

SYNCHRONOUS ELECTRIC MOTOR DRIVE TRAINS

It is becoming increasingly attractive (due to energy considerations) to drive centrifugal or axial compressors and blowers with large synchronous electric motors. The shaft speed step-up is usually accomplished with a parallel-shaft helical gear set enclosed in a gearbox. Fig. 2.8 shows such a train.

Synchronous motors of the salient-pole type are started as induction motors, with the field short-circuited or discharged through a resistor. This produces a pulsating torque, during start-up, with a frequency which varies from twice line frequency[5] initially down to zero at synchronous speed. Any natural frequencies of torsional vibration that lie within this range will be excited during the start-up. The magnitude of the driving torque pulsation varies with speed and varies from motor to motor, but can be quite large, with a peak-to-peak amplitude often greater than the average torque.

Mruk et al. [16] have measured the pulsations and found them to be even larger (by a factor of 5) for an abnormal shaft with malfunctioning exciter circuits. Under the latter condition the predominant frequency of the pulsations during start-up was equal to the motor slip frequency (one-half its usual value under normal conditions).

When this type of excitation becomes resonant with a natural frequency, even if for only a few seconds, it is not unusual for the gears in the train to clatter or even break as a result of tooth separation and impact. Sohre [17] published photographs of a pinion shaft failure due to this phenomenon, and Brown [18] has shown a compressor shaft failure.

The rational technical approaches to avoiding or solving this problem are as follows.

1. If there is only one natural frequency of torsional vibration within the range of excitations, change it to coincide with the motor speed at which the excitation magnitude is the lowest. This requires an accurate prediction capability for the natural frequencies and information from the motor manufacturer regarding direct axis and quadrature axis torques during start-ups. The normal excitation frequency during start-ups, in terms of motor speed, is

$$f_e = 2f\left(\frac{N_s - N}{N_s}\right) \quad \text{Hz,} \tag{2-14}$$

[5]That is, $2 \times 60 = 120$ Hz in the United States.

where

f = line frequency, Hz

N_s = synchronous motor speed, rpm

N = actual motor speed, rpm

2. If the natural frequency being excited cannot be changed, redesign the motor so that its speed of minimum excitation coincides with the natural frequency. This requires the same technical capability and information as approach 1, as well as a capability to modify and predict electric motor characteristics.

3. If there are a number of torsional natural frequencies within the range of excitation frequencies, incorporate adequate damping into the system to reduce the magnification factor at resonance to an acceptable value. The most practical way of accomplishing this is with a special coupling, such as the Holset or Houdaille type, but the effect of the coupling mass on lateral rotordynamics should be checked.

4. Increase the average starting torque or reduce the total drivetrain inertia so that the start-up takes less time. The passage dwell time through each resonance will be reduced proportionably.

TORSIONAL STABILITY CONSIDERATIONS

Some turboshaft-powered drivetrains, notably in helicopters and in constant-frequency electric power sets, have automatic speed control. Figures 2.9 and 2.10 show the basic concept, in which a tachometer signal ω is compared to a preset

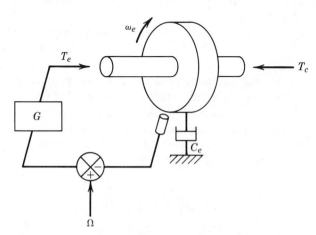

Figure 2.9. Schematic of feedback for turbine speed control.

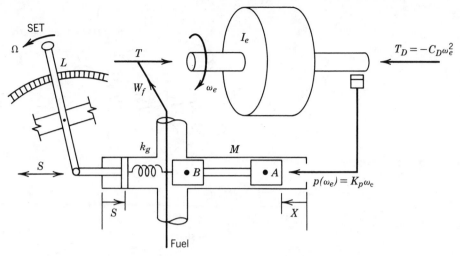

Figure 2.10. Schematic of speed control governor.

derived speed Ω and the error is fed back to change the fuel flow as required. In the language of control theory this is said to be a "closed-loop system." The feedback signal (measured speed) closes the loop. One characteristic of closed-loop systems is a potential for instability, i.e., self-excited oscillations that grow larger with time.

In the helicopter application, the automatic speed governor relieves the pilot of the requirement to maintain constant rotor speed during changes of load and flight conditions. A special problem, which has occurred in several helicopter development programs, is that of maintaining torsional stability of the engine and drivetrain while providing a sufficiently rapid response to demands for power and speed changes.

Figure 2-11 shows the system schematic. The major elements of the system are the gas generator (which produces torque T_e), the drivetrain (which includes the power turbine), and the speed governor. Figure 2.12 is a block diagram as used in control systems analysis. The drivetrain is a three-inertia (power turbine rotor, gearbox, and rotary wing) torsional system with two nonzero natural frequencies. The effective torsional stiffness of the rotary wing mast is greatly reduced by the square of the gear ratio [19], typically 80:1, which produces a very low natural frequency.

To make this system torsionally stable, it has been shown [20] that either special features must be provided in the design of the governor or the gain of the system must be reduced, which will increase the steady-state droop. The droop of a closed-loop system is the difference between the desired output and the actual output obtained. The specific requirements for stability depend on the mechanical parameters of the drivetrain such as torsional damping and stiffness. It has been suggested

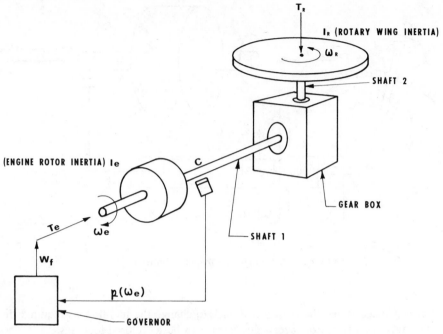

Figure 2.11. Model for a two-inertia drivetrain with speed control.

that locating the speed sensor on the rotary-wing mast, rather than on the power turbine shaft, might solve this problem. This, however, is apparently incompatible with engine–airframe interfacing requirements.

THE ROLES OF ANALYSIS AND TESTING

Most of the rotordynamic considerations presented in this chapter have provided questions, rather than answers, to the turbomachinery design engineer. How many stages? Overhung impeller? Shaft diameter? Rotational speed? Type of bearing?

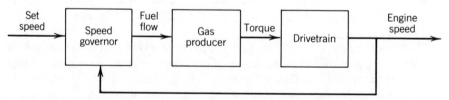

Figure 2.12. Block diagram of helicopter drivetrain with speed control.

Number and location of bearings? Bearing support stiffness? Squeeze film damper? Special shaft coupling? These are all questions which can and should be answered with the aid of rotordynamics analysis and testing.

The design of anything as complex as a turbomachine is an iterative process. Beginning with concepts and configurations synthesized from experience, the preliminary design is analyzed with mathematical models using the best parametric data available. At every point where it is economically feasible, the data are verified or improved by testing. Redesign of components during the process is the rule rather than the exception, and the process is repeated until acceptable performance and reliability are obtained.

Turbomachinery troubleshooting is, in many cases, an extension of this process in which part of the cost is borne by the user. Information gained is often fed back to the manufacturer, who may use it to retrofit other machines or to influence future designs.

The ensuing chapters present methods and information for design analysis, mathematical modeling, troubleshooting, and measuring turbomachinery rotordynamics.

The reader may find it helpful to look through the next chapter on torsional vibrational analysis, even if his/her primary interest is confined to problems of rotor whirling and lateral vibration. For each mathematical modeling technique or method of solution used in the analysis of lateral rotordynamics, there is an analogous but simpler procedure for torsional vibration analysis. Thus, an ancillary function of Chapter III will be to introduce these techniques and solution methods with the simplest possible models.

REFERENCES

1. Pennink, H., "The State of the Art of High Speed Overhung Centrifugal Compressors for the Process Industry," *Proceedings of the Seventh Turbomachinery Symposium, Texas A&M University, December 1978*, pp. 35–46.

2. Kimball, A. T., "Internal Friction as a Cause of Shaft Whirling," *Philosophical Magazine*, Ser. 6, **49,** 724–727 (1925).

3. Marmol, R. A., *Engine Rotor Dynamics: Synchronous and Nonsynchronous Whirl Control*, USARTL-TR-79-2, U.S. Army Research and Technology Laboratories, Ft. Eustis, VA, pp. 41–47 (February 1979).

4. Darlow, M., and Zorzi, E., *Mechanical Design Handbook for Elastomers*, Mechanical Technology Inc., Report 81TR5, NASA Contract NA53-21623 (January 1981).

5. Barrett, L. E., Gunter, E. J., and Allaire, P. E., "Optimum Bearing and Support Damping for Unbalance Response and Stability of Turbomachinery," ASME Paper No. 77-6T-27, *Journal of Engineering for Power*, pp. 89–94 (1978).

6. Simmons, H. R., "Vibration Energy: A Quick Approach to Rotor Dynamic Optimi-

zation," ASME Paper No. 76-PeT-60, presented at the Joint Petroleum/Mechanical Engineering Conference, Mexico City, September 19–24, 1976.

7. Vance, J. M., and Royal, A. C., "High-Speed Rotor Dynamics: An Assessment of Current Technology for Small Turboshaft Engines," *Journal of Aircraft*, **12**(4), (April 1975).

8. Baker, W. E., Westine, P., and Dodge, F. T., *Similarity Methods in Engineering Dynamics*, Hayden, Rochelle Park, NJ, 1973.

9. Hibner, D. H., Kirk, R. G., and Buono, D. F., "Analytical and Experimental Investigation of the Stability of Intershaft Squeeze Film Dampers, Part I: Demonstration of Instability," *Journal of Engineering for Power*, pp. 47–52 (January 1977).

10. Gunter, E. J., Li, D. F., and Barrett, L. E., "Dynamic Characteristics of a Two-Spool Gas Turbine Helicopter Engine," *Proceedings of the Conference on Stability and Dynamic Response of Rotors with Squeeze Film Bearings, Charlottesville, VA May 8–10, 1979*, sponsored by the U.S. Army Research Office.

11. Koppers Co., Baltimore, can supply descriptive information of a commercial nature.

12. Pfeifer, P., "Torsional Vibration Damper Experience," *Diesel and Turbine Progress Worldwide* (December 1977).

13. Vance, J. M., and Brown, R. A., "Suppression of Torsional Vibration With Zero Torsional Stiffness Couplings," *The Shock and Vibration Bulletin*, **44**, Pt. 5, 43–54 (August 1974).

14. Vance, J. M., Brown, R. A., and Darlow, M. S., *Feasibility Investigation of Zero Torsional–Stiffness Couplings for Suppression of Resonance and Instability in Helicopter Drive Trains*, U.S. Army Air Mobility Research and Development Laboratory Report TR-73-103 (June 1974).

15. Nestorides, E. J., *Handbook of Torsional Vibration*, British Internal Combustion Research Association (BICERA), 1958.

16. Mruk, G. K., Halloran, J. D., and Kolodziej, R. M., "New Method Predicts Startup Torque," *Hydrocarbon Processing*, pp. 229–234 (May 1978).

17. Sohre, J. S., "Transient Torsional Criticals of Synchronous Motor Driven, High-Speed Compressor Units," ASME Paper 65-FE-22, presented at the Applied Mechanics Conference, Washington, DC, June 7–9, 1965.

18. Brown, R. N., "A Torsional Vibration Problem as Associated With Synchronous Motor Driven Machines," *Journal of Engineering for Power*, pp. 215–220 (July 1960).

19. Seirig, A., *Mechanical Systems Analysis*, International Textbook Co., Scranton, PA, 1969, pp. 93–96.

20. Peczkowski, J. L., "Automatic Control of Free Turbine Engines," ASME Paper 63-WA-159, presented at the ASME Winter Annual Meeting, Philadelphia, November 17–22, 1963.

Chapter III

Torsional Vibration Analysis

As described in the preceding chapter, torsional vibration may be a problem in drivetrains where several rotating machines are connected together by drive shafting and couplings. Problems occur when the natural frequencies of torsional vibration are excited by pulsating torques with matching frequency components or when automatic speed control produces unstable torsional oscillations.

The typical engineering objectives of torsional vibrations analysis are as follows:

1. Predict the natural frequencies.

2. Evaluate the effect on the natural frequencies and vibration amplitudes of changing one or more design parameters (i.e., "sensitivity analysis").

3. Compute the vibration amplitudes and peak torques under steady-state torsional excitation.

4. Compute the dynamic torques and gear tooth loads under transient conditions (e.g., during a machine start-up).

5. Evaluate the torsional stability of drivetrains with automatic speed control.

From the standpoint of the reader of this book, torsional vibration analysis also provides a description of the mathematical methods referred to later for analyzing rotor critical speeds, response to imbalance, and whirl stability.

MATHEMATICAL MODELING

Figure 3.1 shows a lumped parameter model of a drivetrain. All of the rotary inertia is lumped into N discrete disks, each with a polar mass moment of inertia

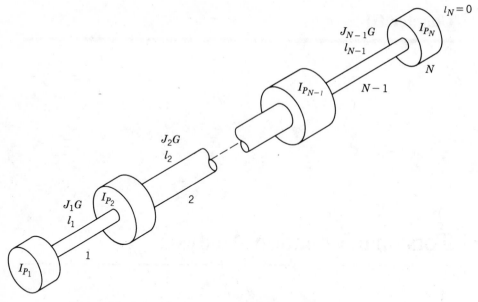

Figure 3.1. Lumped parameter model for torsional vibration analysis.

I_{P_n}, $1 \le n \le N$. For example, I_{P_1} might be the inertia of a power turbine driving the train, I_{P_2} might represent a shaft coupling, I_{P_3} the gears in a speed reducer, and so on.

The inertias are connected by massless torsional springs of stiffness K_n, $1 \le n \le N - 1$. The springs represent the torsional flexibilities of shafts and couplings. Usually, large-diameter wheels and rotors are represented by inertias, and long slender shafts by springs, but even a long shaft of constant diameter can be lumped into a number of "stations" if the higher vibration frequencies are of interest. The inertia of each shaft section is usually divided into equal parts and lumped into the "disks" at each end of the section. Figure 3.2 shows the nth station, which consists of an inertia connected to a massless shaft. The torsional stiffness of the shaft section is given by

$$K_n = \frac{J_n G}{l_n}, \tag{3-1}$$

where

$J_n = \pi d_n^4 / 32$

G = shear modulus

l_n = shaft section length

d_n = shaft section diameter

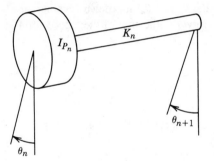

Figure 3.2. The nth station in the model.

In torsional vibration analysis, the number of stations N should be at least one more than the number of natural frequencies to be calculated.

To represent the dissipation of vibration energy, viscous dampers are incorporated into the model. The viscous characteristic produces a torque opposing and linearly proportional to the angular velocity across the damper. The dampers denoted C_n, $1 \leq n \leq N - 1$, represent the energy dissipated in the relative twisting motion of the shafts and couplings, and the dampers denoted B_n, $1 \leq n \leq N$, represent energy dissipated in bearings, fluid impellers, or other elements with an absolute velocity-dependent torque (i.e., the damper is to "ground").

The purpose of lumping the parameters into discrete stations is to describe the dynamics of each inertia mathematically by an ordinary differential equation, thus avoiding the more difficult partial differential equations required for distributed-mass models. If θ_n is the angular displacement of the nth inertia, the system has N degrees of freedom described by the N generalized coordinates θ_n, $1 \leq n \leq N$. The drivetrain is then modeled by N second-order ordinary differential equations of motion with constant coefficients.

If the driver is inertia I_{P_1}, then the driving torque $T(t)$ is imposed on the first differential equation only. If I_{P_N} is the load, then its torque must be imposed on the last differential equation.

The differential equations can be derived from Newton's laws [1], or from Lagrange's equation [2, 3]. The latter method is especially powerful for the three dimensional rotational dynamics of whirling rotors (to be treated in later chapters), so will be illustrated here.

Each coordinate will have a motion $\theta_n(t)$ that satisfies Lagrange's equation:

$$\frac{d}{dt}\left(\frac{\partial L}{\partial \dot{\theta}_n}\right) - \frac{\partial L}{\partial \theta_n} = Q_n, \quad 1 \leq n \leq N, \tag{3-2}$$

where

$L = T - V$ is the "Lagrangian"
$T =$ kinetic energy of the system
$V =$ potential energy of the system
$Q_n =$ the nonconservative torque on the nth coordinate

The Lagrangian L is written in terms of the θ_n and $\dot{\theta}_n$ in order to evaluate the derivatives in equation (3-2). The kinetic energy T is the sum of the $\frac{1}{2}I_{P_n}\dot{\theta}_n^2$ for all the inertias, so

$$T = \tfrac{1}{2} \sum_{n=1}^{N} I_{P_n}\dot{\theta}_n^2. \tag{3-3}$$

The potential energy V is the sum of the strain energy in all the torsional springs,

$$V = \tfrac{1}{2} \sum_{n=1}^{N-1} K_n(\theta_n - \theta_{n+1})^2. \tag{3-4}$$

The nonconservative torques are obtained from the virtual work δW of the damping torques, the driving torque $T(t)$, and the load torque. That is,

$$\delta W = \sum_{n=1}^{N} Q_n \delta\theta_n \tag{3-5}$$

gives

$$Q_1 = T_1(t) - B_1\dot{\theta}_1 - C_1(\dot{\theta}_1 - \dot{\theta}_2), \tag{3-6}$$

$$Q_n = T_n(t) - \sum_{n=2}^{N-1} \left[B_n\dot{\theta}_n + C_n(\dot{\theta}_n - \dot{\theta}_{n+1}) - C_{n-1}(\dot{\theta}_{n-1} - \dot{\theta}_n) \right], \tag{3-7}$$

$$Q_N = T_N(t) - B_N\dot{\theta}_N + B_N'\dot{\theta}_N|\dot{\theta}_N| - C_{N-1}(\dot{\theta}_{N-1} - \dot{\theta}_N). \tag{3-8}$$

The term involving B_N' in equation (3-8) is the load torque, assumed here to be proportional to the speed squared.

Substitution of L from equations (3-3) and (3-4), and the Q from (3-6) through (3-8), into Lagrange's equation (3-2) yields the N differential equations of motion as

$$I_{P_1}\ddot{\theta}_1 + B_1\dot{\theta}_1 + C_1(\dot{\theta}_1 - \dot{\theta}_2) + K_1(\theta_1 - \theta_2) = T_1(t), \qquad n = 1, \tag{3-9}$$

$$
\begin{aligned}
I_{P_n}\ddot{\theta}_n &+ B_n\dot{\theta}_n + C_n(\dot{\theta}_n - \dot{\theta}_{n+1}) \\
&+ C_{n-1}(\dot{\theta}_n - \dot{\theta}_{n-1}) + K_n(\theta_n - \theta_{n+1}) \\
&+ K_{n-1}(\theta_n - \theta_{n-1}) = T_n(t), \qquad n = 2, 3, \cdots, N-1;
\end{aligned}
\tag{3-10}
$$

$$
\begin{aligned}
I_{P_N}\ddot{\theta}_N &+ B_N\dot{\theta}_N + C_{N-1}(\dot{\theta}_N - \dot{\theta}_{N-1}) \\
&+ K_{N-1}(\theta_N - \theta_{N-1}) = -B_N'|\dot{\theta}_N|\dot{\theta}_N + T_N(t), \qquad n = N.
\end{aligned}
\tag{3-11}
$$

If there are speed reducers or increasers (e.g., gears) in the drivetrain, they impose constraints on the shaft speeds of each station, which modify the effective inertia and stiffness parameters in the model. All parameters are typically referred to the shaft speed of the driver at station 1, with the inertia and stiffness of every other station calculated as [4]

$$I_{P_n} = g_n^2 I_{P_n}' \tag{3-12}$$

$$K_n = g_n^2 K_n', \tag{3-13}$$

where

g_n = speed ratio of station n to station 1
I_{P_n}' = actual inertia of station n
I_{P_n} = effective inertia referred to the speed of station 1
K_n' = actual stiffness of station n
K_n = effective stiffness referred to the speed of station 1

Accurate values for the damping coefficients are usually difficult to obtain. The author has found from experiments that an all-steel rotor/shaft assembly with tight fits and no gear backlash will have about 1.5–2.0 percent of critical damping, not including friction between the rotor and ground. Since the damping in a steel shaft is proportional to its stiffness, that is, $C_n = \beta K_n$, then the modal damping ratio ξ_i [5] for the ith torsional mode is

$$\xi_i = \frac{\beta}{2}\omega_i \approx 0.0175, \tag{3-14}$$

where ω_i = the natural frequency of the ith mode.
The station–station coefficients C_n are then given by

$$C_n = \beta K_n = \frac{0.035}{\omega_i} K_n. \tag{3-15}$$

Since the natural frequency ω_i is generally not known a priori, an iterative calculation procedure is required when using equation (3-15).

A simpler and faster method of approximating the C_n is to consider a two-inertia model for the nth station. The single nonzero natural frequency is

$$\omega_n = \sqrt{\frac{K_n(I_{P_n} + I_{P_{n+1}})}{I_{P_n}I_{P_{n+1}}}}. \tag{3-16}$$

The station–station damping coefficient is then approximated by

$$C_n = 0.035 \sqrt{\frac{K_n I_{P_n} I_{P_{n+1}}}{I_{P_n} + I_{P_{n+1}}}}. \tag{3-17}$$

For stations that include a shaft coupling, the factor 0.035 should be replaced by the coupling manufacturer's data.

The station–ground coefficients B_n can be calculated from estimated horsepower losses in bearings, assuming a viscous torque–speed relationship, and from the slopes of the nonlinear torque–speed characteristics of the wheels that do work. For example, the friction torque of a bearing at station n may be approximated by

$$T_n = -B_n \dot{\theta}_n. \tag{3-18}$$

The dissipated power is then

$$P_n = -B_n \dot{\theta}_n^2 \tag{3-19}$$

If the power loss in the bearing can be estimated at a constant shaft speed Ω, the coefficient B_n is given by

$$B_n = P_n / \Omega^2. \tag{3-20}$$

Many types of wheels that do work (e.g., fans or pump impellers) have a hydrodynamic torque–speed characteristic which is approximately quadratic. An example is shown in equation (3-11), where B'_N is the nonlinear (quadratic) load coefficient. In this case the power absorbed by station N at speed $\dot{\theta} = \Omega$ is

$$P_N = -B'_N \Omega^3. \tag{3-21}$$

For a linearized eigenvalue analysis the effective viscous damping coefficient at station N would then be the slope of the torque–speed curve at the operating shaft speed Ω as shown in Fig. 3.3. That is,

$$B_N = 2B'_N \Omega \tag{3-22}$$

is the contribution of the hydrodynamic load at station N to its viscous damping coefficient.

EIGENVALUE ANALYSIS

Usually the first and most important objective of torsional vibration analysis is the determination of natural frequencies and their logarithmic decrements. In mathe-

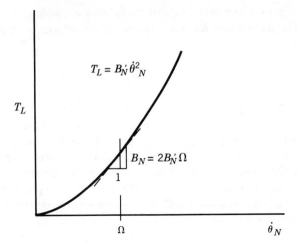

Figure 3.3. Viscous damping coefficient of the load torque.

matical parlance, this is called eigenvalue analysis. The natural frequencies of the train modeled in the previous section are the imaginary parts[1] of the eigenvalues of the linearized differential equations (3-9) through (3-11).

The differential equations are linearized by setting the driving torque $T_n(t) = T^*$, a constant, and letting $\dot{\theta}_n = \Omega + \dot{\epsilon}_n$, where Ω is the steady-state shaft speed and $\dot{\epsilon}_n$ is a small angular perturbation of the speed at station n. With these substitutions, the linearized equations turn out to have the same form as equations (3-9) through (3-11), except with zero right-hand side (the driving and load torques cancel) and with all the θ_n replaced by ϵ_n. It is typical, however, to write the ϵ_n as θ_n, with the tacit understanding that they are small oscillations away from the steady-state angle Ωt at each station. The resulting equations are not only linear but also can be characterized as ordinary, second-order, homogeneous differential equations with constant coefficients. The solution to all such equations has the form [6]

$$\theta_n(t) = \bar{a}_n e^{st}, \qquad n = 1, 2, \cdots, N, \qquad (3\text{-}23)$$

where $\bar{a}_n$ is the (complex) amplitude of the oscillation at the nth station and s is the eigenvalue. The amplitudes are arbitrary here; the problem at hand is to find the eigenvalues.

COMPUTATION OF UNDAMPED NATURAL FREQUENCIES

With no damping ($B_n = C_n = 0$, $n = 1, 2, \cdots, N$), equations (3-9) through (3-11) have $N - 1$ purely imaginary nonzero eigenvalues $s_j = \pm i\omega_j, j = 1, 2, \cdots,$

---

[1] In general, the eigenvalues are complex numbers.

$N - 1$, which are the undamped natural frequencies of torsional vibration. From equation (3-23), the solution for free vibration of the nth station at the jth natural frequency is then

$$\theta_n(t) = a_n \cos{(\omega_j t)}, \tag{3-24}$$

where a_n is the real amplitude.

Computation of the ω_j without damping is a much simpler task than with damping included; consequently the earliest published methods were for the undamped case. The most well known and widely used for these are Holzer's procedure and the method of Stodola (now called matrix iteration). They are well documented in the literature; see, for example, Refs. [7] and [8]. Holzer's method with damping included will be described in a later section as a simple illustration of the general transfer matrix methods now in common use for rotordynamic analysis.

COMPUTATION OF DAMPED NATURAL FREQUENCIES

When the damping is nonzero, the eigenvalues are complex numbers. Then with

$$s_j = \lambda_j \pm i\omega_j, \quad i = \sqrt{-1},$$

$$j = 1, 2, \cdots, N - 1, \tag{3-25}$$

the solution (3-23) can be written as

$$\theta_n(t) = \sum_{j=1}^{N-1} a_{nj} e^{\lambda_j t} \cos(\omega_j t - \beta_{nj}), \tag{3-26}$$

where a_{nj} is the real amplitude at station n with frequency ω_j and phase lag β_{nj}. It can be seen that $N - 1$ nonzero natural frequencies[2] are given by the imaginary parts of the eigenvalues (3-25), and that the sign of the real part λ_j determines whether the free vibration given by equation (3-26) will grow exponentially (unstable) or die out (stable). The torsional vibration of the drivetrain modeled by equations (3-9) through (3-11) will always be stable ($\lambda_j < 0$, all j). Some type of feedback from the angular velocity to the driving torque $T_n(t)$ would be required for instability to occur. Such a system is described in a later section on speed control stability.

The various methods used for computing the damped natural frequencies are simply different ways of evaluating the complex eigenvalues (3-25) of the linearized differential equations.

---

[2]There will be a greater number of natural frequencies of rotor whirling (discussed in later chapters).

EXAMPLE 1

Figure 3.4 shows a schematic of a helicopter drivetrain. The three major inertias are the power turbine rotor I_{P_1}, the gears I_{P_2}, and the rotary wing I_{P_3}. The engine drive shaft has torsional stiffness K_1, and the rotary wing mast has torsional stiffness K_2. The largest sources of torsional damping are from the aerodynamic load on the rotary wing, and from internal friction or dampers in the rotary-wing hub assembly.

Typical numerical values, taken from Ref. [9], are

$$I_{P_1} = 6 \text{ in-lb-sec}^2$$
$$I_{P_2} = 2.3 \text{ in-lb-sec}^2$$
$$I'_{P_3} = 298{,}598 \text{ in-lb-sec}^2$$
$$K_1 = 257{,}004 \text{ in-lb/rad}$$
$$K'_2 = 61.44 \times 10 \text{ in-lb/rad}$$

$$\text{Operating power} = 5270 \text{ hp}$$
$$\text{Turbine speed} = 13{,}820 \text{ rpm}$$
$$\text{Gearbox speed ratio} = 80:1$$

The model can be rendered like Fig. 3.1 by referring all parameters to the turbine shaft. The effective inertia and stiffness of the rotary wing and mast are calculated from equations (3-12) and (3-13) to be

$$I_{P_3} = 298{,}598/80^2 = 46.7 \text{ in-lb-sec}^2$$
$$K_2 = 61.44 \times 10^6/80^2 = 9600 \text{ in-lb/rad}$$

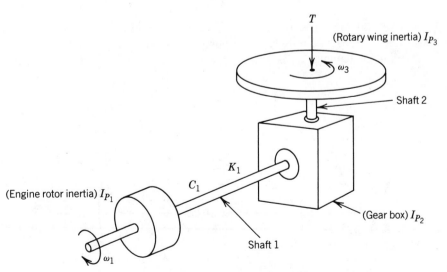

Figure 3.4. A helicopter drivetrain.

Since $K_1 \gg K_2$, it can be seen that I_{P_1} and I_{P_2} will oscillate together as a rigid inertia against the rotary wing in the fundamental mode. To model the fundamental mode, Fig. 3.5 shows a simplified two-inertia system in which K_1 is the torsional stiffness K_2 calculated above. Thus the parameter values for the simplified system are

$$I_{P_1} = 6 + 2.3 = 8.3 \text{ in-lb-sec}^2$$
$$I_{P_2} = 46.7 \text{ in-lb-sec}^2$$
$$K_1 = 9600 \text{ in-lb/rad}$$

The steady-state load torque of the rotary wing is

$$B_2'\dot\theta_2^2 = \frac{(5270)(550)(12)}{(13{,}820)(2\pi/60)} = 24{,}034 \text{ in-lb}$$

at the operating speed $\dot\theta_2 = 13{,}820(2\pi/60) = 1447 \text{ rad/sec} = \Omega$.

The equivalent viscous damping coefficient of the load is calculated from equation (3-22):

$$B_2 = 2B_2'\dot\theta_2 = 33.2 \text{ in-lb/(rad-sec)}.$$

The mast bearing friction is small by comparison.
The minimum shaft damping, from equation (3-17), is

$$C_1 = 0.035 \sqrt{\frac{(9600)(8.3)(46.7)}{8.3 + 46.7}} = 9.1 \text{ in-lb/(rad-sec)}.$$

Rotary wing hinge friction or a specially designed lead-lag damper can increase this value to at least 26 in-lb-sec, depending on its design and condition.

The turbine damping B_1 is neglected here, since a driving torque–speed curve

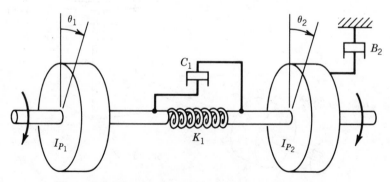

Figure 3.5. Two-inertia model of the fundamental mode.

of positive slope will diminish the net turbine damping coefficient and can even make it negative in extreme cases [10].

The two differential equations of motion (3-9) and (3-11) [$N = 2$ so (3-10) disappears] are linearized by substituting

$$
\begin{aligned}
T_1 &= T^*, & T_2 &= 0, \\
\dot{\theta}_1 &= \Omega + \dot{\epsilon}_1, & \dot{\theta}_2 &= \Omega + \dot{\epsilon}_2, \\
\theta_1 &= \Omega t + \theta_1^* + \epsilon_1, & \theta_2 &= \Omega t + \theta_2^* + \epsilon_2, \\
\ddot{\theta}_1 &= \ddot{\epsilon}_1, & \ddot{\theta}_2 &= \ddot{\epsilon}_2,
\end{aligned}
$$

to give

$$
\begin{aligned}
I_{P_1}\ddot{\epsilon}_1 &+ C_1(\Omega + \dot{\epsilon}_1 - \Omega - \dot{\epsilon}_2) \\
&+ K_1(\Omega t + \theta_1^* + \epsilon_1 - \Omega t - \theta_2^* - \epsilon_2) = T^*, \\
I_{P_2}\ddot{\epsilon}_2 &+ B_2(\Omega + \dot{\epsilon}_2) + C_1(\Omega + \dot{\epsilon}_2 - \Omega - \dot{\epsilon}_1) \\
&+ K_1(\Omega t + \theta_2^* + \epsilon_2 - \Omega t - \theta_1^* - \epsilon_1) = -B_2'(\Omega^2 + 2\Omega\dot{\epsilon}_2 + \dot{\epsilon}_2^2).
\end{aligned}
\tag{3-27}
$$

With no perturbation ($\epsilon_1 = \epsilon_2 = 0$), the steady-state driving torque is seen to be

$$
T^* = K_1(\theta_1^* - \theta_2^*) = B_2'\Omega^2 + B_2\Omega,
$$

where $\theta_1^* - \theta_2^*$ is the steady-state (average) twist in the shaft, and $B_2\Omega$ is bearing friction only.

Assuming the perturbation is small so $\dot{\epsilon}_2^2 \rightarrow 0$, and substracting the steady-state terms from both sides of both equations, leaves the linearized differential equations as

$$
\begin{aligned}
I_{P_1}\ddot{\epsilon}_1 + C_1(\dot{\epsilon}_1 - \dot{\epsilon}_2) + K_1(\epsilon_1 - \epsilon_2) &= 0, \\
I_{P_2}\ddot{\epsilon}_2 + (\bar{B}_2 + C_1)\dot{\epsilon}_2 - C_1\dot{\epsilon}_1 + K_1(\epsilon_2 - \epsilon_1) &= 0,
\end{aligned}
\tag{3-28}
$$

where $\bar{B}_2 = B_2 + 2B_2'\Omega$.

Substituting in the solution $\epsilon_1 = \bar{a}_1 e^{st}$, $\epsilon_2 = \bar{a}_2 e^{st}$, from (3-23), and dividing out e^{st}, yields (3-29):

$$
\begin{bmatrix}
[I_{P_1}s^2 + C_1 s + K_1] & [-C_1 s - K_1] \\
[-C_1 s - K_1] & [I_{P_2}s^2 + (\bar{B}_2 + C_1)s + K_1]
\end{bmatrix}
\begin{Bmatrix}
\bar{a}_1 \\
\bar{a}_2
\end{Bmatrix}
=
\begin{Bmatrix}
0 \\
0
\end{Bmatrix}.
\tag{3-29}
$$

Cramer's rule requires that the determinant of the 2 × 2 matrix be zero if $\bar{a}_1$, $\bar{a}_2$ are nonzero. Expansion of the determinant gives the characteristic polynomial as

$$I_{P_1} I_{P_2} s^4 + \left[I_{P_1}(\bar{B}_2 + C_1) + I_{P_2} C_1 \right] s^3$$
$$+ \left[K_1(I_{P_1} + I_{P_2}) + \bar{B}_2 C_1 \right] s^2 + \left[K_1 \bar{B}_2 \right] s = 0. \qquad (3\text{-}30)$$

Factoring out s shows that one of the eigenvalues (roots) is $s = 0$. This result, which gives a zero natural frequency, occurs for all torsional systems with no spring to ground. Physically, the zero root shows that the drive train is free to rotate as a rigid body. The roots of the remaining third-order polynomial are found numerically by the Siljak method [11], using a digital computer. The complete set of eigenvalues for this example are thus found to be

$$s_0 = 0 + i0, \qquad s_1 = -0.6036 - i0, \qquad s_2 = -0.699 \pm i36.9. \quad (3\text{-}31)$$

The root s_1 is not oscillatory since its imaginary part is zero. The complex conjugate roots s_2 represent a motion of the form

$$\theta_1 = e^{-0.699t} \bar{a}_1 e^{i36.9t}$$
$$\theta_2 = e^{-0.699t} \bar{a}_2 e^{i36.9t}. \qquad (3\text{-}32)$$

The absolute value of $\bar{a}_1$ or $\bar{a}_2$ depends on initial conditions, but the ratio $\bar{a}_2/\bar{a}_1$ can be found from the differential equations, once the value of s_2 is known. The first equation of (3-29), with $s = s_2 = -0.699 + 36.9i$ gives

$$\frac{\bar{a}_2}{\bar{a}_1} = \frac{I_{P_1} s^2 + C_1 s + K_1}{C_1 s + K_1} = 0.178 e^{i3.1608}, \qquad (3\text{-}33)$$

which means that the larger inertia I_{P_2} oscillates with an amplitude 0.178 times that of I_{P_1} and leads by an angle of 3.1608 rad, or 181°.
Thus

$$\theta_1 = e^{-0.699t} \cos(36.9t)$$
$$\theta_2 = e^{-0.699t} (0.178) \cos(36.9t + 3.1608) \qquad (3\text{-}34)$$

represents the free vibration executed by the two inertias.
With the higher value of shaft damping $C_1 = 26.0$ in-lb-sec, the oscillating eigenvalue is

$$s_2 = -1.90 \pm i36.9. \qquad (3\text{-}35)$$

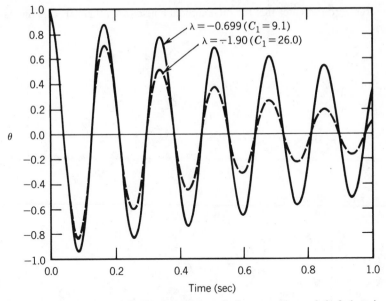

Figure 3.6. Free vibration of the fundamental mode for two values of shaft damping C.

The increased damping changes the frequency almost imperceptively (there is a change in the second decimal place), but substantially increases the rate of decay. Figure 3.6 is a plot of the free vibration for the two different values of shaft damping.

FORCED RESPONSE TO STEADY EXCITATION

If the driving torques in equation (3-9) are harmonic with constant amplitude and frequency, that is, if

$$T_n(t) = T^* + T_{e_n} \cos(\omega t), \qquad (3\text{-}36)$$

then the linearization procedure yields the nonhomogeneous set of equations

$$I\{\ddot{\theta}_n\} + D\{\dot{\theta}_n\} + K\{\theta_n\} = \{T_{e_n} \cos(\omega t)\}, \qquad (3\text{-}37)$$

where I is the inertia matrix:

$$I = \begin{bmatrix} I_{P_1} & & & 0 \\ & I_{P_2} & & \\ & 0 & \ddots & \\ & & & I_{P_N} \end{bmatrix} \qquad (3\text{-}38)$$

(the large zero indicates that the off-diagonal entries are all zero), D is the damping matrix:

$$
D = \begin{bmatrix}
(B_1 + C_1) & -C_1 & & & & 0 \\
-C_1 & (B_2 + C_2 + C_1) & -C_2 & & & \\
& -C_2 & (B_3 + C_3 + C_2) & \ddots & & \\
& & -C_3 & & -C_{N-1} & \\
0 & & & \ddots & \ddots & \\
& & & -C_{N-1} & (B_N + C_{N-1})
\end{bmatrix} , \quad (3\text{-}39)
$$

and K is the stiffness matrix

$$
K = \begin{bmatrix}
K_1 & -K_1 & & & 0 \\
-K_1 & (K_1 + K_2) & -K_2 & & \\
& -K_2 & (K_2 + K_3) & \ddots & \\
0 & & -K_3 & & -K_{N-1} \\
& & & \ddots & \ddots \\
& & & -K_{N-1} & K_{N-1}
\end{bmatrix} ; \quad (3\text{-}40)
$$

θ_n is the vector of N angular displacements:

$$
\{\theta_n\} = \begin{Bmatrix} \theta_1 \\ \theta_2 \\ \vdots \\ \theta_N \end{Bmatrix} ; \quad (3\text{-}41)
$$

and $\{T_{e_n} \cos(\omega t)\}$ is the vector of exciting torques:

$$
\{T_{e_n} \cos(\omega T)\} = \begin{Bmatrix} T_{e_1} \cos(\omega t) \\ T_{e_2} \cos(\omega t) \\ \vdots \\ T_{e_N} \cos(\omega t) \end{Bmatrix} \quad (3\text{-}42)
$$

Note that, in general, an exciting torque $T_{e_n} \cos(\omega t)$ can act on any of the N inertias, thus rendering any or all of the elements in (3-42) nonzero.

The frequency of vibration in this case is the exciting frequency ω; the problem is to find the responding amplitudes of each of the θ_n. The solution procedure is to substitute the expression

$$\theta_n = \theta_{c_n} \cos(\omega t) + \theta_{s_n} \sin(\omega t), \qquad \text{for } n = 1, 2, \cdots, N, \qquad (3\text{-}43)$$

into the differential equations (3-37), performing the indicated derivatives. For this solution to be valid for all time t requires that all the $\cos(\omega t)$ terms add to T_{e_n} in each equation, and that the $\sin(\omega t)$ terms add to zero independently. The result is a set of $2N$ algebraic equations, linear in the unknowns θ_{c_n} and θ_{s_n}. Cramer's rule, Gaussian elimination, or any other method can be used to solve for the unknowns [12].

Resonance will produce large amplitudes of vibration when the exciting frequency is near any of the natural frequencies unless the effective damping on that mode is high. Damping keeps the resonant amplitudes bounded. Note that the exciting frequency ω for torsional vibration is not generally equal to shaft speed and that the exciting torque amplitude is not directly related to rotor imbalance.[3]

EXAMPLE 2

When the helicopter of example 1 (Fig. 3.4) flies with a forward velocity, each rotary wing (blade) is subjected to a cyclic variation of aerodynamic load as its tangential velocity becomes directly opposed to the airstream once during each revolution. The result is a cyclic variation of load torque on inertia I_{P_2} in Fig. 3.5, with the fundamental excitation frequency ω equal to $N_B \omega_W$, where N_B is the number of blades and ω_W is the angular velocity of the rotary wing. Even though the model of Fig. 3.5 is referred to the turbine shaft, the excitation frequency is unaffected by the gear ratio of the speed reducer (except that it determines the speed of the rotary wing). Thus, for this example, if $N_B = 4$ we have the excitation frequency

$$\omega = 4(13,820/80)(2\pi/60) = 72.36 \text{ rad/sec} \quad (11.52 \text{ Hz}),$$

or about twice the 5.87 Hz natural frequency of the drivetrain.

When the excitation torque is included in the differential equations, the term

---

[3]Dynamic coupling between lateral and torsional modes of vibration may occur, especially when gears are present, thus producing a component of torsional vibration synchronous with shaft speed.

$T_{e_2} \cos(\omega t)$ appears on the right-hand side of the second equation of (3-27). The resulting two equations to be used in place of (3-28) are

$$\begin{bmatrix} I_{P_1} & 0 \\ 0 & I_{P_2} \end{bmatrix} \begin{Bmatrix} \ddot{\theta}_1 \\ \ddot{\theta}_2 \end{Bmatrix} + \begin{bmatrix} C_1 & -C_1 \\ -C_1 & (B_2 + C_1) \end{bmatrix} \begin{Bmatrix} \dot{\theta}_1 \\ \dot{\theta}_2 \end{Bmatrix}$$

$$+ \begin{bmatrix} K_1 & -K_1 \\ -K_1 & K_1 \end{bmatrix} \begin{Bmatrix} \theta_1 \\ \theta_2 \end{Bmatrix} = \begin{Bmatrix} 0 \\ T_{e_2} \cos(\omega t) \end{Bmatrix}. \tag{3-44}$$

Substitution of (3-43) with the associated procedure yields (3-45):

$$\begin{bmatrix} (K_1 - I_{P_1}\omega^2) & C_1\omega & -C_1\omega & -C_1\omega \\ -C_1\omega & (K_1 - I_{P_1}\omega^2) & C_1\omega & -K_1 \\ -K_1 & -C_1\omega & (K_1 - I_{P_2}\omega^2) & (B_2 + C_1)\omega \\ C_1\omega & -K_1 & -(B_2 + C_1)\omega & (K_1 - I_{P_2}\omega) \end{bmatrix}$$

$$\cdot \begin{Bmatrix} \theta_{c1} \\ \theta_{s1} \\ \theta_{c2} \\ \theta_{s2} \end{Bmatrix} = \begin{Bmatrix} 0 \\ 0 \\ T_{e_2} \\ 0 \end{Bmatrix}. \tag{3-45}$$

Figure 3.7 shows the solution for excitation torque $T_{e_2} = 7200$ in-lb (30 percent of the load torque).

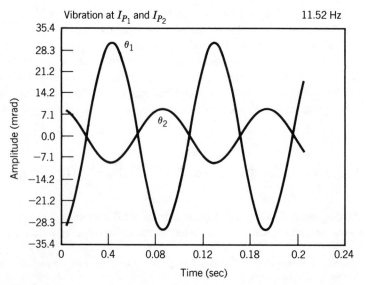

Figure 3.7. Torsional vibration of helicopter drivetrain-forced response.

In this example a more meaningful result is the oscillatory torque in the drive shaft and transmission, which is given by

$$T_s(t) = K_1[\theta_2(t) - \theta_1(t)] + C_1[\dot{\theta}_2(t) - \dot{\theta}_1(t)]$$

$$= K_1[(\theta_{c_2} - \theta_{c_1})\cos(\omega t) + (\theta_{s_2} - \theta_{s_1})\sin(\omega t) \tag{3-46}$$

$$+ C_1\omega[(\theta_{s_2} - \theta_{s_1})\cos(\omega t) + (\theta_{c_1} - \theta_{c_2})\sin(\omega t).$$

The magnitude of this torque is

$$|T_s| = \sqrt{\begin{array}{c}[K_1(\theta_{c_2} - \theta_{c_1}) + C_1\omega(\theta_{s_2} - \theta_{s_1})]^2 \\ + [K_1(\theta_{s_2} - \theta_{s_1}) + C_1\omega(\theta_{c_1} - \theta_{c_2})]^2\end{array}} \tag{3-47}$$

Figure 3.8 shows the frequency response of (3-47) obtained by computing solutions to (3-45) for 50 frequencies over the range 0–12 Hz ($0 < \omega < 24\pi$ rad/sec). It can be seen that the oscillatory torque at the excitation frequency of 11.52 Hz is quite small.

Note, however, what would happen if the excitation frequency was near resonance (5.87 Hz). The magnitude of the oscillatory torque would then be greater than the load torque of 24,000 in-lb, so that the net transmitted torque would go negative during each cycle as shown in Fig. 3.9. This causes the gear teeth to separate and recontact with an impact on each reversal of torque. The author has seen several cases of gear tooth breakage, and all were found to be subjected to this type of torque reversal.

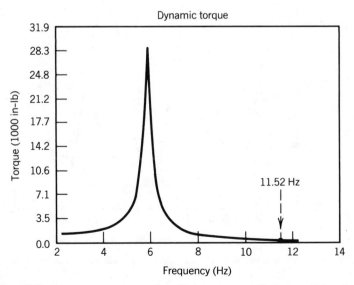

Figure 3.8. Frequency response of oscillatory torque in helicopter drive shaft.

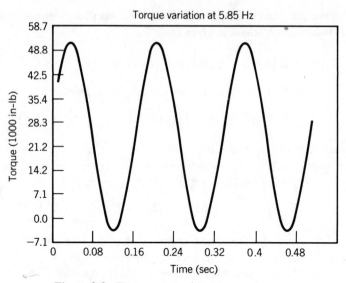

Figure 3.9. Torque versus time near resonance.

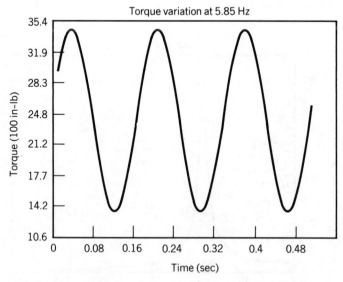

Figure 3.10. Torque versus time near resonance with increased shaft damping.

In such a case, a possible solution may be to increase shaft damping by use of special devices (e.g., a lead-lag damper). Figure 3.10 shows the resonant torque variation for this example with C_1 increased to 26.0 in-lb-sec. Note that the transmitted torque is positive for all time, which should be a required condition except for short transients.

TRANSIENT RESPONSE

If the frequency or amplitude of any of the exciting torques (3-42) vary with time, then the solution just presented is invalid. Except for some special simple cases, the engineering analyst must then resort to a numerical integration of differential equations (3-9) through (3-11), marching out an approximate solution for the $\theta_n(t)$ in small incremental steps of time Δt.

In this case there is no good reason to linearize the differential equations, so one of the advantages of this method is that the nonlinearities are retained in the model. In fact, the parameters K_n, C_n, and I_{P_n} no longer need be constant in the model.

To organize the marching algorithm, the N second-order equations (3-9) through (3-11) are expanded to $2N$ first-order equations [13]. The N new coordinates are defined by

$$\{\dot{\theta}_n\} = \{v_n\}, \qquad n = 1, 2, \ldots, N, \tag{3-48}$$

which also defines N of the first-order equations.

The second N equations are obtained by substituting (3-48) into (3-9) through (3-11) and solving for $\dot{v}_n$ $(= \ddot{\theta}_n)$:

$$\{\dot{v}_n\} = \left\{\frac{f_n}{I_{P_n}}\right\}, \tag{3-49}$$

where each element f_n represents all terms but the first in the nth equation of (3-9) through (3-11), moved to the right-hand side. For example,

$$f_1 = T_1(t) - B_1 v_1 - C_1(v_1 - v_2) - K_1(\theta_1 - \theta_2). \tag{3-50}$$

The complete set of $2N$ first-order equations are now given by (3-48) and (3-49) taken together. Alternatively, the momenta p_n can be used as coordinates in place of the v_n, and the entire set of equations can be derived by a numerical interpretation of Hamilton's principle. See Ref. [14]. Since each of these equations is an expression for a first derivative (slope) in time, the solution can be marched out from time t_{i-1} to time t_i by using Euler's formula

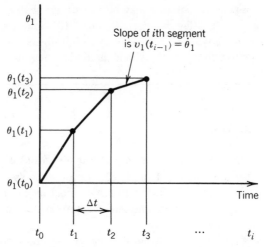

Figure 3.11. Numerical marching scheme for θ_1.

$$
\left\{ \begin{array}{c} \theta_n \\ v_n \end{array} \right\}_{t_i} = \left\{ \begin{array}{c} \theta_n \\ v_n \end{array} \right\}_{t_{i-1}} + \left\{ \begin{array}{c} v_n \, \Delta t \\ \dfrac{f_n}{I_{P_n}} \, \Delta t \end{array} \right\}_{t_{i-1}} , \tag{3-51}
$$

where $\Delta t = t_i - t_{i-1}$ is small.

To get started, some initial conditions for all the θ_n, v_n at time t_0 must be known or assumed. Figure 3.11 shows graphically how the numerical solution proceeds.

The solution is expedited with a high speed digital computer, and numerical integration schemes much more sophisticated than (3-51) have been devised [15].

At first look, numerical integration appears to be an almost omnipotent method for dynamics analysis, but the method has three serious disadvantages: (1) large-order systems require impractically large amounts of computation time; (2) numerical integration errors and round-off errors are difficult to identify, evaluate, and control; and (3) the solutions obtained apply only to the special case represented by the numerical input data and initial conditions for each run of the computer. If these limitations are recognized and dealt with, numerical integration can yield useful simulations of transient response.

EXAMPLE 3

This example is based on the drivetrain of Ref. [16], consisting of a salient-pole synchronous electric motor driving an axial-flow compressor through a gearbox. A schematic and three-inertia torsional model are shown in Fig. 3.12, with all parameters referred to the motor shaft. The numerical values of the parameters are

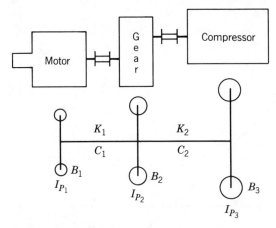

Figure 3.12. Industrial drivetrain with three-inertia torsional model.

I_{P_1} = 4192 in-lb-sec^2, I_{P_2} = 4907 in-lb-sec^2, I_{P_3} = 10,322 in-lb-sec^2
K_1 = 73.59 × 10^6 in-lb/rad, K_2 = 351.7 × 10^6 in-lb/rad
$B_1 = B_2 = B_3$ = 22.99 in-lb-sec (1% bearing loss)
C_1 = 16,663 in-lb-sec, C_2 = 39.411 in-lb-sec

   In this case at least three degrees of freedom must be retained to model the transient excitation of two resonances during start-up. As described in Chapter II, the salient-pole synchronous electric motor produces an oscillating torque during start-up, with an excitation frequency of twice line frequency times the motor slip ratio. Figure 3.13 is a torsional interference diagram, showing how the excitation frequency passes through the two natural frequencies during the start-up from 0–1200 rpm (motor speed). The motor slip ratio is given by

$$S_R = \frac{N_s - N_m}{N_s},$$ (3-52)

where

N_s = synchronous speed = 1200 rpm
N_m = motor speed, rpm (varies 0–1200 rpm during the start-up)

   Thus, the motor torque at any instant of time during the start-up can be described by

$$T_1(t) = T_{avg} + T_{osc} \cos(\omega_e t)$$ (3-53)

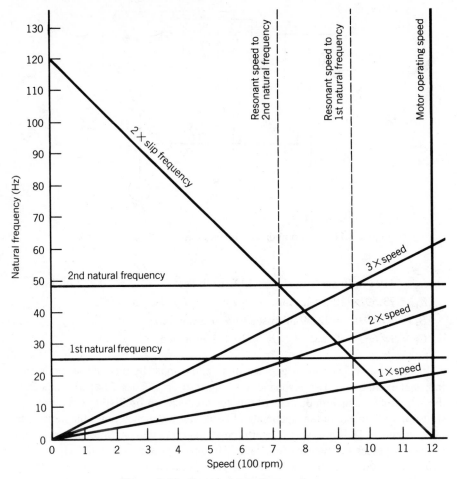

Figure 3.13. Torsional interference diagram.

and

$$\omega_e = 4\pi L_f \left(\frac{N_s - N_m}{N_s} \right) \quad \text{rad/sec}, \tag{3-54}$$

where L_f = line frequency = 60 Hz in the United States.

The angular speed ω_e of the rotating vector $\mathbf{T}_0$ that generates the oscillating torque should be integrated to give correct values of torque at each time value, for use in any computer program for numerical transient analysis. Note that the frequency ω_e is not a constant, but if $\alpha(t)$ denotes the angular position of $\mathbf{T}_0$, then

$$\alpha = \int_0^t \omega_e \, dt = 4\pi L_f \int_0^t \left(1 - \frac{N_m}{N_s}\right) dt$$

$$= 4\pi L_f \left(t - \frac{\theta_1}{\Omega_s}\right),$$

(3-55)

where $\Omega_s = (2\pi/60) N_s$.

Thus the final expression for the driving torque of a synchronous electric motor during start-up is

$$T_1(t) = T_{avg} + T_{osc} \cos\left[4\pi L_f\left(t - \frac{\theta_1}{\Omega_s}\right)\right]$$

(3-56)

where

T_{avg} = average value of starting torque at time t

T_{osc} = magnitude of the oscillating torque at time t

θ_1 = angular position of the motor inertia at time t

Figure 3.14 shows the torque–speed characteristics for the motor of this example.

Motor speeds at which the torsional natural frequencies are encountered can be

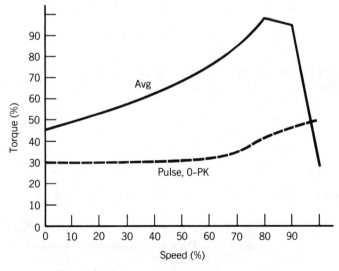

Figure 3.14. Motor torque–speed characteristics.

obtained from equation (3-54) by solving for N_m and setting ω_e equal to the torsional natural frequency ω_j. This gives the encounter speeds in rpm as

$$N_{\text{res}} = N_s\left(1 - \frac{\omega_j}{4\pi L_f}\right), \qquad j = 1, 2,$$

$$\omega_j \sim \text{rad/sec.} \tag{3-57}$$

A computer code for transient response should include computation of the eigenvalues at the beginning, so the values of ω_j are made available to precompute the expected encounter speeds. It is therefore convenient to separate the differential equations into linear and nonlinear terms so the linear part can be used to compute the eigenvalues and eigenvectors. This will be described further below.

To formulate the complete set of nonlinear differential equations in first-order form for efficient numerical integration, the alternative use of the momenta p_n as coordinates (in place of the velocities v_n) will be illustrated here. The momenta are defined as

$$p_n = \frac{\partial L}{\partial \dot{\theta}_n} = I_{P_n}\dot{\theta}_n, \qquad n = 1, 2, 3, \tag{3-58}$$

where $L = T - V$ from equations (3-3) and (3-4).

These three equations are inverted to give

$$\begin{Bmatrix} \dot{\theta}_1 \\ \dot{\theta}_2 \\ \dot{\theta}_3 \end{Bmatrix} = \begin{bmatrix} 1/I_{P_1} & 0 & 0 \\ 0 & 1/I_{P_2} & 0 \\ 0 & 0 & 1/I_{P_3} \end{bmatrix} \begin{Bmatrix} p_1 \\ p_2 \\ p_3 \end{Bmatrix}. \tag{3-59}$$

Lagrange's equation (3-2) is solved for $\dot{p}_n$, using the definitions (3-3) through (3-8), to give

$$\dot{p}_1 = -K_1(\theta_1 - \theta_2) - (B_1 + C_1)\dot{\theta}_1 + C_1\dot{\theta}_2 + T_1(t),$$

$$\dot{p}_2 = K_1\theta_1 - (K_1 + K_2)\theta_2 + K_2\theta_3 + C_1\dot{\theta}_1$$
$$- (B_2 + C_2 + C_1)\dot{\theta}_2 + C_2\dot{\theta}_3, \tag{3-60}$$

$$\dot{p}_3 = K_2(\theta_2 - \theta_3) + C_2\dot{\theta}_2 - (B_3 + C_2)\dot{\theta}_3 - B_3'\dot{\theta}_3|\dot{\theta}_3|.$$

The expressions (3-59) for the $\dot{\theta}_n$ are substituted into (3-60), so that the right-hand side is in terms of θ_n, p_n, and t. The resulting equations, together with (3-59), comprise the six first-order differential equations to be integrated in time to give $\theta_n(t)$, $p_n(t)$, for $n = 1, 2, 3$.

Equations (3-59) and (3-60) are the first integrals (in time) of equations (3-9) through (3-11) for $N = 3$ and are therefore more efficient than the latter for transient analysis [14].

For numerical integration, both the Runge–Kutta and the Newmark [15] method were found satisfactory in this example, with Newmark about 40 percent faster. Some computer systems designed for engineering analysis have software packages for numerical integration readily available, and this may influence the choice of method.

For precomputation of the eigenvalues, the linearization procedure illustrated in Example 1 eliminates the terms in $T_1(t)$ and B_3' from equations (3-60). These linearized equations together with (3-59) define the characteristic matrix described later in this chapter as a method for eigenvalue analysis of large-order systems. The eigenvalues of the characteristic matrix are computed to be as shown in the accompanying table.

Real Part		Frequency	
1/sec	log dec.	Hz	rad/sec
−7.49	(0.1369)	54.72	(343.82)
−7.49	(0.1369)	−54.72	(−343.82)
−2.12	(0.0943)	22.51	(141.43)
−2.12	(0.0943)	−22.51	(−141.43)
0.0	(0.0)	0.0	0.0
−0.0	(0.0)	0.0	0.0

Figure 3.15 shows the two mode shapes, plotted from the eigenvectors of the nonzero eigenvalues.

Using $\omega_1 = 141.43$ rad/sec and $\omega_2 = 343.82$ rad/sec in equation (3-57) gives the expected encounter speeds as 652 rpm and 975 rpm, respectively.

Figure 3.16 shows the motor speed versus time during the start-up, obtained by numerically integrating equations (3-59) and (3-60) with all (linear and nonlinear) terms retained. The time and amplitude scales are too compressed on this figure to clearly detail the torsional oscillations. Figure 3.17 shows the oscillations of motor speed on an expanded time scale during the second resonance, which begins at about 11 sec. Figure 3.18 shows the torque in the motor shaft versus time, in which the first resonance can also be seen, beginning at 7.5 sec. Inspection of the eigenvalues reveals why the first resonance (54.7 Hz) has a much smaller amplitude. The real part of the eigenvalue shows that this mode is more highly damped than the 22.5 Hz mode. Figure 3.19 shows the difference in decay rates of free vibration for the two modes.

The negative torques during the second resonance do not necessarily imply gear

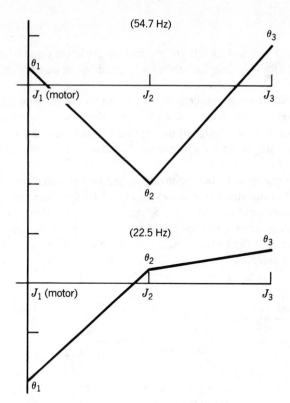

Figure 3.15. Torsional mode shapes for the industrial drivetrain.

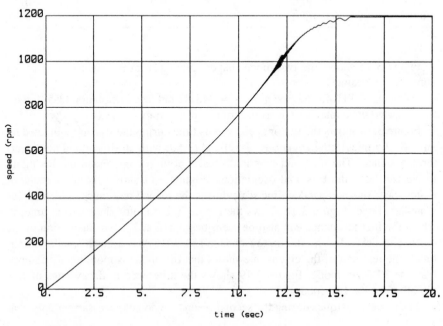

Figure 3.16. Motor speed versus time in start-up of industrial drivetrain.

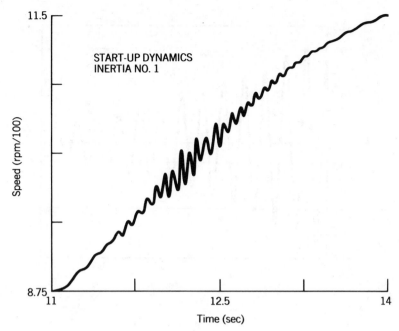

Figure 3.17. Motor speed oscillations during start-up.

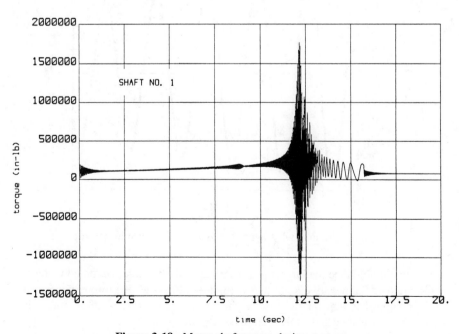

Figure 3.18. Motor shaft torque during start-up.

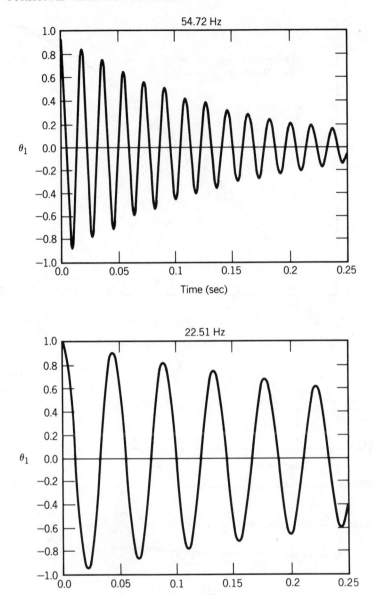

Figure 3.19. Free vibration of the two modes excited by start-up.

tooth separation, since the largest gear is attached to the motor shaft and its inertia could produce the negative amplitudes. Figure 3.20 shows the torque versus time in the compressor shaft. This curve verifies the gear tooth separation, since only a small pinion is attached to the compressor shaft.

At the time of commissioning, microphone measurements on this train recorded

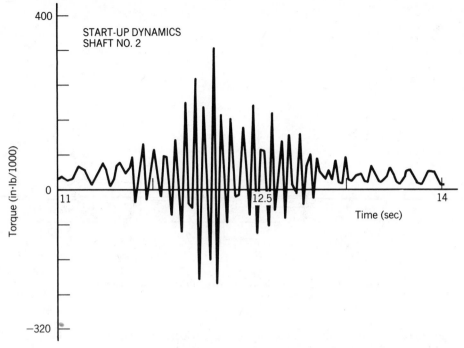

Figure 3.20. Compressor shaft torque during start-up.

a slight clattering of tooth separation for a brief instant during the start-up. It should be pointed out that the drivetrain was designed with a great amount of attention to this problem, and a long life is predicted. Since the resonance is only a transient condition, the life of the machine can be expressed in terms of numbers of start-ups, using a cumulative fatigue analysis [17].

SPEED CONTROL STABILITY

Some of the potential problems associated with automatic speed governors were described in Chapter II. The unstable oscillations of speed that sometimes occur are seen as torsional vibration and usually have a frequency close to one of the natural frequencies of torsional vibration.

To analyze torsional stability with automatic speed control, the dynamics of the governor must be added to the set of differential equations (3-9) through (3-11). The coupling of governor dynamics with drivetrain dynamics occurs at the driving torque [for example, T_1 in equation (3-9)] since it is this quantity which is ultimately controlled by the governor.

Figure 3.21 illustrates a simple mechanical governor that uses the feedback signal from a tachometer (engine speed ω_e) to control fuel flow. Modern turbine

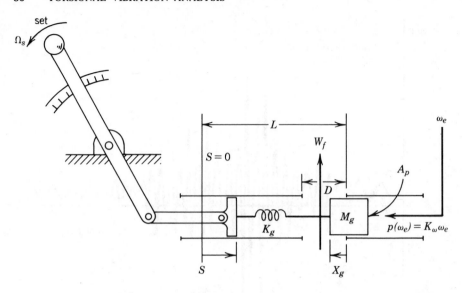

Figure 3.21. Automatic speed governor (fuel flow control).

governors are electronic or electrohydraulic devices, but they all have dynamics which can be modeled as internal inertia (M_g), stiffness (K_g), and damping (C_g). In control theory, these parameters are usually grouped and expressed as "time constants," to be defined below.

In Fig. 3.21 the equilibrium value X_g^*, and hence the steady-state value of fuel flow W_f^* is determined by the set speed Ω_s and the "pressure" p generated by engine speed ω_e. It is clear that oscillations of X_g will produce oscillations of fuel flow and engine speed. The question to be answered by stability analysis is whether these oscillations will die out rapidly, as desired. In the worst case they will grow larger with time.

With proper sizing of the pressure coefficient K_ω and the piston area A_p, the dynamics of the governor in Fig. 3.21 can be described by the differential equation.

$$M_g\ddot{X}_g + C_g\dot{X}_g + K_gX_g = K_gh(\omega_e - \Omega_s), \qquad (3\text{-}61)$$

where $h = S/\Omega_s$ is determined by the design of the mechanism to adjust the set speed. The fuel flow rate is determined by

$$W_f = Q(D - X_g), \qquad (3\text{-}62)$$

where Q in general will be a function of operating conditions. The linearity of equation (3-62) will hold only for small changes of X_g and W_f. Taking two time derivatives of (3-62) and substituting the result into (3-61) yields

$$M_g \ddot{W}_f + C_g \dot{W}_f + K_g W_f = K_g Q h (\Omega_s - \omega_e),   \tag{3-63}$$

Equation (3-63) is the differential equation for fuel flow dynamics. The fuel flow changes are driven by the speed error ($\Omega_s - \omega_e$) on the right-hand side of (3-63). It is not practical to have the speed error, or "droop," equal to zero, but it can be made small by making the sensitivity Q large. Dividing by K_g and factoring the coefficients allows equation (3-63) to be written as [18]

$$\tau_1 \tau_2 \ddot{W}_f + (\tau_1 + \tau_2) \dot{W}_f + W_f = K_p (\Omega_s - \omega_e),   \tag{3-64}$$

where $K_p = Qh$ is called the gain; τ_1 and τ_2 are called the time constants of the governor. The steady-state value of the fuel flow rate is

$$W_f^* = K_p (\Omega_s - \omega_e^*),   \tag{3-65}$$

where ω_e^* is the steady engine speed, hopefully close to Ω_s.

For readers familiar with classical control theory, equation (3-64) can be Laplace transformed to produce the transfer function of fuel flow rates W_f to speed error ϵ

$$\frac{\Delta W_f}{\Delta \epsilon} = \frac{K_p}{(\tau_1 s + 1)(\tau_2 s + 1)}.   \tag{3-66}$$

This transfer function describes the dynamics of a "direct fuel flow govenor." It, and transfer functions for several other types of governors, can be found in Ref. [19].

The most direct approach for stability analysis is to simply add equation (3-64) to the set (3-9) through (3-11), substituting $\dot{\theta}_1$ for ω_e [since inertia I_{P_1} is assumed to be the driver in (3-9)].

EXAMPLE 4

Figure 3.22 shows feedback speed control applied to the two-inertia drive train model of Example 1 (Fig. 3.5). It should be noted that helicopters designed in the United States usually have a two-spool gas turbine engine, whereas the model shown here would be applicable to a drive train with a single-spool engine (sometimes used in European helicopters). Recall that the inertia I_{P_1} represents the power turbine and gears; I_{P_2} represents the rotary wing.

The closed-loop system is described by three ordinary differential equations:

$$I_{P_1} \ddot{\epsilon}_1 + (B_1 + C_1) \dot{\epsilon}_1 - C_1 \dot{\epsilon}_2 + K_1 (\epsilon_1 - \epsilon_2) - K_f \epsilon_3 = 0,$$

$$I_{P_2} \ddot{\epsilon}_2 + (\bar{B}_2 + C_1) \dot{\epsilon}_2 - C_1 \dot{\epsilon}_1 + K_1 \epsilon_2 - K_1 \epsilon_1 = 0,   \tag{3-67}$$

$$\tau_1 \tau_2 \dddot{\epsilon}_3 + (\tau_1 + \tau_2) \ddot{\epsilon}_3 + \epsilon_3 + K_p \dot{\epsilon}_1 = 0,$$

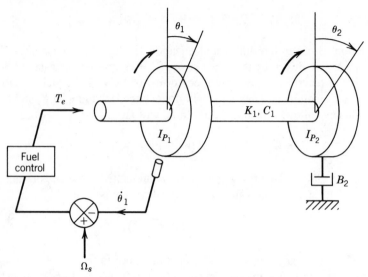

Figure 3.22. Two-inertia drivetrain model with feedback speed control.

where ϵ_1, ϵ_2, and ϵ_3 are small perturbations of θ_1, θ_2, and fuel flow W_f, respectively, away from their steady-state (constant speed) values.

Taking drivetrain values from Example 1 and speed governor values for the direct fuel flow governor of Ref. [19], we obtain the following results:

$$I_{P_1} = 8.3 \text{ in-lb-sec}^2, \quad I_{P_2} = 46.7 \text{ in-lb-sec}^2$$
$$K_1 = 9600 \text{ in-lb/rad}, \quad C_1 = 9.1 \text{ in-lb-sec}$$
$$B_1 = 0, \quad \bar{B}_2 = 33.2 \text{ in-lb-sec (damping of the load only)}$$
$$\tau_1 = 0.02 \text{ sec}, \quad \tau_2 = 0.03 \text{ sec}.$$
$$K_f = 24{,}624 \text{ in-lb/(lb-sec)}, \quad K_p = 0.0112 \text{ lb/(sec-rad-sec)}$$

An eigenvalue analysis of equations (3-67) will determine torsional stability. Substitution of the solution $\epsilon_n(t) = \bar{a}_n e^{st}$ yields (3-68):

$$
\begin{bmatrix}
[I_{P_1}s^2 + (B_1 + C_1)s + K_1] & [-C_1 s - K_1] & [-K_f] \\
[-C_1 s - K_1] & [I_{P_2}s^2 + (\bar{B}_2 + C_1)s + K_1] & [0] \\
[K_p s] & [0] & [\tau_1 \tau_2 s^2 + (\tau_1 + \tau_2)s + 1]
\end{bmatrix}
$$
$$
\cdot \begin{Bmatrix} \bar{a}_1 \\ \bar{a}_2 \\ \bar{a}_3 \end{Bmatrix} e^{st} = \begin{Bmatrix} 0 \\ 0 \\ 0 \end{Bmatrix}. \qquad (3\text{-}68)
$$

For nonzero amplitudes $\bar{a}_n$, Cramer's rule requires the determinant of the 3 × 3 D matrix in (3-68) to be zero. Expansion of the determinant yields the characteristic polynomial, the roots of which are the eigenvalues.

Expansion of the determinant can be programmed on a digital computer to yield the coefficients of the characteristic polynomial. In the program, each element of the matrix is represented by an array of three polynomial coefficients, for example: d_{12} and d_{21} are represented by the numerical values of $-K_1$, $-C_1$, 0. A BASIC (HP 3.0) program listing to compute the eigenvalues is presented here.

PROGRAM "TORSTAB2"

```
10      ! Program "TORSTAB2"--computes eigenvalues of a 2-inertia drive train
20      ! with "direct fuel flow" feedback speed governor
30      OPTION BASE 0
40      DIM A(4),B(2),C(6),D22(2),D33(2),D(6),Icoef(6),Rroot(6),Iroot(6)
50      LOADSUB ALL FROM "Siljak"
60      READ Ip1,Ip2,K1
70      DATA  8.3,46.7,9600.0
80      READ B1,B2eff,C1
90      DATA 0.0,33.2,26.
100     READ T1,T2,Kf,Kp
110     DATA .02,.03,24624.,.0112
120     PRINT "IP1=",Ip1," IP2=",Ip2," K1=",K1
130     PRINT
140     PRINT "B1=",B1," B2eff=",B2eff," C1=",C1
150     PRINT
160     PRINT "T1=",T1," T2=",T2," Kf=",Kf," Kp=",Kp
161     !  T1 and T2 are the time constants of the speed governor
170     PRINT
180     PRINT
190     ! THE D MATRIX ELEMENTS ARE DEFINED IN TERMS OF POLYNOMIAL COEFFICIENTS
200     A(0)=K1
210     A(1)=B1+C1
220     A(2)=Ip1
230     B(0)=K1
240     B(1)=B2eff+C1
250     B(2)=Ip2
255     FOR I=0 TO 2
256     D22(I)=B(I)        ! SAVE D22
257     NEXT I
260     CALL Polymul(2,2,A(*),B(*),C(*))
270     FOR I=0 TO 4
280     A(I)=C(I)          ! D11*D22
290     NEXT I
330     B(0)=1
340     B(1)=T1+T2
350     B(2)=T1*T2
355     FOR I=0 TO 2
356     D33(I)=B(I)           ! SAVE D33
357     NEXT I
360     MAT C= (0)
370     CALL Polymul(4,2,A(*),B(*),C(*))
390     MAT D= C            !    D11*D22*D33
440     MAT B= (0)
450     B(1)=Kp             ! D31
460     MAT C= (0)
470     CALL Polymul(1,2,B(*),D22(*),C(*))
480     MAT C= C*(Kf)    !!  -D31*D22*D13
490     MAT D= D+C          !  D11*D22*D33-D31*D22*D13
500     MAT B= (0)
```

```
510    B(0)=-K1
520    B(1)=-C1
530    MAT C= (0)
540    CALL Polymul(1,1,B(*),B(*),C(*))
550    !   C(*) IS NOW D21*D12
560    MAT A= (0)
570    FOR I=0 TO 2
580    A(I)=C(I)              !              D21*D12
590    NEXT I
600    MAT C= (0)
610    CALL Polymul(2,2,A(*),D33(*),C(*))
620    MAT D= D-C         !   D(*) are the polynomial coefficients
621    PRINT
622    PRINT "THE POLYNOMIAL COEFFICIENTS ARE:",D(*)
623    PRINT
630    MAT Icoef= (0)
640    CALL Siljak(6,D(*),Icoef(*),1.E-8,1.E-6,20,Rroot(*),Iroot(*))
650    PRINT
651    PRINT "THE EIGENVALUES ARE:"
660    PRINT "           REAL                         IMAG          "
670    FOR I=1 TO 6
680    PRINT "           ",Rroot(I),"           ",Iroot(I)
690    NEXT I
700    PRINT
710    PRINT
720    PRINT
730    DELSUB Siljak TO END
740    END
750    SUB Polymul(Na,Nb,A(*),B(*),C(*))
760    FOR I=0 TO Na
770    FOR J=0 TO Nb
780    C(I+J)=A(I)*B(J)+C(I+J)
790    NEXT J
800    NEXT I
810    SUBEND
```

Subprogram "Polymul" multiplies any two polynomials and so is used to expand the characteristic polynomial. Subprogram "Siljak" (not listed) finds the complex roots that are the eigenvalues. The program output, for three different values of shaft (mast) damping C_1, is presented here.

```
THE INPUT DATA IS:
IP1=        8.3         IP2=      46.7        Ki=       9600

B1=         0           B2eff=    33.2        C1=       9.1

T1=         .02         T2=       .03         Kf=       24624      Kp=      .0112


THE POLYNOMIAL COEFFICIENTS ARE:              0       2.96629248E+6
555903.98624        40261.73496        743.394272        19.846136
.232566


THE EIGENVALUES ARE:
          REAL                        IMAG
          -.116910820433              -44.2063616356
          -.116910820433              44.2063616356
          -68.6893040453              5.87511179833E-16
          -8.20618738696              -5.26084827452
          -8.20618738696              5.26084827452
          1.8231900806E-23            -1.27138629502E-23
```

```
THE INPUT DATA IS:
IP1=        8.3        IP2=      46.7        K1=        9600

B1=         0          B2eff=    33.2        C1=        4.55

T1=         .02        T2=       .03         Kf=        24624        Kp=        .0112


THE POLYNOMIAL COEFFICIENTS ARE:              0           2.96629248E+6
  554498.0872         40003.93196         730.791136             19.695986
  .232566


THE EIGENVALUES ARE:
        REAL                          IMAG
          .187501180983              -44.1656915548
          .187501180983               44.1656915548
         -68.6234369071                9.64987904443E-16
         -8.2207213685               -5.26338015415
         -8.2207213685                5.26338015415
          1.69378155117E-23          -1.16508066161E-23


THE INPUT DATA IS:
IP1=        8.3        IP2=      46.7        K1=        9600

B1=         0          B2eff=    33.2        C1=        26

T1=         .02        T2=       .03         Kf=        24624        Kp=        .0112


THE POLYNOMIAL COEFFICIENTS ARE:              0           2.96629248E+6
  561125.89696        41219.28896         790.20592              20.403836
  .232566


THE EIGENVALUES ARE:
        REAL                          IMAG
         -1.24433291733             -44.3395528783
         -1.24433291733              44.3395528783
         -68.940151292               5.97169391903E-16
         -8.15235614004             -5.25072146907
         -8.15235614004              5.25072146907
          2.38961492974E-23         -1.75420450998E-23
```

Note that the 36.9 rad/sec natural frequency of the drivetrain has been raised
to 44.2 rad/sec by the addition of speed control. Also, there is now a second
natural frequency at 5.26 rad/sec, but it is highly damped.

The most important result is that the real part of the 44.2 rad/sec eigenvalues
is positive when $C_1 = 4.55$ (half its nominal value of 9.1), indicating the torsional
oscillations will grow with time (unstable). Increasing C_1 stabilizes the system,
showing that shaft damping is a very important parameter in this system.

It is remarkable that the closed-loop system can be unstable even with a significant amount of damping in the drivetrain and an overdamped governor.

ANALYSIS OF LARGE-ORDER SYSTEMS

When the drivetrain model is large (many stations), the following formulations are
useful for efficient computation of eigenvalues or response using a digital computer.

The Characteristic Matrix

By defining N new coordinates v_n to represent the oscillating angular velocities $\dot{\theta}_n$, the N linearized and homogeneous differential equations can be expanded to $2N$ first-order equations as described earlier for transient response analysis. With the equations linearized, the result can be expressed as (3-69):

$$\begin{Bmatrix} \dot{\theta}_n \\ \dot{v}_n \end{Bmatrix} = [A] \begin{Bmatrix} \theta_n \\ v_n \end{Bmatrix}, \qquad n = 1, 2, \cdots, N, \qquad (3\text{-}69)$$

where A is known as the characteristic matrix of order $2N$.

Alternatively, the canonical variables consisting of the angular displacements θ_n and the momenta p_n [20] can be used as coordinates to obtain the similar form (3-70):

$$\begin{Bmatrix} \dot{\theta}_n \\ \dot{p}_n \end{Bmatrix} = [A'] \begin{Bmatrix} \theta_n \\ p_n \end{Bmatrix}, \qquad n = 1, 2, \cdots, N, \qquad (3\text{-}70)$$

where A' is also a characteristic matrix of order $2N$.

Either equation (3-69) or (3-70) can be written as

$$\dot{\mathbf{x}} = A\mathbf{x}, \qquad (3\text{-}71)$$

where the prime has been dropped from A if equation (3-70) is denoted, and $\mathbf{x}$ represents the vector of $2N$ coordinates in equation (3-69) or (3-70).

Substitution of the solution (3-23) yields

$$se^{st}\mathbf{a} = A\mathbf{a}e^{st} \qquad (3\text{-}72)$$

where the first N elements of $\mathbf{a}$ are the amplitudes of the θ_n, and the second N elements are the amplitudes of either the velocities v_n or the momenta p_n.

After we divide out e^{st}, equation (3-72) can be rewritten as

$$(A - s\mathcal{I})\mathbf{a} = 0, \qquad (3\text{-}73)$$

where $\mathcal{I}$ is the identity matrix. Equation (3-73) represents a system of $2N$ linear algebraic equations.

Since $\mathbf{a} = 0$ represents no oscillation, the only meaningful satisfaction of equation (3-73) is to require [21]

$$|A - s\mathcal{I}| = 0. \qquad (3\text{-}74)$$

The complex numerical values of s which make the determinant in (3-74) zero are the eigenvalues (3-25). Iteration schemes have been devised to improve trial values of s until the determinant is close enough to zero to satisfy the accuracy requirements.

If the elements of the determinant are treated as polynomials in s (not substituting a numerical complex value for s), then expansion of the determinant yields the characteristic polynomial of order $2N$ in s. When N is large, the coefficients of the characteristic polynomial can be evaluated from the determinant with a digital computer by using a polynomial multiplier subprogram to expand the determinant. The eigenvalues are the roots of this polynomial and can be computed by various rootfinder algorithms [11,22].

The characteristic polynomial can also be generated by multiplication of the transfer matrices. Some mathematical procedures that have been found useful in generating the polynomial and finding its roots are described below.

The Transfer Matrix: Holzer's Method

Holzer's method was the first algorithm devised to compute the natural frequencies of large-order torsional systems. It was developed before the age of high speed electronic computers, so it is especially efficient in terms of required digital storage (memory) and computation time. This makes it well suited for the contemporary personal computer or minicomputer.

Early users of Holzer's method did not formulate it in terms of matrices, as it can be implemented quite nicely with scalar operations. Also, damping was usually omitted from the model to avoid dealing with complex numbers.

Here, the method will be presented in matrix form, with damping included, as the simplest possible illustration of the general transfer matrix method that has been used so successfully for rotordynamic eigenvalue analysis. For the more complex case of rotor whirling, implementation with purely scalar operations becomes impractical, so the matrix formulation is necessary. With appropriate modifications, the method can also be used for forced response analysis.

The basic idea of a transfer matrix is to express the boundary values (e.g., torque, angular displacement) at one end of a station in terms of the boundary values at the other end. Multiplying the matrices of all the stations together, we can express the boundary conditions at the right end of the rotor in terms of the left-end boundary conditions. The elements of the transfer matrices contain the eigenvalue, so if it is the correct value the true boundary conditions at the right end will be obtained. Trial and error iteration schemes are used to converge on the correct value.

For torsional vibration, the boundary condition for torque at each end of the rotor is usually zero; even if not it can be made zero by putting any external torque acting on the end inertia ''inside'' the outer face of the station. For eigenvalue analysis, the boundary condition for angular displacement is arbitrary since the eigenvalues are independent of vibratory amplitude. However, the *ratio* of amplitudes at each end is determined by the mode shape (eigenvector) associated with each eigenvalue.

Figure 3.23 shows the nth station of the model illustrated in Fig. 3.1, broken into two parts: the inertia element and the stiffness element. On the inertia element the left-end angular displacement and torque is unprimed; the right-end values are primed. Using the notation of the figure, the transfer equation for the torque on the nth inertia is determined by Newton's Second Law or by Lagrange's equation as follows:

$$T'_n = I_{P_n}\ddot{\theta}_n + B_n\dot{\theta}_n + T_n \tag{3-75}$$

Substituting the solution (3-23) and adding the identity $\theta'_n = \theta_n$, we obtain transfer equations for the nth inertia in terms of the eigenvalue s:

$$\begin{Bmatrix} \theta'_n \\ T'_n \end{Bmatrix} = \begin{bmatrix} 1 & 0 \\ [I_{P_n}s^2 + B_n s] & 1 \end{bmatrix} \begin{Bmatrix} \theta_n \\ T_n \end{Bmatrix}. \tag{3-76}$$

The 2×2 inertia transfer matrix defined by (3-76) will be denoted as $[T_I]_n$.

Since the torsional stiffness element is massless, its end torques are equal. That is,

$$T_{n+1} = T'_n. \tag{3-77}$$

The shaft torque is resisted by the elastic torsional stiffness and internal damping, so

$$T'_n = (K_n + C_n s)(\theta_{n+1} - \theta_n). \tag{3-78}$$

Solving (3-78) for θ_{n+1} allows this equation and (3-77) to be written as

$$\begin{Bmatrix} \theta_{n+1} \\ T_{n+1} \end{Bmatrix} = \begin{bmatrix} 1 & \dfrac{1}{K_n + C_n s} \\ 0 & 1 \end{bmatrix} \begin{Bmatrix} \theta'_n \\ T'_n \end{Bmatrix}. \tag{3-79}$$

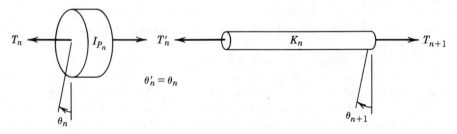

Figure 3.23. Torsional inertia and stiffness elements of the nth station.

Equation (3-79) defines the shaft transfer matrix for the nth shaft to be denoted as $[T_s]_n$.

Substitution of (3-76) into (3-79) gives

$$\begin{Bmatrix} \theta_{n+1} \\ T_{n+1} \end{Bmatrix} = [T_{sI}]_n \begin{Bmatrix} \theta_n \\ T_n \end{Bmatrix},$$ (3-80)

where

$$[T_{sI}]_n = [T_s]_n [T_I]_n$$ (3-81)

is evaluated by the rules of matrix multiplication. Here $[T_{sI}]_n$ is the transfer matrix of the nth station.

Now let $n = N$ in equation (3-76) (note that the Nth station has no shaft element), let $n + 1 = N$ in equation (3-80), so $n = N - 1$, etc. Successive substitutions are made until the left end of the drivetrain is reached where $n = 1$. The result is (3-82):

$$\begin{Bmatrix} \theta'_N \\ T'_N \end{Bmatrix} = [T_0] \begin{Bmatrix} \theta_1 \\ T_1 \end{Bmatrix},$$ (3-82)

where

$$[T_0] = [T_I]_N [T_{sI}]_{N-1} \cdots [T_{sI}]_1$$ (3-83)

is the overall transfer matrix for the drivetrain. For torsional vibration it is of order 2, independently of how many stations are incorporated in the model.

In Holzer's original scheme damping was not included, so $B_n = C_n = 0$ and $s^2 = -\omega_n^2$ was always a real number. The algorithm started with a trial value for the natural frequency ω_n, $T_1 = 0$, and $\theta_1 = 1$. The multiplications (3-83) in (3-82) gave a value for T'_N that would be zero only if ω_n was correct. A plot of T'_N versus incremented values of ω_n identified the natural frequencies where the graph crossed the zero axis ($T'_N = 0$).

With damping included, s is a complex number, so the elements of the transfer matrices are complex numbers in the typical digital computer algorithm. When s is an eigenvalue of the system, both the real and imaginary parts of T'_N will compute to be zero. Iteration schemes have been developed to converge on the eigenvalue, but the author's experience is that they are generally unreliable when s is complex. Consequently, it has been found better to treat the elements of the transfer matrices as polynomials in s, storing the coefficients as arrays in the computer. For example, element 2,1 in the matrix of (3-76) is a second-order polynomial and can be represented by the array of numerical values 0, B_n, I_n. A BASIC subroutine which multiplies any two polynomials to give a resultant polynomial was given in Example 4. The polynomials are of the form $a_0 + a_1 s + a_2 s^2 \cdots$.

When the polynomial multiplications in (3-83) are carried out to give (3-82), the 2,1 element of $[T_0]$ is the characteristic polynomial of the system. That is $T_N' = T_{0_{2,1}}(s) = 0$ since $\theta_1 = 1$ and $T_1 = 0$. The complex roots of this polynomial are the eigenvalues.

The associated eigenvector (mode shape) is obtained by computing successive values of θ_N' from (3-82) and (3-83), substituting 1, 2, 3, $\cdots$ for N, and with s equal to the previously computed eigenvalue. Note that the θ_n will be complex numbers with meaning defined by equations (3-23) and (3-26).

The Modal Method

The modal method is useful to minimize the number of coordinates. For example, a detailed definition of mode shapes may be desired for more accurate calculation of forces dependent on rotor motion without a large increase in the number of coordinates.

In this method the "generalized modal coordinates" may be thought of as the amplitudes of the individual modes associated with the natural frequencies.

The equations of motion (3-9) through (3-11) for the entire system may be expressed in matrix form as

$$[I]\{\ddot{\theta}\} + [C]\{\dot{\theta}\} + [K]\{\theta\} = \{T(t)\}, \qquad (3\text{-}84)$$

where

$[I]$ = diagonal matrix of polar moment of inertia

$[C]$ = triple band diagonal damping matrix (including the B coefficients)

$[K]$ = triple band diagonal stiffness matrix

$\{T(t)\}$ = external torque vector

The global equations of motion, equation (3-84), can be reduced to a smaller set by assuming the angular displacement vector for n inertias is represented as a linear combination of the lowest r undamped normal modes [23] as follows:

$$\theta(x, t) = \sum_{j=1}^{r} \phi_j(x) \, q_j(t), \qquad (3\text{-}85)$$

or, in matrix form,

$$\{\theta\} = [P]\{q\},$$

where

$\phi_j(x)$ = the jth undamped normal mode (x is the axial space coordinate here)

$\{\theta\}$ = vector of angular displacement

$\{q\}$ = vector of the generalized modal coordinates

$[P] = [\{\phi_1(x)\}, \{\phi_2(x)\}, \cdots, \{\phi_r(x)\}]$, modal matrix of the eigenvectors

The $n \times r$ modal matrix $[P]$ is composed of the r undamped normal modes (eigenvectors), one in each column. The undamped normal modes and natural frequencies can be obtained from the Holzer method as described above, where exclusion of damping greatly simplifies the algorithm.

As a general rule, the undamped modes associated with all the natural frequencies below 4 times the highest excitation frequency should be used to get good accuracy.

Substituting equation (3-85) into equation (3-84) and premultiplying it by $[P]^T$, the modal equations of motion are given by

$$[DI]\{\ddot{q}\} + [CC]\{\dot{q}\} + [DK]\{q\} = \{\overline{T}\} \tag{3-86}$$

where

$[DI] = [P]^T [I] [P]$, diagonal modal inertia matrix

$[CC] = [P]^T [C] [P]$, cross coupled modal damping matrix

$[DK] = [P]^T [K] [P]$, diagonal modal stiffness matrix

$\{\overline{T}\} = [P]^T \{F(t)\}$, vector of modal external torque

The orthogonal property of normal modes is employed in the derivation of the modal equations of motion that are expressed in terms of the generalized coordinates q_r of the undamped modes. Equation (3-86) is coupled through the modal cross-coupled damping matrix. It is also important to note that there are only $r\,(<n)$ coupled second-order differential equations of motion to be solved if only r modes are used to express the torsional mode shapes.

Damped frequencies and damped mode shapes are determined by linearizing the system to eliminate external torques, as in Example 1. Equation (3-86) then is written as

$$[DI]\{\ddot{q}\} + [CC]\{\dot{q}\} + [DK]\{q\} = 0. \tag{3-87}$$

In order to reduce the r second-order differential equations to $2r$ first-order differential equations, let

$$\{\dot{q}\} = \{v\} \tag{3-88}$$

Combination of equations (3-87) and (3-88) produces

$$
\begin{Bmatrix} \dot{q} \\ \dot{v} \end{Bmatrix} = \begin{bmatrix} 0 & [\mathcal{I}] \\ -[DI]^{-1}[DK] & -[DI]^{-1}[CC] \end{bmatrix} \begin{Bmatrix} q \\ v \end{Bmatrix} , \quad (3\text{-}89)
$$

where $[\mathcal{I}]$ is the identity matrix. Assume the solution is of the form (3-23); that is,

$$
\{q\} = \{\psi\} e^{st}, \quad (3\text{-}90)
$$

where $\{\psi\}$ and s are in complex form.
   Then,

$$
\begin{aligned} \{\dot{q}\} &= s\{\psi\} e^{st}, \\ \{\dot{v}\} &= s\{\varphi\} e^{st}, \end{aligned} \quad (3\text{-}91)
$$

where $\{\varphi\} = s\{\psi\}$.
   Therefore, the modal damped equations of motion are transformed to

$$
s \begin{Bmatrix} \psi \\ \varphi \end{Bmatrix} = \begin{bmatrix} 0 & [\mathcal{I}] \\ -[DI]^{-1}[DK] & -[DI]^{-1}[CC] \end{bmatrix} \begin{Bmatrix} \psi \\ \varphi \end{Bmatrix} . \quad (3\text{-}92)
$$

Equation (3-92) is an eigenproblem that can be solved by standard eigensolver routines available in computer numerical libraries. Finally, the vector of the generalized modal coordinates, $\{q\}$, is transformed back to the physical coordinates, $\{\theta\}$, by the use of equation (3-85).

   An assumption that the damping matrix is proportional to the mass or stiffness matrix is often employed to facilitate the solution of equation (3-87). That is, all modal cross-coupled damping terms are neglected. The assumption of proportional damping in the analysis serves to uncouple the modal equations of motion. However, this simplification is valid only if the damping is extremely light, only 1 or 2 percent of critical damping. The assumption cannot generate accurate results in the analysis of rotating machinery with significant damping.

   Furthermore, even if modal cross-coupled damping terms are retained in the modal analysis, considerable errors in the first torsional damped eigenvalue will occur if the torsional system has a large ground damping (B terms). This situation is corrected by introducing the rigid body modes into the modal matrix. That is, the angular displacement of all masses is assumed to be a sum of the angular displacement of the rigid body modes plus the oscillatory modes as follows:

$$
\begin{aligned} \{\theta\} &= \{\theta_r\} + \{\theta_o\} \\ &= [P_r]\{q_r\} + [P_o]\{q_o\}, \end{aligned} \quad (3\text{-}93)
$$

where

$[P_r]$ = modal matrix consisted of rigid body modes

$[P_o]$ = modal matrix consisted of oscillatory modes

We substitute equation (3-93) into equation (3-84) and premultiply by $[P_r]^T$ for the damped eigenvalues analysis, so that the equations of motion become

$$[P_r]^T [I] [P_r] \{\ddot{q}_r\} + [P_r]^T [C] [P_r] \{\dot{q}_r\} \tag{3-94}$$
$$+ [P_r]^T [K] [P_r] \{q_r\} = 0,$$

where the orthogonal property of normal modes has been used. The last term in equation (3-94) is zero, because the rigid body modes cannot produce any torque in the system. Also, recognizing that the torsional system has only a rotational rigid body mode, equation (3-94) is reduced to

$$\sum_{i=1}^{n} I_i \ddot{q}_r + \sum_{i=1}^{n} B_i \dot{q}_r = 0, \tag{3-95}$$

where

$$q_r = \theta_i, \qquad i = 1, 2, \cdots, n.$$

In applying numerical integration to determine the transient response, the Newmark method (or the average acceleration method) is recommended, since it is unconditionally stable.

The modal equations of motion, equation (3-86), at time t_{i+1} are

$$[DI] \{\ddot{q}_{i+1}\} + [CC] \{\dot{q}_{i+1}\} + [DK] \{q_{i+1}\} = \{\overline{T}_{i+1}\}. \tag{3-96}$$

The acceleration in the time interval t_i to t_{i+1} is taken to be the average of the initial and final values of acceleration, i.e.,

$$\{\ddot{q}\} = \tfrac{1}{2}\{\{\ddot{q}_i\} + \{\ddot{q}_{i+1}\}\}. \tag{3-97}$$

Integration of equation (3-97) twice gives

$$\{\dot{q}_{i+1}\} = \{\dot{q}_i\} + \frac{\tau_i}{2}\{\{\ddot{q}_i\} + \{\ddot{q}_{i+1}\}\}, \tag{3-98}$$

$$\{q_{i+1}\} = \{q_i\} + \tau_i\{\dot{q}_i\} + \frac{\tau_i^2}{4}\{\{\ddot{q}_i\} + \{\ddot{q}_{i+1}\}\}. \tag{3-99}$$

where $\tau_i = t_{i+1} - t_i$ is the time step.

Substituting the recursion equations (3-98) and (3-99) into (3-96) and solving for $\{\ddot{q}_{i+1}\}$, we obtain the following relation:

$$\{\ddot{q}_{i+1}\} = [D]^{-1} \{\{\overline{T}_{i+1}\} - [CC]\{G\} - [DK]\{H\}\}, \quad (3\text{-}100)$$

where

$$[D] = [DI] + (\tau_i/2)[CC] + (\tau_i^2/4)[DK],$$
$$\{G\} = \{\dot{q}_i\} + (\tau_i/2)\{\ddot{q}_i\},$$
$$\{H\} = \{q_i\} + \tau_i\{\dot{q}_i\} + (\tau_i^2/4)\{\ddot{q}_i\}.$$

Equations (3-98), (3-99), (3-100) define the values of velocity, displacement, and acceleration at time t_{i+1} in the generalized coordinates. The modal torque $\{\overline{T}_{i+1}\}$ can be calculated from the motor start-up torque–speed data.

To start computation, the initial acceleration $\{\ddot{q}_o\}$ is obtained from

$$\{\ddot{q}_o\} = [DI]^{-1} \{\{T_o\} - [CC]\{\dot{q}_o\} - [DK]\{q_o\}\}. \quad (3\text{-}101)$$

The total angular displacement of the motor $\theta_1(t)$, required in calculating the driving torque, is approximated in finite difference form as by

$$\theta_m(t) \cong \theta_m(t - \tau_i) + \tau_i\dot{\theta}_m(t - \tau_i). \quad (3\text{-}102)$$

The time step size is determined by the pulsating torque frequency and natural frequencies below twice slip frequency. As a general rule, the time step size is typically one-tenth to one-twentieth of the period of the highest pulsating torque or natural frequency in the range between 0 and 120 Hz in a 60-Hz electrical power system.

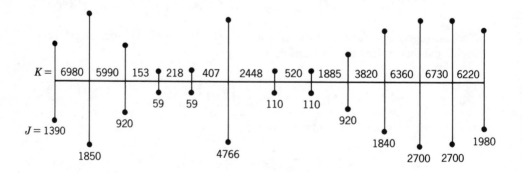

$$J = (\text{in.-lb-sec2})$$
$$K = (\text{in.-lb/rad}) * 10E\text{-}6$$

Figure 3.24. Thirteen-inertia model for the drivetrain of Example 3.

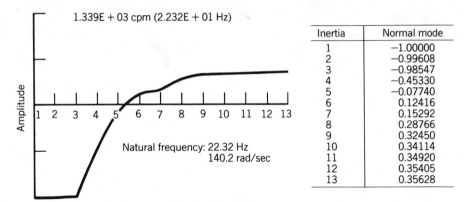

Figure 3.25. The first damped mode of the model in Fig. 3.24.

Figure 3.24 shows a 13-inertia model of the drivetrain analyzed in Example 3. Figures 3.25 and 3.26 show a much better definition of the first two mode shapes, using the 13-inertia model in a modal analysis as described above. (Compare with Fig. 3.15.) Note that the natural frequencies are only slightly different from those computed from the three-inertia model in Example 3.

Figures 3.27 and 3.28 show the results of transient analysis using four modal coordinates including the rigid body mode. These are practically identical to the results obtained in Example 3 using a three-inertia model. Furthermore, they are practically identical to results presented in Ref. [23] from an analysis using all 13 angular displacements as degrees of freedom (i.e., 13 second-order differential equations). Thus it can be seen that the main advantage of this method is a better

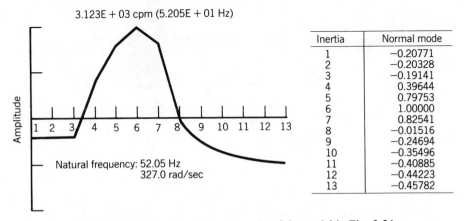

Figure 3.26. The second damped mode of the model in Fig. 3.24.

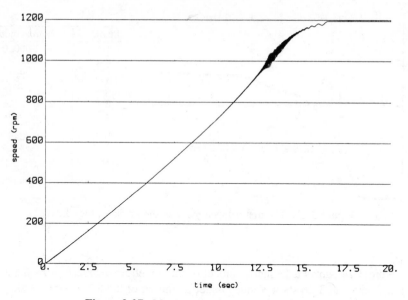

Figure 3.27. Motor start-up by modal analysis.

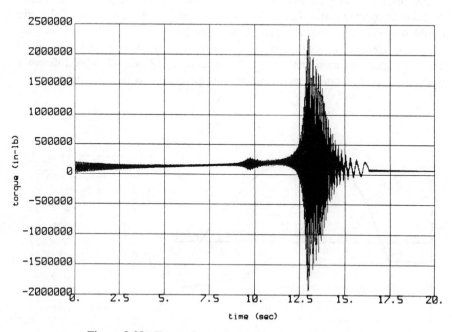

Figure 3.28. Torque in the motor shaft by modal analysis.

definition of the mode shapes, but with only one additional coordinate required for the transient analysis.

BRANCHED SYSTEMS

A new method, developed by Dr. B. T. Murphy,[4] for calculating damped eigenvalues of branched torsional systems is presented here. The method easily handles any branched[5] torsional system regardless of the number of branches or their respective locations within the system. Branches on top of other branches are permitted. It is also easily applied to systems with redundant, or closed-loop, branches. Damping to ground or between adjacent inertias can be included without any additional refinements.

The general system depicted in Fig. 3.29 represents a general open-loop torsional system. Any conceivable combination of branches is permitted. The Holzer method is difficult to apply to such complex multibranched systems and even more difficult to apply to closed-loop (or redundant) gear trains such as that shown in Fig. 3.30.

The characteristic polynomial for any of these torsional systems can be generated by a procedure which starts with the polynomial for a simple system of just one or two inertias. The system is then built up by adding a single inertia/spring combination at a time. Each time an inertia/spring combination is added, the characteristic polynomial of the resulting system is easily calculated from the previous polynomial. The procedure is continued, adding inertia/spring combinations, until the entire system is obtained.

To begin, Fig. 3.31 shows a two-inertia torsional system. The P subscript has been dropped from the inertias for brevity. The two second-order differential equations are (note that K_0 and/or K_2 could be zero)

$$I_1 \ddot{\theta}_1 + K_0 \theta_1 + K_1 \theta_1 - K_1 \theta_2 = 0, \tag{3-103}$$

$$I_2 \ddot{\theta}_2 + K_1 \theta_2 - K_1 \theta_1 + K_2 \theta_2 = 0. \tag{3-104}$$

The solution to these equations, from (3-23), is

$$\theta_1 = \bar{\theta}_1 \, e^{st}, \qquad \theta_2 = \bar{\theta}_2 \, e^{st}, \tag{3-105}$$

where s is the complex eigenvalue and the $\bar{\theta}_i$ define the eigenvectors (i.e., mode shapes). Substituting (3-105) into (3-103) and (3-104) results in (3-106):

---

[4] Now employed at Rocketdyne Division of Rockwell Corp., Canoga Park, California.
[5] One shaft driving two or more shafts—e.g., two marine propeller shafts from one gearbox.

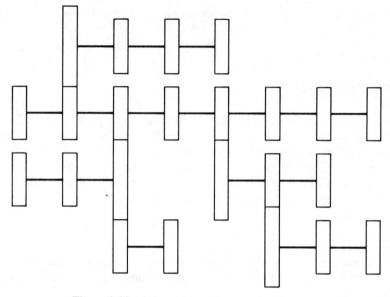

Figure 3.29. A general open-loop torsional system.

$$\begin{bmatrix} (I_1 s^2 + K_0 + K_1) & -K_1 \\ -K_1 & (I_2 s^2 + K_1 + K_2) \end{bmatrix} \begin{Bmatrix} \bar{\theta}_1 \\ \bar{\theta}_2 \end{Bmatrix} = 0. \qquad (3\text{-}106)$$

The eigenvalues s are found by equating the determinant of the square matrix equal to zero to give

$$D_2 = (I_1 s^2 + K_0 + K_1)(I_2 s^2 + K_1 + K_2) - K_1^2 = 0. \qquad (3\text{-}107)$$

Figure 3.32 shows a three-inertia torsional system that is obtained by attaching a single inertia/spring combination to the spring K_2. Note that the spring to which the new inertia is connected, K_2, was going to ground, and the newly added spring,

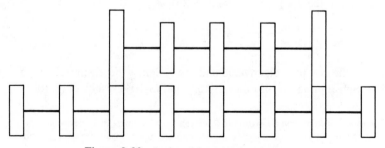

Figure 3.30. A closed-loop torsional system.

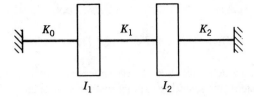

Figure 3.31. A two-inertia torsional system.

K_3, now goes to ground. The three equations for this system in matrix form are given in (3-108):

$$\begin{bmatrix} (I_1 s^2 + K_0 + K_1) & -K_1 & 0 \\ -K_1 & (I_2 s^2 + K_1 + K_2) & -K_2 \\ 0 & -K_2 & (I_3 s^2 + K_2 + K_3) \end{bmatrix} \begin{Bmatrix} \bar{\theta}_1 \\ \bar{\theta}_2 \\ \bar{\theta}_3 \end{Bmatrix} = 0.$$

(3-108)

Expansion of the determinant yields the characteristic polynomial as

$$D_3 = (I_3 s^2 + K_2 + K_3) D_2 - K_2^2 (I_1 s^2 + K_0 + K_1). \qquad (3\text{-}109)$$

Just as before, the eigenvalues are the roots of the polynomial D_3. Now attach another inertia/spring combination onto the spring K_3, with the newly added spring K_4 going to ground. The four differential equations in matrix form are given in (3-110):

$$\begin{bmatrix} (I_1 s^2 + K_0 + K_1) & -K_1 & 0 & 0 \\ -K_1 & (I_2 s^2 + K_1 + K_2) & -K_2 & 0 \\ 0 & -K_2 & (I_3 s^2 + K_2 + K_3) & -K_3 \\ 0 & 0 & -K_3 & (I_4 s^2 + K_3 + K_4) \end{bmatrix}$$

$$\cdot \begin{Bmatrix} \theta_1 \\ \theta_2 \\ \theta_3 \\ \theta_4 \end{Bmatrix} = 0. \qquad (3\text{-}110)$$

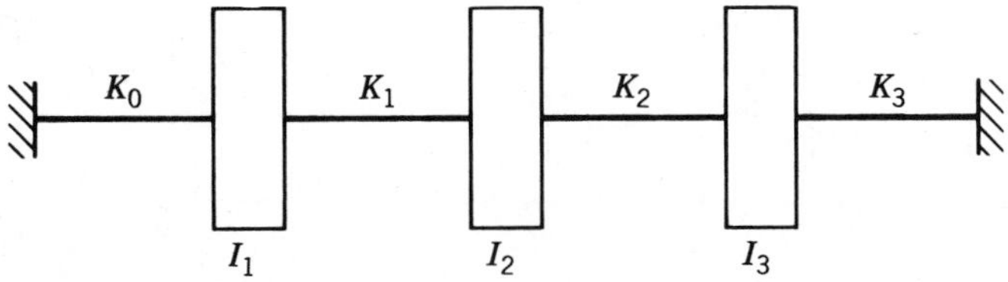

Figure 3.32. A three-inertia torsional system.

Row–column expansion gives the characteristic polynomial for this system as

$$D_4 = (I_4 s^2 + K_3 + K_4) D_3 - K_3^2 D_2. \qquad (3\text{-}111)$$

A pattern can now be established. Every time a new inertia/spring combination is added to the system, the resulting characteristic polynomial can be obtained from the previous one as

$$D_i = (I_i s^2 + K_{i-1} + K_i) D_{i-1} - K_{i-1}^2 D_{i-2}. \qquad (3\text{-}112)$$

The polynomial D_{i-1} is the characteristic polynomial of the previous system (i.e., before adding the new inertia/spring combination). The factor multiplying D_{i-1} is seen to be the diagonal element that appears in the system matrix due to the newly added inertia. The polynomial D_{i-2} is the characteristic polynomial of the previous system minus the inertia/spring combination to which the new one is being attached. The factor multiplying D_{i-2} is the square of the spring rate for the spring onto which the new inertia/spring combination is attached. Thus,

$$D_i = (\text{diag. elem.}) D_{i-1} - (\text{conn. spring})^2 D_{i-2}. \qquad (3\text{-}113)$$

Note that the new inertia/spring combinations must be attached onto previously grounded springs and that the newly added spring is grounded.

A very simple method is thus defined for finding the characteristic polynomial for the single branch system of Fig. 3.33. Using $D_0 = 1$ and $D_{-1} = 0$, start with $i = 1$ and apply equation (3-113) repeatedly to build up the system one inertia/spring at a time. After the last inertia is added (the last spring will most likely be zero), the result is the characteristic polynomial for the entire system. Any polynomial rootfinder algorithm can then be used to find the eigenvalues.

To add a side branch to the system, consider the system of Fig. 3.34. The only difference between the matrix equation for this system with and without the extra inertia $\bar{I}_3$ and stiffness $\bar{K}_3$ is the diagonal element of the third row: $(I_3 s^2 + K_2 + K_3)$ versus $(I_3 s^2 + \bar{I}_3 s^2 + K_2 + K_3 + \bar{K}_3)$. This being the only difference, the method of equation (3-113) can be used to find its characteristic polynomial D_7. It is possible to build up the side branch by adding another inertia/spring onto the

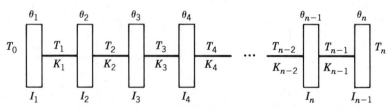

Figure 3.33. A single-branch torsional system.

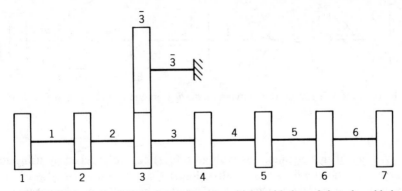

Figure 3.34. A single-branch system onto which a side branch is to be added.

grounded spring $\overline{K}_3$ (see Fig. 3.35). An attempt to apply equation (3-113) would result in

$$D_8 = (I_8 s^2 + \overline{K}_3 + K_8) D_7 - \overline{K}_3^2(?). \qquad (3\text{-}114)$$

The polynomial D_{i-2} was said to be obtained by removing the interconnecting spring $\overline{K}_3$ and its associated inertia (in this case I_3 and $\overline{I}_3$). Performing this removal gives the arrangement shown in Fig. 3.36. The polynomial D_{i-2} is to be the characteristic polynomial of this system. Therefore we use the product of the two separate parts shown here:

$$D_8 = (I_8 s^2 + \overline{K}_3 + K_8) D_7 - \overline{K}_3^2 D_2 D_7^4, \qquad (3\text{-}115)$$

where D_7^4 designates the polynomial for the subsystem of inertias 4 through 7.

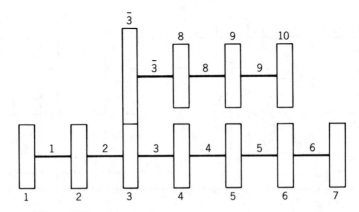

Figure 3.35. A torsional system model with one side branch.

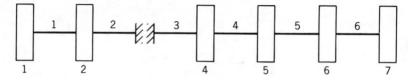

Figure 3.36. The result of removing inertia I_3 from the main branch of Fig. 3.34.

The validity of this equation (as with all others) can be checked by using row–column expansion on the system matrix. Recall that the values of $\bar{I}_3$, $\bar{K}_3$, etc. must be scaled by the gear ratio squared. Also, had there been other side branches already coming off of I_3, then they too would be included in the product (see, for example, Fig. 3.37).

The process is continued by adding the remaining inertia/springs of the side branch (see Fig. 3.35) onto K_8 by once again using equation (3-113).

The method, as described thus far, is sufficient to handle any system that fits the form of Fig. 3.29. Any conceivable combination of branches is permitted as long as there are no closed loops.

Now consider the system shown in Fig. 3.38. The characteristic polynomial for this system can be found since it is open loop. The loop can be closed by adding the last inertia/spring onto K_{12}. The last inertia I_{13} is to be attached to the grounded

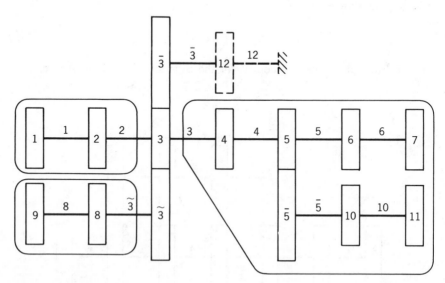

Figure 3.37. When inertia I_{12} is added, D_{i-2} of equation (3-113) is taken as the product of the subsystems that are circled.

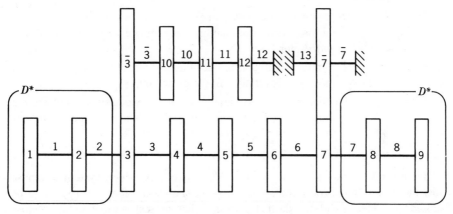

Figure 3.38. An open-loop system ready to be closed.

spring K_{12}, but its associated spring K_{13} is already present being attached to $\bar{I}_7$. The equation to close the loop is

$$D_{13} = (I_{13}s^2 + K_{12} + K_{13}) D_{12} - K_{12}^2 D_{11} - K_{13}^2 D^*$$
$$- 2(K_3 K_4 \cdots K_6 \bar{K}_3 K_{10} \cdots K_{13}) D_2 D_9^8. \qquad (3\text{-}116)$$

The polynomial D^* is the product of the characteristic polynomials of the systems circled in Fig. 3.38 with K_2 and K_7 to ground. Equation (3-116) can be written in a general form that is used whenever a loop is to be closed:

$$D_i = (\text{diag. elem.}) D_{i-1} - K_l^2 D_l - K_r^2 D_r - 2(K_{\text{loop}}) D^*. \qquad (3\text{-}117)$$

The terms of this equation are defined as follows:

diag. elem.: the diagonal element that appears in the system matrix due to the newly added inertia; this will consist of all the inertia being added here and every spring that will be attached to it in the final system

D_{i-1}: the characteristic polynomial of the previous system, or just before adding the new inertia

K_l: the stiffness of the spring on the immediate left of the newly added inertia

D_l: the characteristic polynomial of the previous system minus the inertia/spring combination to the immediate left of the newly added inertia

K_r: the stiffness of the spring on the immediate right of the newly added inertia

D_r: the characteristic polynomial of the previous system minus the
                inertia/spring combination to the immediate right of the newly
                added inertia

K_{loop}:   the product of all spring rates that are part of the loop itself (i.e.,
                those springs that form the closed curve)

$D*$:   the characteristic polynomial of the previous system outside the
                loop (i.e., all the springs of the product K_{loop} and all inertias
                to which they are attached)

Figure 3.39 shows a slightly more complicated closed-loop system. Suppose
that the system has been built up to the point shown and the loop is now to be
closed. The inertia with the dashed outline is the one that is to be added to close
the loop. For application of equation (3-117), the springs that are part of the closed
curve mentioned above are marked with an asterisk (*). The polynomial D_{i-1}
should be self-evident. The other three polynomials are for the systems circled in
the figure. Note that $D*$ will be the product of the separate parts that are circled

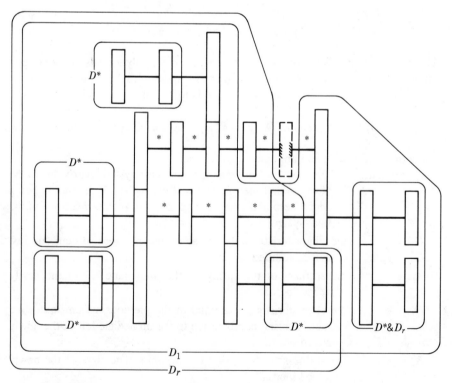

Figure 3.39. A general combination closed- and open-loop torsional system.

and labeled as D^*. Also, D_r is the product of the two parts circled and labeled as D_r.

Calculation of eigenvalues by this method does not automatically yield the mode shapes as with Holzer-based methods. The relative amplitudes must be calculated using Holzer (transfer-matrix) calculations or by deflating the characteristic matrix. For multibranched open-loop systems, iteration is not required because the eigenvalue s is known beforehand. For closed-loop systems, the calculations are slightly more involved. References [24] and [25] should be consulted for more information on performing Holzer calculations for closed-loop systems, keeping in mind the fact that the eigenvalue s will be known.

Viscous torsional damping either to ground or between adjacent inertias is easily accounted for in this method. Should an inertia I_n have damping between itself and ground then the inertia and damper are in series. So replace $I_n s^2$ with $I_n s^2 + B_n s$, where B_n is the viscous damping coefficient to ground. Should there be damping between adjacent inertias, then the damper must be in parallel with some spring, say K_n. Replace all occurrences of K_n with $C_n s + K_n$, where C_n is the viscous shaft damping coefficient.

NUMERICAL COMPUTATION OF CHARACTERISTIC POLYNOMIAL COEFFICIENTS AND ROOTS (EIGENVALUES)

The following discussion, from Ref. [26], should be helpful in programming a digital computer to use any of the methods described earlier involving the characteristic polynomial.

Scaling

The magnitudes of the polynomial coefficients are found to follow a definite trend. The coefficient of the term s^i tends to get smaller as i gets larger. The coefficient of the term s^0 will be of roughly the same order of magnitude as the product of all the spring rates in the system. The coefficient of the last term (i.e., the largest power of s) will be roughly of the same order of magnitude as the product of all the inertias in the system. For systems with large numbers of inertias the coefficients of the polynomial may become prohibitively large (or small) for the computer being used, resulting in numerical overflow and underflow. An easy way to combat this problem is to scale the polynomial by making the following substitution:

$$s = f\bar{s}. \tag{3-118}$$

The scale factor is f, an arbitrary constant (preferably real), and the polynomial originally in powers of s will now be in powers of $\bar{s}$. Each coefficient of the original polynomial is now multiplied by f raised to the same power as $\bar{s}$. The unscaled polynomial

$$P = a_0 + a_1 s^1 + a_2 s^2 + a_3 s^3 + \cdots + a_n s^n \qquad (3\text{-}119)$$

becomes

$$P = a_0 + a_1 f \bar{s}^1 + a_2 f^2 \bar{s}^2 + a_3 f^3 \bar{s}^3 + \cdots + a_n f^n \bar{s}^n, \qquad (3\text{-}120)$$

or

$$P = \bar{a}_0 + \bar{a}_1 \bar{s}^1 + \bar{a}_2 \bar{s}^2 + \bar{a}_3 \bar{s}^3 + \cdots + \bar{a}_n \bar{s}^n. \qquad (3\text{-}121)$$

Scaling in this manner, with f chosen greater than 1, will increase the magnitude of the coefficients for the higher order terms. Thus f is chosen to make all the coefficients as close as possible to the same order of magnitude as the coefficient a_0, which is not changed by the scaling. Then divide through the entire polynomial by the coefficient a_0. The resulting polynomial will now have all its coefficients as close to 1 as possible. After solving for the scaled roots $\bar{s}$, use equation (3-118) to obtain the desired unscaled eigenvalues s.

There is an interesting consequence of scaling in this manner. Consider the case of an undamped rotor. In this case the unscaled eigenvalues s have magnitudes which are equal to the natural frequencies in radians per second. Suppose that the only natural frequencies of interest are those that are less than 10,000 rad/sec (or roughly 1500 Hz, or 100,000 rpm). If we then set the scale factor f equal to 10,000, all the scaled eigenvalues $\bar{s}$ that we are interested in will have magnitudes less than 1 by virtue of equation (3-118). So when finding the roots of the scaled characteristic polynomial, equation (3-121), we need only be concerned with roots which have magnitudes less than 1.

Now focus attention on the coefficients $\bar{a}_i$ of the scaled polynomial. Consider any two coefficients $\bar{a}_i$ and $\bar{a}_j$, where j is greater than i. The sum of these two terms of the polynomial is

$$\text{sum} = \bar{a}_i \bar{s}^i + \bar{a}_j \bar{s}^j. \qquad (3\text{-}122)$$

Remember that j is always greater than i and for the problem at hand $\bar{s}$ will always be less than 1. If $\bar{a}_j$ happens to be less than $\bar{a}_i$, then the second term will always be smaller than the first term. In fact, we can say that if

$$\bar{a}_i \gg \bar{a}_j$$

then

$$\text{sum} = \bar{a}_i \bar{s}^i.$$

In this way one can selectively eliminate terms of the scaled polynomial that will always be insignificant so long as $\bar{s}$ is less than 1. By throwing out the insignificant terms, the degree of the polynomial can be cut drastically. This procedure can be performed even while the system is being built up, thereby saving computation effort in the calculation of the coefficients as well as in solving for the roots.

An easy way to incorporate the scale factor f into the above calculations is to modify the input data. Multiply all inertia values by the scale factor squared and all damping values by the scale factor. All polynomials that are calculated will then be scaled automatically (except for dividing through the a_0). At any time during the buildup of the system all polynomials can be reduced in size by throwing away the smaller terms.

The scaling and reducing procedure just described shares a remarkable resemblance to condensation, which is an important part of the finite element method [27]. For torsional models with springs to ground, the maximum number of natural frequencies that can be calculated is equal to the number of inertias in the system model. If estimations for the first three natural frequencies of a machine are desired, the system must then be broken up into a minimum of three inertias. Of course, it is always recommended that more than the minimum number of inertias be used in order to increase accuracy. Increasing the number of inertias from 3 to 60 will increase the accuracy of the first three modes. But the increased accuracy is paid for with increased execution time since the model now has 60 degrees of freedom instead of just 3. In finite element analysis, condensation is a method where the accuracy of many degrees of freedom is achieved without all the additional computation effort. A large set of degrees of freedom is reduced to a smaller set in such a way as to retain the extra accuracy of the degrees of freedom that are eliminated. This is precisely the effect that is obtained by the scaling and reduction procedure described above.

Diagonal Elements

When programming the method described for branched systems, it was found convenient to construct arrays of variables which represent the diagonal elements of the system matrix. This is because equations (3-113) and (3-117) are always the basic equations which are to be applied, and both require the matrix's diagonal elements.

Subprograms

It is usually convenient to have two basic subroutines from which to build the main program: One that adds two polynomials and one that multiplies two polynomials (see the subprogram on page 90).

Rootfinders

A variety of polynomial rootfinders are available. The Newton–Raphson method, the Siljak method, and the Bairstow method of quadratic factoring have all been used and found satisfactory. The Bairstow method was found to be superior for large-order systems, averaging out to be about twice as fast as the Newton–Raphson method.

REFERENCES

1. Den Hartog, J. P., *Mechanical Vibrations*, McGraw-Hill, New York, 1956, pp. 26–30, 131.
2. Lanczos, C., *The Variational Principles of Mechanics*, University of Toronto Press, Toronto, 1966.
3. Wells, D. A., *Theory and Problems of Lagrangian Dynamics*, Schaum's Outline Series, McGraw-Hill, New York, 1967.
4. Seireg, A., *Mechanical Systems Analysis*, International Textbook Co., Scranton, PA, 1969, p. 95.
5. Thomson, W. T., *Theory of Vibration with Applications*, 2nd ed., Prentice-Hall, Englewood Cliffs, NJ, 1981, pp. 196–197.
6. Rainville, E. D., *Elementary Differential Equations*, Macmillan, New York, 1958, p. 115.
7. Den Hartog [1], pp. 187, 229, 156, 162.
8. Thomson [5], pp. 296, 300, 285.
9. Vance, J. M., Brown, R. A., and Darlow, M. S., *Feasibility Investigation of Zero Torsional–Stiffness Couplings for Suppression of Resonance and Instability in Helicopter Drive Trains*, U.S. Army Air Mobility Research and Development Laboratory Report TR-73-103, Ft. Eustis, VA, pp. 113–114 (June 1974).
10. Wahl, A. M., and Fischer, E. G., "Investigation of Self-Excited Torsional Oscillations and Vibration Damper for Induction Motor Drives," *Journal of Applied Mechanics (ASME Transactions)*, pp. A-175–A-183 (December 1942).
11. Moore, J. B., "A Convergent Algorithm for Solving Polynomial Equations," *Journal of the Association for Computing Machinery*, **14**(2), 311–315 (April 1967).
12. Pipes, L. A., *Matrix Methods for Engineering*, Prentice-Hall, Englewood Cliffs, NJ, 1963, pp. 34–35.
13. D'Souza, A. F., and Garg, V. K., *Advanced Dynamics, Modeling and Analysis*, Prentice-Hall, Englewood Cliffs, NJ, 1984, pp. 162–164.
14. Vance, J. M., and Sitchin, A., "Derivation of First-Order Difference Equations for Dynamical Systems by Direct Application of Hamilton's Principle," *Journal of Applied Mechanics*, pp. 276–278 (June 1970).
15. D'Souza and Garg [13], pp. 198–213.
16. Jackson, C., and Leader, M. E., "Design, Testing, and Commissioning of a Synchronous Motor–Gear–Axial Compressor," *Proceedings of the Twelfth Turbomachinery Symposium, Texas A&M University, November 15–17, 1983*, pp. 97–111.
17. Evans, B. F., Smalley, A. J., and Simmons, H. R., "Startup of Synchronous Electric Motor Drive Trains: The Application of Transient Torsional Analysis to Cumulative Fatigue Assessment," ASME Paper 85-DET-122, presented at the ASME Conference on Vibration and Noise, Cincinatti, September 10–13, 1985.

18. Vance, Brown, and Darlow [9], p. 101.

19. Swick, R. M., and Skarvan, C. A., *Investigation of Coordinated Free Turbine Engine Control Systems for Multiengine Helicopters*, USAAVLABS Technical Report 67-73, U.S. Army Aviation Material Laboratories, Ft. Eustis, VA, AD 666796 (December 1967).

20. D'Souza and Garg [13], pp. 147–151.

21. Thomas, G. B., Jr., *Calculus and Analytic Geometry*, 3rd ed., Addison-Wesley, Reading, MA, 1960, pp. 426–428.

22. Carnahan, B., Luther, H. A., and Wilkes, J. O., *Applied Numerical Methods*, Wiley, New York, 1969.

23. Jung, S. Y., *A Torsional Vibration Analysis of Synchronous Motor Drive Trains by the Modal Method*, M.S. Thesis in Mechanical Engineering, Texas A&M University (August 1986).

24. Rao, D. K., "Torsional Frequencies of Multi-Stepped Shafts With Rotors," *International Journal of Mechanical Science*, **20**, 415–422 (1978).

25. Sankar, S., "On the Torsional Vibration of Branched Systems Using Extended Transfer Method," ASME Paper No. 77-WA/DE-4, presented at the ASME Winter Annual Meeting, Atlanta, November 27–December 2, 1977.

26. Murphy, B. T., *Eigenvalues of Rotating Machinery*, Doctoral Thesis in Mechanical Engineering, Texas A&M University (May 1984).

27. Zienkiewicz, O. C., *The Finite Element Method*, McGraw-Hill, New York, 1956.

Chapter IV

Critical Speeds and Response to Imbalance

Every rotor–bearing system has a number of discrete natural frequencies of lateral vibration. Associated with each natural frequency is a mode shape, which can be thought of as a snapshot of the rotor deflection curve at the instant of maximum strain during the vibration.

When one of the natural frequencies is excited by rotor imbalance rotating at shaft speed, the shaft speed which coincides with that natural frequency is called a critical speed. In this case the rotor does not vibrate, but rather is bowed into the mode shape associated with the particular natural frequency and whirls about its bearing centerline. To a stationary external observer the rotor appears to vibrate, but this is simply the planar projection of the whirl orbit as seen from one side.

In mathematical terms, the natural frequencies are called eigenvalues and the mode shapes are called eigenvectors. Theoretically, a distributed mass–elastic system has an infinite number of eigenvalues and associated eigenvectors. In practice, only the lowest three or four critical speeds and associated whirl modes are excited in the operating speed range of a typical turbomachine.

The mode shapes are determined by the distribution of mass and stiffness along the rotor, as well as by the bearing support stiffnesses. The first three modes, associated with the lowest three natural frequencies of a uniform shaft, change with increasing support stiffness as shown in Fig. 4.1. Note that the first two modes with low support stiffness ($K \approx 0$) involve a negligible amount of shaft bending. The same two rigid-rotor modes for a rotor with two massive disks are shown in Fig. 4.2. If the disks are identical and spaced equidistant from the rotor midspan, and if the (soft) bearing supports have equal stiffnesses, then the first rigid-rotor mode will trace a cylinder and the second will trace two cones with a common apex at midspan, as shown in Fig. 4.2. These cylindrical and conical modes will

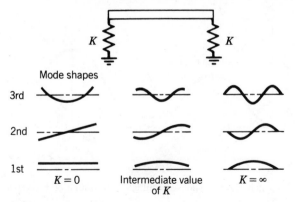

Figure 4.1 Effect of bearing support stiffness K on the modes of lateral vibration for a uniform shaft.

be somewhat modified by rotor asymmetry or by unequal bearing stiffnesses, but the terminology persists.

With a moderate amount of damping, the synchronous response plot for the rotor–bearing system of Fig. 4.2 would appear as shown in Fig. 4.3. This is a plot of the whirling amplitude measured at the bearings, caused by imbalance, as a function of shaft speed. The critical speeds are identified by the peaks of the whirl amplitude for each of the two modes.

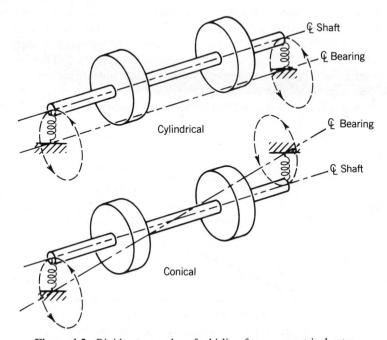

Figure 4.2. Rigid-rotor modes of whirling for a symmetrical rotor.

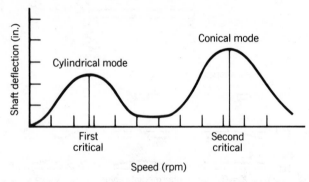

Figure 4.3. Synchronous response to unbalance through both rigid-rotor modes.

If the shaft speed is increased beyond the conical critical, a third critical speed will be reached that involves shaft bending, as shown on the top left of Fig. 4.1. If the supports are very soft, this mode is essentially the same as a free–free (unsupported) mode.

The first three critical speeds typically vary with support stiffness, as shown in Fig. 4.4. This type of plot is called a critical speed map. In this case, the insensitivity of the third critical speed to support stiffness permits a range of operating speeds that does not traverse any of the critical speeds, as shown by the vertical arrow of Fig. 4.4. From a rotordynamics standpoint, this is good turbomachinery design practice, but other considerations and the modern trend toward higher speeds makes it difficult to avoid approaching or traversing the third critical speed.

For machines with oil-film bearings, there can be enough support damping to make one or both of the rigid-rotor critical speeds disappear, i.e., the peaks illustrated on Fig. 4.3 may not be observed. In such a case, the third calculated critical speed may in actuality become the first or second observed critical speed. This is sometimes referred to as the first bender because of the mode shape.

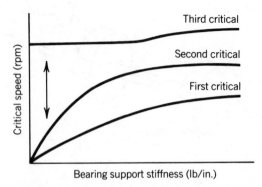

Figure 4.4. Critical speed map for three modes.

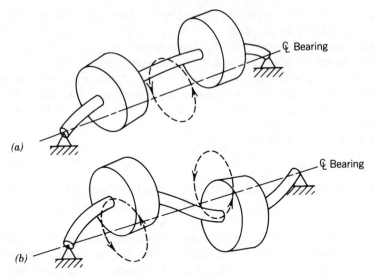

Figure 4.5. First two rigid-support modes of whirling for a symmetric elastic two-disk rotor.

If for some reason the bearing supports are stiffened, the modes become more like those shown on the right side of Fig. 4.1. For a symmetric rotor with two identical disks on an elastic shaft, the first two modes of operation on rigid supports are shown in Fig. 4.5. Mode shape (a) is approximated by a half sine wave; mode shape (b) by a full sine wave. Since rigid bearing supports cannot dissipate energy, and since internal damping in a rotor is destabilizing at high speeds, this type of

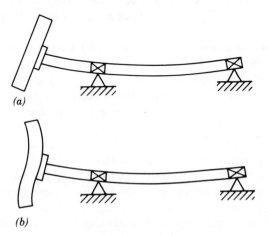

Figure 4.6. Whirl modes of an overhung rotor with (a) a rigid disk and (b) a flexible disk.

design can create severe rotordynamic problems. Almost all of the flexing is in the rotor, rather than in the supports.

Mode shapes associated with the first critical speed of an overhung (cantilevered) disk are shown in Fig. 4.6. The disk participates in the mode as a rigid body (Fig. 4.6a), or the disk may flex (Fig. 4.6b). Another possibility is that the disk may remain rigid but its attachment to the shaft may flex. These flexing or floppy disk effects can be observed on any rotor (not necessarily overhung) that has a large wheel diameter to shaft diameter ratio, but usually they are significant only at very high rotor speeds.

METHODS OF ANALYSIS AND THE EQUATIONS OF ROTOR MOTION

Critical speeds can be computed as the imaginary parts of the complex eigenvalues of the rotor–bearing system, under the constraint that the whirling frequency ω_d equals shaft speed ω (synchronous whirl). A detailed description of eigenvalue analysis is given in Chapter III for torsional vibration. For lateral whirling, it should be noted that the imaginary parts of the eigenvalues (natural frequencies) can change with shaft speed due to gyroscopic effects and speed-dependent bearing properties. Thus the constraint $\omega_d = \omega$ must be specified and incorporated in the analysis for synchronous whirl.

Until recently, the eigenvalue approach to computing critical speeds was often carried out without including damping in the model. This was based on the concept of "indifferent equilibrium," in which the undamped rotor can whirl in dynamic equilibrium with any amplitude at the critical speed.

Even with damping included, the differential equations for eigenvalue analysis are always homogeneous, since there is no imbalance, or any other forcing function, in the model. Thus, eigenvalue analysis is "free vibration" analysis. Eigenvalue analysis becomes critical speed analysis only under the constraint $\omega_d = \omega$.

Critical speeds can also be identified from a computation of imbalance response, as noted above (Fig. 4.3). This is consistent with the definition of critical speeds given in Chapter I, and graphically shows the effects of damping and of the distribution of imbalance along the rotor. The differential equations for imbalance response analysis are nonhomogeneous, since they must include the imbalance (forcing) terms.

For either approach the differential equations can be assembled either as a characteristic matrix or as transfer matrices (see Chapter III). The latter formulation has certain advantages peculiar to the rotor–bearing problem, and so has been developed and refined as the principal method for critical speed and imbalance response analysis.

All of the modern computational methods require the differential equations of motion for a rigid disk, since any rotor–bearing system can be modeled as an assembly of such disks connected by flexible shafts and supported by flexible bearings (and sometimes on flexible foundations or housing structures).

The rigid disk equations of motion can be derived by Lagrange's equation (see Chapter III) from the Lagrangian of a spinning disk, which is [1]

$$L = T = \tfrac{1}{2} m(\dot{X}^2 + \dot{Y}^2) + \tfrac{1}{2} I_T(\dot{\alpha}^2 + \dot{\beta}^2)$$

$$+ \tfrac{1}{2} I_P(\omega^2 - 2\omega\dot{\alpha}\beta) \tag{4-1}$$

where

ω = constant shaft speed

m = mass of the disk

I_T = transverse moment of inertia of the (symmetric) disk about x or y

I_P = polar mass moment of inertia of the disk about z

Equation (4-1), which is simply the kinetic energy of the disk, is expressed in terms of the Euler angle coordinates defined on Fig. 4.7. The order of rotations is (1) α about Y, (2) β about x', (3) ψ about z. The capital subscripts denote space-

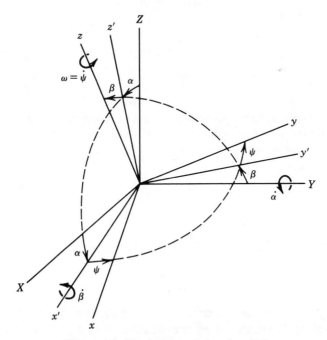

Figure 4.7. The Euler angles α, β, and ψ.

fixed axes, the lower case is for body-fixed principal axes, and the primes denote intermediate positions. The body-fixed xyz and space-fixed XYZ axes are initially coincident. The angles α and β are required to be small for (4-1) to be valid, a requirement which is generally satisfied by turbomachinery rotor–bearing systems. Under this restriction, α and β can be regarded as rotations about the space-fixed Y and X axes, respectively, and so cease to be proper Euler angles.

The shaft elastic forces, plus any conservative bearing forces, acting on the disk could be computed from the Lagrangian by including their potential energy in the Lagrangian if desired. (This was done in Chapter III for torsional vibration.) Instead, all forces acting on the disk will be considered as part of the virtual work and so will appear on the right-hand side of Lagrange's equation (3-2). The coordinates θ_n in (3-2) are here replaced by X, Y, α, β. The virtual work of all the forces acting on the disk is

$$\delta W = \Sigma F_X \, \delta X + \Sigma F_Y \, \delta Y$$

$$+ \Sigma M_Y \, \delta \alpha + \Sigma M_X \, \delta \beta \qquad (4\text{-}2)$$

Referring to Lagrange's equation (3-2), the generalized forces and moments are

$$Q_1 = \Sigma F_X,$$

$$Q_2 = \Sigma F_Y,$$

$$Q_3 = \Sigma M_Y,$$

$$Q_4 = \Sigma M_X. \qquad (4\text{-}3)$$

Thus, ΣF_X represents the X component of all forces acting on the disk, etc.

Substituting (4-1) and (4-3) into Lagrange's equation (3-2) yields the four differential equations of motion for the disk as

$$m\ddot{X} = \Sigma F_X, \qquad (4\text{-}4)$$

$$m\ddot{Y} = \Sigma F_Y, \qquad (4\text{-}5)$$

$$I_T\ddot{\beta} + I_P\omega\dot{\alpha} = \Sigma M_X, \qquad (4\text{-}6)$$

$$I_T\ddot{\alpha} - I_P\omega\dot{\beta} = \Sigma M_Y. \qquad (4\text{-}7)$$

The terms involving the polar moment of inertia I_P in equations (4-6) and (4-7) are the gyroscopic moments. They tend to stiffen the rotor in forward whirl and destiffen it in backward whirl, thus raising and lowering the natural frequencies and critical speeds of forward and backward modes, respectively. Since the gyroscopic terms contain angular velocities, they are grouped with damping terms in the usual matrix formulations of the equations. It can be seen that they are cross-coupled, since angular velocity $\dot{\alpha}$ about y produces a moment about x and vice versa.

A rotor–bearing model with N disks will have $4N$ equations of motion with the form of equations (4-4) through (4-7). This would produce a characteristic matrix of order $4N$ using second-order equations or order $8N$ using first-order equations (see Chapter III). Conversely, the maximum size of the transfer matrices, even with X–Y asymmetry, is 8×8 regardless of the number of disks in the model.

The forces and moments on the right-hand side of equations (4-4) through (4-7) can be classified as follows:

1. *Elastic and Internal Friction Forces Exerted on the Disk by the Shaft.* One way to include the internal friction is to use a complex modulus of elasticity for the shaft [2]. Expressions for the shaft shear components V_x V_y and shaft moment components M_y, M_x on each side of the disk are taken from Euler–Timoshenko beam theory [3]. In the transfer matrix formulation of the equations for multiple disks connected by elastic shafts, the shear and moment equations are used to express the shaft deflection and slope at each disk in terms of the values at an adjacent disk. A description of the symmetric case ($V_x = V_y$; $M_y = M_x$) is given later in this chapter. Although the symmetric case provides a simplified and concise illustration of the transfer matrix computational procedure, it should be pointed out that a principal advantage of the method is the relative ease with which nonrotating (e.g., bearing or support structure) asymmetries can be modeled. Conversely, rotating (with the rotor) asymmetries, such as a noncircular shaft or a keyway effect, present difficulties.

2. *Mass Imbalance Forces.* The distribution of mass imbalance along the rotor is represented by a center of gravity offset u in each disk. There is no loss of generality provided the flexible model is divided into a large enough number of disks.[1] The imbalance force components U_X and U_Y at each disk can be derived from D'Alembert's principle or by including the center of gravity offset in the Lagrangian (4-1). Either way, they turn out to be

$$U_X = m\omega^2 u \cos(\omega t + \psi_u)$$

$$U_Y = m\omega^2 u \sin(\omega t + \psi_u), \tag{4-8}$$

[1]This modeling philosophy is contrary to the prevailing (but sometimes tacit) assumption used for balancing, which is that the entire rotor can be considered as rigid so that the rotor imbalance can be specified as "static" and "dynamic" quantities. See Chapter V.

where ψ_u is the angle measured from the rotor-fixed x axis to the imbalance vector (see Chapter V). It is often convenient to let $\psi_u = 0$, unless the distribution of imbalance in the rotor is known. The imbalance forces are included only for the computation of response to imbalance. For eigenvalue analysis, $U_X = U_Y = 0$, so the differential equations are homogeneous as required.

3. *Fluid Forces.* The working fluid around a turbine wheel, a pump impeller, or a shaft seal can exert significant forces on the rotor. For rotordynamic analysis, they are expressed in terms of stiffness coefficients (force $\sim$ displacement), damping coefficients (force $\sim$ velocity), and in some cases inertia coefficients (force $\sim$ acceleration). For example, the Y component of a fluid force can be expressed as

$$F_Y = -K_{YX}X - K_{YY}Y - K_{Y\alpha}\alpha - K_{Y\beta}\beta$$
$$- C_{YX}\dot{X} - C_{YY}\dot{Y} - C_{Y\alpha}\dot{\alpha} - C_{Y\beta}\dot{\beta}$$
$$- D_{YX}\ddot{X} - D_{YY}\ddot{Y} - D_{Y\alpha}\ddot{\alpha} - D_{Y\beta}\ddot{\beta}, \qquad (4\text{-}9)$$

where

K is stiffness

C is damping

D is inertia

The first subscript on each coefficient denotes the direction of force, and the second subscript denotes the direction of motion producing the force. The negative signs follow the traditional expressions for a mechanical spring force and viscous damper, $F_Y = -KY - C\dot{Y}$.

If the K, C, and D coefficients are constant, then the force is linear. In any case, the force model can be linearized by restricting the rotor motions to be small so that the changes in force are small. This is useful for eigenvalue analysis even when the forces are nonlinear, as discussed in Chapters VI and VII.

The complete set of 32 stiffness and damping coefficients for a single disk are presented in matrix form on page 277, Chapter VII.

The most important characteristic of fluid forces is that they are typically cross-coupled. That is, the off-diagonal terms in the stiffness and damping matrices are nonzero. The destabilizing effect of cross-coupled stiffness K_{XY} was described in Chapter I. The effects of cross-coupled stiffness and damping on response to imbalance and critical speeds are described later in this chapter.

4. *Bearing Forces.* Fluid-film bearing forces are a special case of classification 3, above, and are treated in Chapter VI along with seal forces. In constructing a rotor–bearing model for transfer matrix analysis, a disk should be located at every

bearing and seal location along the rotor, even if its mass is small, so that the bearing or seal forces [such as (4-9)] can be included on the right-hand side of equations (4-4) through (4-7).

For rolling-element bearings, a purely elastic force model is usually appropriate.

THE LONG RIGID SYMMETRIC ROTOR

This special case is analyzed here as a simple illustration of gyroscopic effects, forward and backward eigenvalues, synchronous whirl analysis, and the characteristic matrix formulation of the equations of motion. (The transfer matrix method is superfluous with only one rigid mass element in the model.)

Figure 4.8 shows the model, in which the disk may be a long cylinder. The origin of the nonrotating XYZ axes is at the undeflected centroid of the rotor. The two bearings are located at $z = \pm L/2$. The undamped eigenvalues can be found by substituting a purely elastic and symmetric model ($K_{XX} = K_{YY} = K$; $F_X = -KX$; $F_Y = -KY$) for the bearing forces into equations (4-4) through (4-7) to give the equations of motion:

$$m\ddot{X} + 2KX = 0, \tag{4-10}$$

$$m\ddot{Y} + 2KY = 0, \tag{4-11}$$

$$I_T\ddot{\beta} + I_p\omega\dot{\alpha} + \tfrac{1}{2}KL^2\beta = 0, \tag{4-12}$$

$$I_T\ddot{\alpha} - I_p\omega\dot{\beta} + \tfrac{1}{2}KL^2\alpha = 0. \tag{4-13}$$

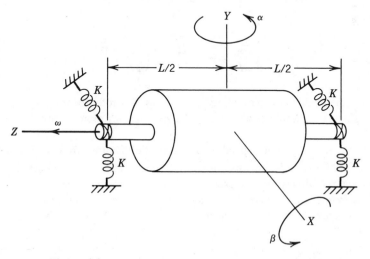

Figure 4.8. The long rigid-rotor model with coordinates.

The characteristic matrix resulting from the homogeneous solution $a_j e^{st}$, for j = 1, 2, 3, 4 (see Chapter III), is

$$
\begin{bmatrix}
(ms^2 + 2K) & 0 & 0 & 0 \\
0 & (ms^2 + 2K) & 0 & 0 \\
0 & 0 & I_T s^2 + \dfrac{KL^2}{2} & I_P \omega s \\
0 & 0 & -I_P \omega s & I_T s^2 + \dfrac{KL^2}{2}
\end{bmatrix}
$$

$$
\cdot \begin{Bmatrix} a_1 \\ a_2 \\ a_3 \\ a_4 \end{Bmatrix} = \begin{Bmatrix} 0 \\ 0 \\ 0 \\ 0 \end{Bmatrix}. \tag{4-14}
$$

The purely imaginary eigenvalues of the system are $s_j = \pm i\omega_j$, for j = 1, 2, 3, 4, where

$$
\omega_1 = \omega_2 = \sqrt{2K/m}, \tag{4-15}
$$

$$
\omega_3 = \left[\frac{I_P}{2I_T}\omega + \sqrt{\frac{KL^2}{2I_T} + \left(\frac{I_P}{2I_T}\omega\right)^2} \right], \tag{4-16}
$$

$$
\omega_4 = \left[\frac{I_P}{2I_T}\omega - \sqrt{\frac{KL^2}{2I_T} + \left(\frac{I_P}{2I_T}\omega\right)^2} \right]. \tag{4-17}
$$

These are the undamped natural frequencies of the rotor–bearing system. The following notes regarding them are instructive, even for more complex systems:

1. If the rotor angular speed ω is zero, the natural frequencies become $\sqrt{2K/M}$ and $\sqrt{KL^2/2I_T}$. In this case the vibration modes are planar: heaving–swaying and pitching–yawing, respectively. A nonzero shaft speed ω changes the ω_3 and ω_4 frequencies but not ω_1 and ω_2. The latter are the natural frequencies of X and Y vibration to produce cylindrical whirl (Fig. 4.2) and are the same eigenvalues shown for the short rigid rotor analyzed in Chapter I.

2. Shaft speed ω raises the ω_3 frequency above the planar pitching vibration value $\sqrt{KL^2/2I_T}$ and lowers the ω_4 frequency. These are the natural frequencies of forward and backward conical whirl, respectively. Inspection of equations

(4-12) and (4-13) shows that the gyroscopic terms prevent either α or β from remaining zero whenever the other is nonzero. Thus, the rotor whirls when perturbed and cannot execute a planar pitching vibration when gyroscopic moments are acting (unless additional forces are acting or the bearing supports are asymmetric).

3. The strength of the gyroscopic moment is determined by the ratio $P = I_P/I_T$, which becomes larger as the rotor is shortened, until $P = 2$ for a thin disk. Figure 4.9 shows how the dimensionless natural frequencies ω_3/ω_T and ω_4/ω_T vary with dimensionless rotor speed ω/ω_T, for four different values of P, where $\omega_T = \sqrt{KL^2/2I_T}$. The value $P = 0$ is a fictitious case with no gyroscopic moment. The forward synchronous excitation frequency of rotating imbalance is shown by the dashed line, and its intersection with the $P = 0$ and $P = 0.5$ curves are the conical critical speeds for these two cases. Note that if P is large enough (e.g., short rotors) there is no intersection and therefore no conical critical speed. In this case the eigenvalue exists but is never excited by rotating imbalance.

The foregoing eigenvalue analysis has determined the critical speeds of the system without including imbalance forces in the equations, even though a critical speed is the "speed at which synchronous response to imbalance is maximum" (Chapter I). This was done by noting the speeds at which the eigenvalues are excited by the synchronous frequency. It is more efficient computationally to preconstrain the frequency in the analysis to be synchronous with shaft speed, as follows:

Forward synchronous whirl in the conical mode results from synchronous vibration of the angular coordinates α and β in equations (4-12) and (4-13), described by

$$\alpha = A \cos \omega t, \qquad \beta = -B \sin \omega t, \tag{4-18}$$

where ω is the rotor speed and $A = B$ with symmetric bearing supports. Substitution of (4-18) into (4-12) or (4-13) gives

$$(I_T - I_P)\,\omega^2 A = \tfrac{1}{2} KL^2 A,$$

or

$$\omega_{con}^2 = \frac{\omega_T^2}{1 - P}; \tag{4-19}$$

where

ω_{con} = the conical critical speed
$\omega_T^2 = KL^2/2I_T$, as above
$P = I_P/I_T$, as above

Equation (4-19) gives the locii of the intersections of synchronous excitation with the eigenvalue curves on Fig. 4.9. It is thus referred to as a "critical speed analysis," as opposed to an eigenvalue analysis (which can also give the critical speeds as special cases).

None of the analyses presented to this point for the long rigid rotor are capable

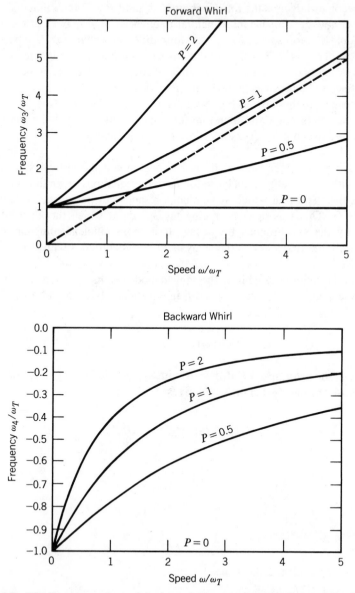

Figure 4.9. Whirling eigenvalues of the long rigid-rotor versus shaft speed, for four mass inertia ratios.

of predicting amplitudes of synchronous whirl. Damping, which has been omitted from the model so far, has a strong effect on the amplitudes near critical speeds but not on the critical speeds. Likewise the rotor imbalance, also omitted so far, affects the amplitude but not the critical speed.[2] An analysis of the response to imbalance requires both imbalance and damping to be included in the equations of motion. When this is done, equations (4-6) and (4-7) can be decoupled (only for the present case of symmetric bearing supports) and expressed as

$$(I_T - I_P)\ddot{\alpha} + \frac{CL^2}{2}\dot{\alpha} + \frac{KL^2}{2}\alpha = (I_T - I_P)\,\omega^2\,\theta\cos(\omega t), \quad (4\text{-}20)$$

$$(I_T - I_P)\ddot{\beta} + \frac{CL^2}{2}\dot{\beta} + \frac{KL^2}{2}\beta = (I_T - I_P)\,\omega^2\,\theta\sin(\omega t), \quad (4\text{-}21)$$

where $C = C_{XX} = C_{YY}$ is the symmetric bearing damping and θ is the couple imbalance angle, defined as the angular misalignment of the principal z axis of inertia. The X and Y response to static imbalance u, that is, response to imbalance in the cylindrical mode, is described in Chapter I (the short rigid rotor), and equations (4-20) and (4-21) for the conical mode are analogous to equations (1-6) and (1-7).

The particular solution to the nonhomogeneous equations (4-20) and (4-21) is

$$\alpha = A\cos(\omega t - \gamma), \qquad \beta = B\sin(\omega t - \gamma),$$

where

$$A = B = \frac{(I_T - I_P)\,\theta\omega^2}{\sqrt{[KL^2/2 - (I_T - I_P)\,\omega^2]^2 + (CL^2/2)^2\,\omega^2}}; \quad (4\text{-}22)$$

$$\gamma = \arctan\left[\frac{(CL^2/2)\,\omega}{KL^2/2 - (I_T - I_P)\,\omega^2}\right]. \quad (4\text{-}23)$$

The response plots of conical whirl amplitude A versus rotor speed ω, and phase angle γ versus ω, look similar to Fig. 1.7 for any given inertia ratio $P = I_P/I_T$. It has already been shown that increasing P moves the amplitude peak to a higher critical speed. The total amplitude response of the long rigid rotor to dynamic imbalance is a linear combination of equations (1-4) and (4-22), which are the amplitudes of the cylindrical and conical whirl modes, respectively. The plots are exemplified by Fig. 4.3, where the total vibration amplitude measured at either bearing would be given by $r_s + (L/2)A$. Note that the couple unbalance θ can be

[2]The location or distribution of imbalance in the rotor can affect the measured critical speeds if the measurement transducer is located so as to respond only to the strongly excited mode. See Chapter VIII.

positive or negative, so that the amplitudes of the two modes can either be additive or subtractive at either bearing.

BACKWARD WHIRL

In the example just presented, the eigenvalues of backward whirl were not excited by the forward rotating rotor imbalance. The actual occurrence of backward whirl in turbomachinery has been doubted in the past [4], but modern instrumentation confirms that it does occur. The author has observed it in his laboratory.

With light damping, backward whirl is excited by rotor imbalance when the rotor speed is between two natural frequencies split by bearing support stiffness asymmetry, for instance, ($\omega_1 < \omega < \omega_2$), where ω_1 is the eigenvalue associated with the lower stiffness and ω_2 is associated with the higher. The speed range $\omega_1 - \omega_2$ is lowered and narrowed by damping. Sufficient damping will make the backward whirl disappear.

The simplest model capable of illustrating this is that of Fig. 1.8, with K_B replaced by $K_{XX} < K_{YY}$. The equations of motion are (dropping the subscripts on $C_B = C_{XX} = C_{YY}$)

$$m\ddot{X} + 2C\dot{X} + 2K_{XX}X = m\omega^2 u \cos(\omega t), \tag{4-24}$$

$$m\ddot{Y} + 2C\dot{Y} + 2K_{YY}Y = m\omega^2 u \sin(\omega t), \tag{4-25}$$

with particular solutions

$$X(t) = X_C \cos(\omega t) + X_S \sin(\omega t), \tag{4-26}$$

$$Y(t) = Y_C \cos(\omega t) + Y_S \sin(\omega t). \tag{4-27}$$

Expressions for the amplitudes X_C, X_S, Y_C, and Y_S can be obtained by substituting (4-26) and (4-27) into (4-24) and (4-25), and solving the resulting set of linear algebraic equations.

Referring to Fig. 1.6, $X(t)$ and $Y(t)$ are the coordinates of the point C and the precession or whirl angle is

$$\phi = \arctan\left(\frac{Y}{X}\right). \tag{4-28}$$

Backward whirl is described by $\dot{\phi} < 0$. Taking the time derivative in (4-28) and using (4-26) and (4-27) gives the condition for backward whirl as

$$X_C Y_S - Y_C X_S < 0. \tag{4-29}$$

A test for this condition can be programmed into any computer code for imbalance response to determine whether the whirl is forward or backward. This should be done at each inertia station in a multidisk model, since the rotor can whirl forward and backward simultaneously at different locations! For the simple example here, X_C, Y_S, Y_C, and X_S are available as functions from the solution described above. Substitution of these functions into (4-29) yields the condition for backward whirl as $\omega_1 < \omega < \omega_2$, where

$$\omega_1^2 = \frac{2K_{XX}}{m} - \frac{1}{2}\left(\frac{2C}{m}\right)^2 + \frac{1}{2}\left(\frac{K_{YY} + K_{XX}}{K_{YY} - K_{XX}}\right)\left(\frac{2C}{m}\right)^2,$$

$$\omega_2^2 = \frac{2K_{YY}}{m} - \frac{1}{2}\left(\frac{2C}{m}\right)^2 - \frac{1}{2}\left(\frac{K_{YY} + K_{XX}}{K_{YY} - K_{XX}}\right)\left(\frac{2C}{m}\right)^2. \tag{4-30}$$

EFFECTS OF CROSS-COUPLED FORCES

The gyroscopic moments in equations (4-6) and (4-7) are an example of cross-coupled force although they are atypical, being moments instead of forces and being inertia induced rather than externally applied by fluid forces on the rotor. However, as was pointed out above, they appear in the damping matrix and raise or lower the natural whirling frequencies depending on the direction of whirl. These are generally the characteristic effects of all cross-coupled damping forces no matter what the source.

The destabilizing effect of cross-coupled stiffness (K_{XY}, $-K_{YX}$) on nonsynchronous whirl is described in Chapters I and VI. In the light of that analysis the effect of K_{XY} on synchronous imbalance response is surprising, since it can either act like direct damping or cancel the existing damping.

Once again, the simplest illustrative model is that of Fig. 1.8, this time with the symmetric bearings retained but with a cross-coupled force acting on the central disk described by $F_X = -K_{XY}Y$, $F_Y = -K_{YX}X$, where K_{XY} and K_{YX} have the same magnitude $\mathcal{K}$, but are of opposite sign, $K_{XY} = \mathcal{K}$, $K_{YX} = -\mathcal{K}$. The resultant force is tangential to the circular whirl orbit and is forward driving, as shown in Fig. 1.16. It is representative of a type of force that can be induced by fluid pressure forces around a turbine wheel, centrifugal impeller, or fluid seal.

The equations of motion are thus modified to be (dropping the subscripts from $K_B = K_{XX} = K_{YY}$ and $C_B = C_{XX} = C_{YY}$)

$$m\ddot{X} + 2C\dot{X} + 2KX + \mathcal{K}Y = m\omega^2 u \cos \omega t, \tag{4-31}$$

$$m\ddot{Y} + 2C\dot{Y} + 2KY - \mathcal{K}X = m\omega^2 u \sin \omega t. \tag{4-32}$$

The solution for $X(t)$ and $Y(t)$ could be obtained by using (4-26) and (4-27) in (4-31) and (4-32), but a 4 × 4 determinant can be avoided by describing the whirl vector r as a complex function $\bar{r} = X + iY$. Equations (4-31) and (4-32) can then be combined to give

$$m\ddot{\bar{r}} + 2C\dot{\bar{r}} + 2K\bar{r} - i\mathcal{K}\bar{r} = m\omega^2 u e^{i\omega t}, \tag{4-33}$$

with particular solution $re^{i\omega t}$, where

$$r = \frac{m\omega^2 u}{(2K - m\omega^2) + i(2\omega C - \mathcal{K})}.$$

The complex amplitude r can be rationalized to give

$$r = |r| \exp\left[i(\omega t - \beta)\right],$$

where

$$|r| = \frac{m\omega^2 u}{\sqrt{(2K - m\omega^2)^2 + (2\omega C - \mathcal{K})^2}}, \tag{4-34}$$

$$\beta = \arctan\left[\frac{2\omega C - \mathcal{K}}{2K - m\omega^2}\right]. \tag{4-35}$$

Comparison with the solution (1-4) and (1-5) without cross-coupling shows that the effect of cross-coupled stiffness on response to imbalance is to modify the effective damping in the system. The new effective damping coefficient is

$$C_e = 2C - \mathcal{K}/\omega. \tag{4-36}$$

Note that there is a value of $\mathcal{K}(= 2\omega C)$ which completely removes damping from the system, but if $\mathcal{K} > 4\omega C$ or if $\mathcal{K} < 0$ (backward driving) the effective damping is increased and the synchronous whirl amplitude is reduced. Figure 4.10 shows plots of the dimensionless whirl amplitude $|r|/u$ versus dimensionless speed ω/ω_n for four different values of forward driving cross-coupled stiffness $\mathcal{K}$. All of the plots have five percent of critical damping (exclusive of $\mathcal{K}$).

This analysis shows that the identification of system damping from measured vibration response plots is really an identification of direct damping and cross-coupled stiffness combined, whenever the latter is present in a turbomachine.

EFFECTS OF LOAD TORQUE

It is quite common to note changes in imbalance response amplitudes when the load torque is varied in a turbomachine. In the past this has usually been attributed to changes in system damping due to changes in fluid flow conditions around the

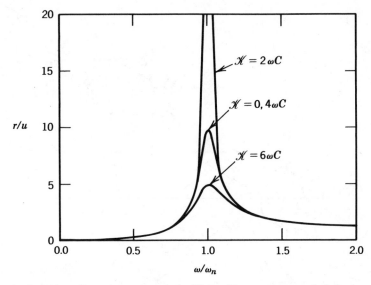

Figure 4.10. Effect of cross-coupled stiffness $\mathcal{K}$ on response to imbalance.

wheels and through the seals. While these changes undoubtedly occur, Ref. [5] shows that shaft torque alone has an effect on imbalance response which is very similar to the effect of cross-coupled stiffness described in the previous section. In fact, there are some special cases in which the torque can be represented in the mathematical model by cross-coupled stiffness coefficients. In the general case, the field transfer matrix for the shaft must be modified to include the effect of shaft torque. This matrix was derived by Yim [6] for the torquewhirl stability analysis described in Chapter VII and is given there.

The closing statement in the preceding section regarding identification of system damping from vibration measurements should be extended to include the effect of shaft torque also.

THE TRANSFER MATRIX (MYKLESTAD–PROHL) METHOD

This method for critical speed analysis was developed independently and almost concurrently by Prohl [7] (for steam turbines) and Myklestad [8] (for airplane wings). It is analogous to Holzer's method for torsional vibration analysis described in Chapter III.

A general flexible rotor is modeled by a number (typically $8 < N < 80$) of lumped inertias (disks) connected by massless elastic shaft sections (see Fig. 4.11). Point and field matrices, for the inertia and shaft elements, respectively, are multiplied together to give expressions for the boundary conditions at the right end ot the rotor in terms of the boundary conditions at the left end. The final matrix of

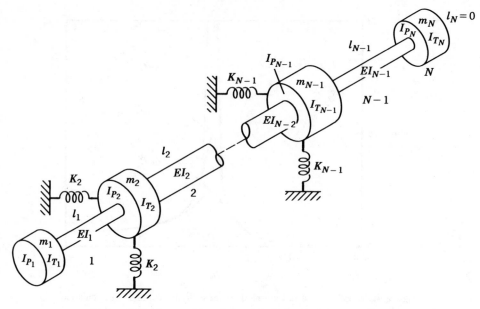

Figure 4.11. Lumped-inertia model of a rotor–bearing system.

coefficients that results from these multiplications is called the "overall transfer matrix." Its elements are determined, in part, by the natural frequency or eigenvalue.

For critical speed analysis, the synchronous assumption (4-18) is used and the elements of the overall transfer matrix are computed based on guesses or iteratively improved values for the natural frequency (critical speed). With zero for the boundary values the linear equations are homogeneous, which requires that the determinant of the overall transfer matrix be zero. Frequency values that make this determinant close enough to zero (tolerance determined by the analyst) are printed out as critical speeds. The eigenvectors (mode shapes) are then determined by multiplication of the transfer matrices to compute the displacements at each station.

The transfer matrix method has been further developed and refined by Lund [9] to accomplish damped eigenvalue analysis with cross-coupled forces included. Critical speeds can be identified as the intersections of synchronous excitation with the damped natural frequencies, but the Lund analysis is even more useful for predicting nonsynchronous whirl instability. Rotordynamic stability analysis, using a modification of Lund's method, is described in Chapter VII.

The transfer matrix method can also be easily modified to compute response to imbalance [5,10]. In this case the frequency is a known input value (shaft speed), and the imbalance forces make the system of equations nonhomogeneous.

As a simple illustration of the transfer matrix procedure, consider critical speed analysis of an undamped, symmetric ($F_X = F_Y$; $M_X = M_Y$) rotor–bearing system with two bearings of stiffness K_2 and K_{N-1}, where N denotes the last inertia station

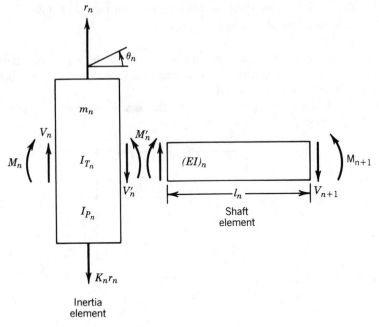

Figure 4.12. Free-body diagram of the elements in the nth rotor station.

at the right end of the rotor. (The bearings can be located at any inertia station in the model. In this example they are located as illustrated in Fig. 4.11.) Figure 4.12 shows free-body diagrams of the inertia (disk) and shaft (elastic) elements constituting the nth station. With X–Y symmetry, the X, Y, α, β coordinates of the disk displacement can be reduced to just $r(= X = Y)$ and $\theta(= \alpha = \beta)$. Rigid-body kinematics and equations (4-4) through (4-7) give the transfer equations across the disk for synchronous whirl (4-37):

$$
\begin{Bmatrix} r'_n \\ \theta'_n \\ V'_n \\ M'_n \end{Bmatrix} = \begin{bmatrix} 1 & 0 & 0 & 0 \\ 0 & 1 & 0 & 0 \\ (m_n\omega^2 - k_n) & 0 & 1 & 0 \\ 0 & (I_{Tn} - I_{Pn})\omega^2 & 0 & 1 \end{bmatrix} \begin{Bmatrix} r_n \\ \theta_n \\ V_n \\ M_n \end{Bmatrix}, \qquad (4\text{-}37)
$$

where

$$k_n = K_n, n = 2, N - 1$$
$$k_n = 0, n \neq 2, N - 1$$
V_n, V'_n are the shears on the left and right faces of the disk

$M_n,\ M_n'$ are the moments on the left and right faces of the disk

ω = shaft speed = whirl frequency

The bracketed matrix of coefficients in (4-37) is called the inertia transfer matrix for the nth station. This matrix will be denoted $[T_I]_n$, following the notation used for Holzer's method in Chapter III.

Statics and beam deflection theory give the transfer equations for the flexible shaft element (4-38):

$$\begin{Bmatrix} r_{n+1} \\ \theta_{n+1} \\ V_{n+1} \\ M_{n+1} \end{Bmatrix} = \begin{bmatrix} 1 & l_n & l_n^3/6EI_n & l_n^2/2EI_n \\ 0 & 1 & l_n^2/2EI_n & l_n/EI_n \\ 0 & 0 & 1 & 0 \\ 0 & 0 & l_n & 1 \end{bmatrix} \begin{Bmatrix} r_n' \\ \theta_n' \\ V_n' \\ M_n' \end{Bmatrix},\qquad (4\text{-}38)$$

where

l_n = shaft station length

E = modulus of elasticity

I_n = area moment of inertia for the shaft station

The matrix defined by (4-38) is the shaft transfer matrix $[T_S]_n$ for the nth station.

Multiplying the transfer matrices from left to right with n = 1, 2, 3, $\cdots$ as described in Chapter III for the Holzer procedure yields the state vector at the Nth (last) station (4-39):

$$\begin{Bmatrix} r_N' \\ \theta_N' \\ V_N' \\ M_N' \end{Bmatrix} = [T_o] \begin{Bmatrix} r_1 \\ \theta_1 \\ V_1 \\ M_1 \end{Bmatrix},\qquad (4\text{-}39)$$

where $[T_o]$ is the overall transfer matrix.

Substitution of the left-end boundary conditions $V_1 = 0$, $M_1 = 0$, gives the right-end boundary conditions as (4-40):

$$\begin{Bmatrix} V_N' \\ M_N' \end{Bmatrix} = \begin{Bmatrix} 0 \\ 0 \end{Bmatrix} = \begin{bmatrix} d_{11} & d_{12} \\ d_{21} & d_{22} \end{bmatrix} \begin{Bmatrix} r_1 \\ \theta_1 \end{Bmatrix}\qquad (4\text{-}40)$$

The D matrix in (4-40) is a submatrix of $[T_o]$.

Nonzero solutions for r_1, θ_1 in (4-40) exist only if the determinant of the D matrix is zero. Inspection of (4-37) shows that the elements d_{ij} of the D matrix depend on ω. The critical speeds computer program is therefore structured as follows:

1. Input numerical values for all station parameters.
2. Input a starting value for ω.
3. Compute the elements of all the individual station transfer matrices.
4. Take the values $r_1 = 1$, $\theta_1 = 0$, and multiply all the transfer matrices to give $V'_N = d_{11}$ and $M'_N = d_{21}$ [see (4-40)].
5. Take the values $r_1 = 0$, $\theta_1 = 1$, and multiply all the transfer matrices to give $V'_N = d_{12}$, $M'_N = d_{22}$.
6. Compute the determinant of the D matrix and check for closeness to zero. If close enough, print out the critical speed $\omega = \omega_{cr}$. If not close enough, increment to a new value for ω and return to step 3.
7. When ω_{cr} is found, use either equation in (4-40) to compute the value of θ_1, with $r_1 = 1$; then use the associated elements of the transfer matrices in (4-37) and (4-38) to compute the values of $r_1, r_2, \cdots, r_N$ by matrix multiplication. This gives the mode shape for the computed critical speed, normalized with respect to $r_1 = 1$.

The addition of damping to the model makes the eigenvalues and eigenvectors complex, as described in Chapter III. The removal of the symmetry assumption doubles the size of the state vectors and the order of the transfer matrices. The transfer matrices without symmetry and an improved computational procedure to find the complex eigenvalues are given in Chapter VII.

REFINEMENTS TO IMPROVE THE ACCURACY OF COMPUTED CRITICAL SPEEDS

The author and coworkers at Shell Westhollow Laboratories conducted an extensive research program to optimize the transfer matrix method for best accuracy, as determined by comparisons with careful experimental measurements [11,12]. The following refinements of the model and of the computational procedure were found to be beneficial. All are applicable to either a critical speed code or an eigenvalue code.

Effect of Disk/Shaft Attachment Flexibility

One of the usual assumptions inherent to critical speed analysis is that disks and wheels attached to the shaft remain rigid in all the calculated modes of interest. A further assumption usually made is that the attachment of the disks to the shaft is also rigid, so that the disks remain normal to the shaft in all modes. There is some evidence in the recent literature that these assumptions are not always valid [13].

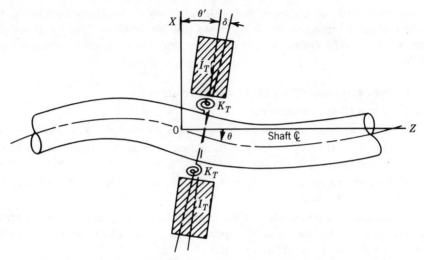

Figure 4.13. Effect of disk attachment flexibility.

Figure 4.13 shows the effect considered. The local slope θ of the shaft is different from the angular deflection θ' of the disk by the amount δ, due to a finite moment stiffness of the attachment, $K_T(\text{in} - \text{lb/rad})$.

The most straightforward analysis of the dynamic effect is by consideration of mechanical mobility and impedance. By definition, the impedance of the rigid disk is

$$Z_d = \frac{M}{\theta} = -I_T \omega^2, \tag{4-41}$$

where ω is the frequency of the local exciting moment M.

For a rigid attachment, the moment transferred across the nth disk in the computer program is

$$M_{n+1} = M_n + Z_{dn}\theta_n$$
$$= M_n - I_{Tn}\omega^2\theta_n. \tag{4-42}$$

For a flexible attachment, the moment stiffness K_T is in series with the inertia I_T. The resulting modified impedance of the disk is

$$Z_d = \frac{1}{\overline{M}_K + \overline{M}_I}, \tag{4-43}$$

where

$$\overline{M}_K = \frac{1}{K_T}$$

and

$$\overline{M}_I = \frac{-1}{I_T\omega^2}$$

are the mobilities of the attachment stiffness and disk inertia respectively.

Thus, we have

$$Z_d = \frac{-K_T I_T \omega^2}{K_T - I_T \omega^2}, \qquad (4\text{-}44)$$

so that the new transfer equation for the nth disk is

$$M_{n+1} = M_n + Z_{dn}\theta_n$$

$$= M_n - \frac{K_T I_T \omega^2}{K_T - I_T \omega^2}\theta_n. \qquad (4\text{-}45)$$

To utilize equation (4-45) in the computer program, a numerical value for the attachment stiffness K_T must be supplied. Three methods for calculating K_T have been explored. Interestingly, all three of the methods investigated have been found to give results of the same order of magnitude. The three methods are as follows:

1. *Calculation of attachment stiffness based on an equation from the theory of elasticity*, given in Ref. [14]. This model is actually based on the bending of the disk itself.

Figure 4.14 shows the model, the equation, and associated values of the deflection δ in the equation.

The variables in the figure and in the equation are defined as

$E =$ Young's modulus, psi
$t =$ disk thickness, in
$\mu =$ Poisson's ratio
$r_0 =$ shaft radius, in
$a =$ radius of the disk, in
$\alpha =$ shape factor

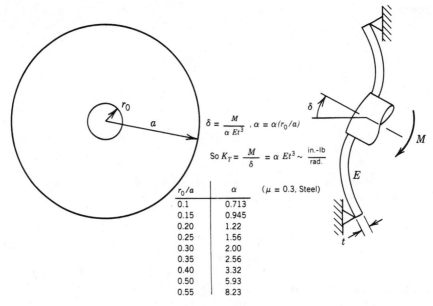

$$\delta = \frac{M}{\alpha\, Et^3}\ , \alpha = \alpha(r_0/a)$$

$$\text{So } K_T = \frac{M}{\delta} = \alpha\, Et^3 \sim \frac{\text{in.-lb}}{\text{rad.}}$$

r_0/a	α	($\mu = 0.3$, Steel)
0.1	0.713	
0.15	0.945	
0.20	1.22	
0.25	1.56	
0.30	2.00	
0.35	2.56	
0.40	3.32	
0.50	5.93	
0.55	8.23	

Figure 4.14. Formula for K_T.

A quadratic curve has been derived to fit the data for the shape factor, α (Fig. 4.15).

2. *Calculations based on an assumption that the disk attachment flexibility is a certain fraction of the local shaft moment flexibility.* For the three-mass laboratory rotor described later in this chapter, accelerometer measurements on the disk indicated that the effective attachment stiffness is about 20 times the local shaft moment stiffness. The disks on this rotor are unusually thick, so this number should be considered an upper limit.

For a concentrated moment midway between two simple supports, the local shaft moment stiffness is

$$K_S = 24EI/l.$$

For any real case of practical interest, the effective length l must be estimated. Then the attachment stiffness would be

$$K_T \le 20K_S \tag{4-46}$$

3. *Selection of the value for K_T that produces calculated natural frequencies closest to the measured values.* In some cases, it may be possible to establish values of K_T for various types of rotor construction through experience. This method was used by the author to verify that methods 1 and 2 give resulting natural frequencies that are more accurate than ignoring the attachment flexibility.

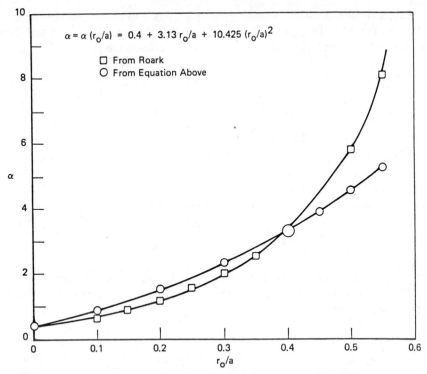

Figure 4.15. Shape factor α for $K_T = \alpha E t^3$.

Effect of Concentrated End Masses Due to the Definition of Rotor Stations in the Transfer Matrix Model

Figure 4.16 shows the transfer matrix model for a uniform shaft of diameter D and length L, using only four stations (three shaft elements). As usual, the shaft element in the last station on the right is of zero length, which puts a mass element on the end.

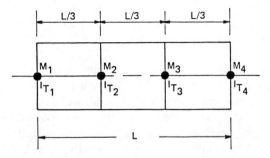

Figure 4.16. Transfer matrix model for uniform shaft.

A typical convention for lumping mass properties is to use one-half of the mass and mass moment of inertia from each of the shaft segments to the left and right of the mass element being considered. Using this convention gives

$$m_1 = m_4 = \bar{\rho}L/6,$$

$$m_2 = m_3 = \bar{\rho}L/3;$$

where $\bar{\rho}$ is the mass/unit length.
Similarly,

$$I_{T_1} = I_{T_4} = \frac{\bar{\rho}L}{6}\left(\frac{D^2}{16} + \frac{L^2}{108}\right),$$

$$I_{T_2} = I_{T_3} = \frac{\bar{\rho}L}{3}\left(\frac{D^2}{16} + \frac{L^2}{108}\right),$$

where moments of inertia are taken about the centroids of the segments.

Using the values above to calculate the total mass and resulting moment of inertia for the shaft gives

$$\sum_1^4 m_i = \bar{\rho}L,$$

which is correct, and

$$I_T = \sum_1^4 I_{T_i} + 2m_2\left(\frac{L}{6}\right)^2 + 2m_1\left(\frac{L}{2}\right)^2$$

$$= \bar{\rho}L\left(\frac{D^2}{16} + \frac{L^2}{9}\right),$$

which is incorrect. The correct moment of inertia for the shaft is

$$I_T = \bar{\rho}L\left(\frac{D^2}{16} + \frac{L^2}{12}\right). \tag{4-48}$$

The discrepancy is due to m_1 and m_4 placed further outboard from the center of gravity of the shaft than they should be. This will cause significant errors in the calculated natural frequencies when a small number of stations is used, especially for "rigid-rotor" modes and for shafts with large masses near the ends. The evolution of a tendency by engineering analysts to use an inordinately large number of stations to model even simple rotors may be partially attributable to this effect.

A correction factor (f) has been derived to move part of the end mass to the next station inboard. If we take the left-end station as an example, the corrected lumped masses m_1' and m_2' are

$$m_1' = fm_1, \tag{4-49}$$

$$m_2' = m_2 + (1 - f)m_1, \tag{4-50}$$

$$f = \frac{1}{1 + l/4a}, \tag{4-51}$$

where

$l = l_1 + l_2$ (the length of segments 1 and 2)
$a = $ distance from station 2 to the center of gravity of the rotor

If $l_1 \approx l_2$ and $a \approx L/2$, where L is the total shaft length, then f is approximated by

$$f = \frac{1}{1 + l_1/l}. \tag{4-52}$$

Effect of Shear Deflection

For some rotating machinery, the geometry of the rotor is such that practically all of the shaft deflection is due to bending. However, for short shafts of large diameter, shear deflection can be a significant effect in the calculation of critical speeds.

The model found to be most accurate is due to Timoshenko and has been used by Lund [9]. The transfer equation for deflection of the nth shaft element due to both bending M and shear V is

$$r_{n+1} = r_n + l_n\theta_n + \left(\frac{l_n^2}{2EI_n}\right)M_n + \frac{l_n^3}{6EI_n}V_n - 1.33\frac{l_n}{A_nG_n}V_n. \tag{4-53}$$

The last term, with the shear modulus G_n in the denominator, represents the shear deflection effect; A_n is the cross-sectional area.

Convergence Criterion

The d_{ij} elements of the D matrix [see equation (4-40)] are functions of frequency. The object of the computer algorithm is to find the natural frequencies that make the determinant $|D(\omega)|$ of the matrix D as close to zero as possible. The previously described criterion for success, which is sometimes used, is to require $|D| \leq$

ϵ, where ϵ is a small predetermined number. If the criterion is not satisfied, an improved estimate for the frequency is calculated from a Newton–Raphson scheme, as follows:

$$\omega = \omega_0 - \frac{D_0 \left(\omega_0 - \omega_1 \right)}{D_0 - D_1}, \qquad (4\text{-}54)$$

where

    ω_0, D_0 are the most recent values of frequency and determinant;

    ω_1, D_1, are previous values of frequency and determinant;

    ω is the improved estimate for natural frequency.

As D_0 gets smaller, the difference between ω_0 and ω_1 also becomes smaller. However, an acceptable value of D_0 for one rotor, or for one mode, will often produce unacceptable accuracy (in ω_0) for another rotor or mode.

It was found that a convergence criterion on $\omega_0 - \omega_1$ was much easier to implement and produced much more consistent and accurate results than a criterion based on D_0.

Rules for Lumped Mass–Elastic Modeling of Shafts and Rotors

Out of this study came three modeling rules or guidelines that were found to produce optimum accuracy. Since the computer programs used in the study were based on the transfer matrix method, with lumped masses and inertias, the rules and guidelines should be used only for this type of program. They are as follows:

1. The optimum length of shaft used for each station is the maximum diameter in the station. Using additional shorter stations does not increase the program accuracy. Furthermore, the number of stations must be at least four times the number of desired modes to be computed.

2. Wherever disks, wheels, or sleeves are present on the shaft (even if pressed on or keyed), the disk or sleeve outside diameter should be used in calculating the bending stiffness of the section. The possible effect of centrifugal growth at high speed was not explored in reaching this conclusion.

3. Whenever disks or wheels of diameter more than twice the shaft diameter are located near a node of a particular mode of natural vibration, the flexible disk effect is likely to be significant and should be included in the model. This is especially true for free-free modes. This will split the computed frequency into two natural frequencies, one below and one above the frequency for a rigid disk attachment.

Machine Foundations and Housings

Closely spaced critical speeds ("split criticals") often indicate that the rotor response is being strongly influenced by the machine foundation or housing. For an accurate computer simulation to be possible, this effect must therefore be included. The transfer matrix method permits any bearing(s) to be modeled as mounted on a foundation mass, which in turn is supported by a linear spring/damper combination. Figure 4.17 shows the model, in which the bearing stiffness becomes frequency dependent. Different foundation parameters can be specified for each bearing, and for each bearing different parameters can be specified for the horizontal and vertical directions. Note that one limitation inherent in this type of foundation model is that there is no direct coupling between the various foundation masses. However, the best way to obtain numerical estimates for the foundation or housing parameters is from impedance or mobility measurements made directly on the foundation when the rotor has been removed, and these measurements will include the cross-bearing mass coupling if it exists.

The foundation parameters are determined from driving point impedance measurements made directly on the bearing housing. With a swept-frequency harmonic force applied to the structure by a shaker, or by using an instrumented hammer to excite the structure over a broad range of frequencies, the response of the structure is measured with an accelerometer or velocity pickup. Mechanical impedance is defined as the ratio of the applied force (magnitude and phase) to the measured velocity (magnitude and phase). The inverse of impedance is called mobility. Should the shaker and velocity pickup be at the same location, then the result is termed "driving point" impedance. Figure 4.18 shows a shaker connected to an idealized foundation model through a force transducer, with a velocity pickup mounted next to the force transducer (an accelerometer with integrated signal can be used if the phase is kept accurate). A convenient measurement procedure is to drive the shaker with a random noise signal and record the force and acceleration on a magnetic tape recorder for later analysis. Using a random noise

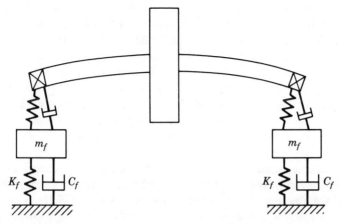

Figure 4.17. Lumped-mass model for the foundation or bearing housing.

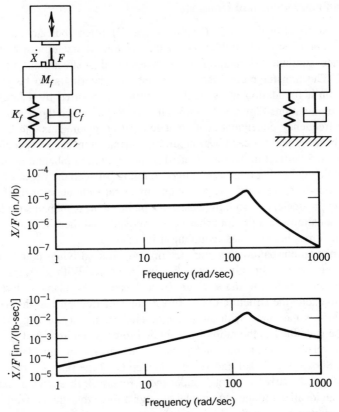

Figure 4.18. Shaker mounted on a foundation and the resulting fast Fourier transform (FFT) analyzer displays (for the ideal case).

signal allows the measurements to be taken for all frequencies simultaneously. Analysis of the recorded data can be performed with a two-channel frequency analyzer that permits integration and differentiation of the signals as well as computing their transfer functions. Figure 4.18 also shows the analyzer displays that would be obtained for the ideal one-degree-of-freedom foundation shown. Foundation stiffness is obtained from the X/F curve by reading off the value where the frequency approaches zero, and taking its reciprocal ($K = F/X$). Foundation damping is obtained from the $\dot{X}/F$ mobility curve by reading off the value at the peak (i.e., at the natural frequency) and taking its reciprocal ($C = F/\dot{X}$). Foundation mass is then calculated from the expression $\omega^2 = K/m$, with the frequency and stiffness known.

As an example, some simple rap tests with a hammer were used to check the natural frequencies of the foundation for a laboratory rotor (described below). It was found that the structure had a horizontal mode around 149 Hz. Vertically, the structure was found to be essentially rigid, with a fundamental frequency in excess of 300 Hz. Measurements were therefore made horizontally only and were made

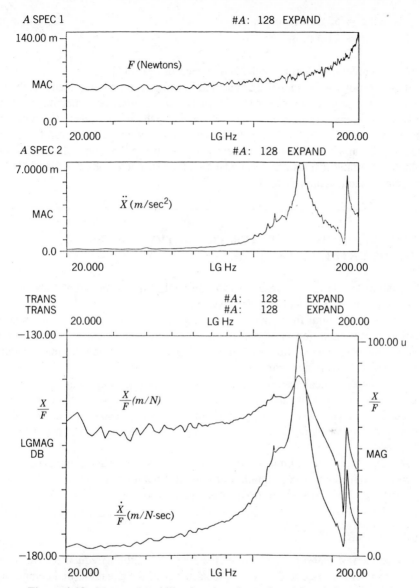

Figure 4.19. Measured mobility functions for outboard foundation of test rig.

on just one housing due to the symmetry of the structure. The plots obtained from the frequency analyzer are shown in Fig. 4.19. The parameters calculated as described above are as follows:

frequency = 149 Hz
stiffness = 208,000 lb-ft/in

damping = 55.6 lb-ft-sec/in
mass = 92.0 lb-m

A more complex foundation model is analyzed in Ref. [15].

It is also possible to model the machine housing or foundation analytically by branching the transfer matrices [15,16], or by the finite-element method [17,18]. Two-spool or three-spool rotors can be treated in these ways also. However, it should be remembered that accurate numerical values will be required for all of the lumped mass and elastic parameters which make up the model. In the case of complex asymmetric housings or foundations, the measured mobilities may actually be a more practical approach.

COMPARISON OF COMPUTER PREDICTIONS WITH EXPERIMENTAL MEASUREMENTS

The principal experimental results of the research program mentioned above and described in Refs. [11] and [12] are given in this section.

"Free–free"natural frequencies were measured by hanging rotors horizontally from long ropes and impacting them in the horizontal plane. Free–free frequency measurements are an excellent way of assessing accuracy of the rotor mass–elastic model without involving errors in bearing properties.

The reported critical speeds on bearings were obtained from measured response plots during coast-down, and damped natural frequencies were obtained from a measured frequency spectrum with the rotor running at constant speed. Details of the measurement techniques can be found in Refs. [11] and [12] and in Chapter VIII.

The "log dec" (logarithmic decrement) predictions were made using the damped eigenvalue transfer matrix program described in Chapter VII.

The following rotors and rotor–bearing systems were tested:

Uniform Shafts

A comparison of measured and calculated free–free natural frequencies for three uniform shafts was used to determine the accuracy of the equations and computing procedure used. Since these shafts had only minute changes in station diameter, errors were eliminated that sometimes may occur in modeling rotors having large changes in station diameters.

Table 4-1 shows the measured versus computed natural frequencies for a 3-in diameter shaft 50.33 in long. Figure 4.20 shows the calculated mode shapes for each frequency, with the measured modal deflections plotted as individual points for comparison.

The measured versus computed natural frequencies for a 2-in diameter shaft 48 in long are shown in Table 4–2.

**TABLE 4-1. Free–Free Natural Frequencies of a Uniform Shaft, 3-in Diameter ×
50.33 in Long**

Measured (Hz)	Computed (Hz)	Error (%)
212.5	213.3	+0.4
581.25	579.9	−0.2
1106.2	1117.6	+1.0
1787.5	1811.9	+1.4
2575	2648.9	+2.9
3731	3612.4	−3.2
		Overall RMS = 1.9%

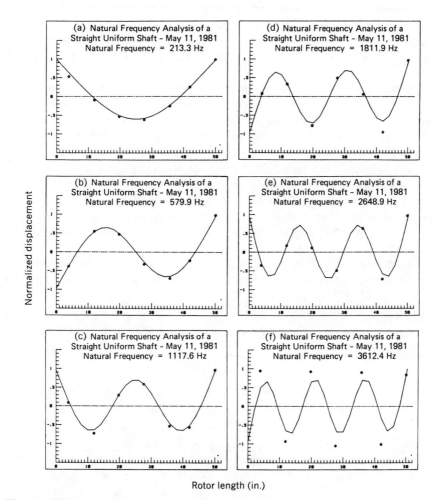

Figure 4.20. A comparison of measured and calculated free–free mode shapes of a 3-in
diameter shaft, 50.33 in long.

**TABLE 4-2.   Free–Free Natural Frequencies of a Uniform Shaft, 2-in Diameter ×
48.0 in Long**

Measured (Hz)	Computed (Hz)	Error (%)
156.05	154.06	−1.3
427.25	418.90	−2.0
827.73	805.53	−2.7
1320.6	1304.22	−1.2
		Overall RMS = 2.5%

The simplicity of this shaft, and the uniform magnitude and consistent sign of the error, suggests that most of the error could be due to a slight variation in Young's modulus from 30 × 10 psi. As with the previous shaft, the errors are quite small, with a root mean square, (RMS) average for five modes of only 2.5 percent.

The frequencies of a uniform shaft 6-in diameter by $11\frac{1}{2}$ in long were measured in order to determine the error associated with small L/D ratios, where shear deflection effects become significant and where the rule for choosing the number of stations in the transfer matrix model must be modified (see the section on rules for computer modeling below).

Table 4-3 shows that the effect of direct shear deflection in this shaft is indeed quite significant, changing the RMS error for the first three modes from 2.0-percent (with shear included) to 33.6-percent (without shear included) when a 24-station model was used. Even with only a 12-station model (four times the number of modes to be computed), the RMS error is less than 4-percent when shear is included.

Some interesting observations were made when making the experimental measurements on this shaft. The frequencies showing up in the spectrum were

**TABLE 4-3.   Free–Free Natural Frequencies of a Uniform Shaft, 6-in Diameter ×
$11\frac{1}{2}$ in Long**

Measured (Hz)	Computed (% Error) 24 Stations With Shear Effect	24 Stations No Shear Effect	12 Stations With Shear Effect	12 Stations No Shear Effect
5252	5292 (+0.8)	5829 (+11.0)	5593 (+6.5)	6160 (+17.3)
9700	9540 (−1.6)	12947 (+33.5)	9758 (+0.6)	13484 (+39.0)
14489	14063 (−2.9)	21217 (+46.4)	14256 (−1.6)	21828 (+50.7)
Overall RMS Error:	2.0%	33.6%	3.9%	38.3%

Figure 4.21. Uniform shaft with pressed-on disk. Photograph courtesy of Westhollow Research Center Graphics Dept., Shell Development Company, a division of Shell Oil Company.

quite sensitive to the measurement technique. For example, the natural frequency at 9700 Hz disappeared completely when one particular combination of shaker and random noise generator was used. These types of anomaly were not observed in the measurement on longer shafts and rotors.

A uniform shaft with a single massive disk was made up to represent one step of complexity away from the uniform shaft. A 9-in diameter by 3-in thick steel disk was hydraulically press-fitted onto a 2-in diameter by $48\frac{3}{16}$-in long steel shaft. Figure 4.21 shows a photograph of the assembly.

Table 4-4 gives the first four modes, measured and calculated, along with the error in percent. The RMS error for all four modes was 1.2-percent.

Three-Disk Laboratory Rotor

This rotor is the heart of a rotor bearing test apparatus purchased from Centritech, Inc. for rotordynamics research at Shell Westhollow Research Center. It has three

TABLE 4-4. Free–Free Natural Frequencies of a Uniform Shaft with a Single Massive Disk

Measured (Hz)	Computed (Hz)	Error (%)
130.76	130.45	−0.24
356.64	353.57	−0.86
632.65	632.19	−0.07
1132.80	1157.26	+2.16
		Overall RMS = 1.2%

Figure 4.22. Three-disk laboratory rotor. Photograph courtesy of Westhollow Research Center Graphics Dept., Shell Development Company, a division of Shell Oil Company.

large disks but was machined from one piece of stock to eliminate uncertainty about the effects of attachments or interfaces. The rotor weighs 369.7 lb and is 52.4 in long. The disks are 9.95-in outer diameter by 5.00-in axial length and are spaced on 10-in centers. Figure 4.22 is a photograph of this rotor.

Table 4-5 gives the measured and computed frequencies and the error in percent,

TABLE 4-5. Free–Free Natural Frequencies of a Three-Disk Laboratory Rotor

Measured (Hz)	Computed	
	No Flexible Disk [Hz (% Error)]	With Flexible Disk [Hz (% Error)]
94	95 (+1.1)	94 (0)
207	207 (0)	204 (−1.4)
356	353 (−0.8)	350 (−1.7)
463	408 (−11.9)	400 (−13.6)
832	850 (+2.2)	838 (+0.7)
1037	968 (−6.7)	938 (−9.5)
	Overall RMS = 5.7%	Overall RMS = 6.8%

both with and without the flexible disk effect included in the computations. The attachment stiffness of the 5-in thick disks is so high that the difference between the two results is insignificant. Figure 4.23 shows the computed mode shapes for each natural frequency, with the measured modal deflections plotted as individual points for comparison. The computed mode shapes have the flexible disk effect included.

Single-Stage Steam Turbine Rotor

This spare rotor, weighing 262 lb, from a single-stage steam turbine at the Shell Deer Park Refinery was made available to the research project for vibration testing. Figure 4.24 is a photograph of the rotor, suspended for free–free tests.

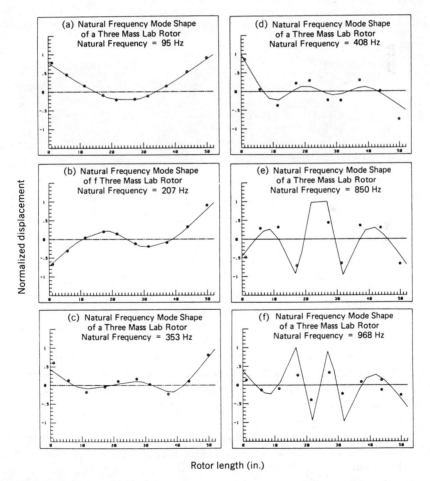

Figure 4.23. A comparison of measured and calculated free–free mode shapes of the three-disk laboratory rotor.

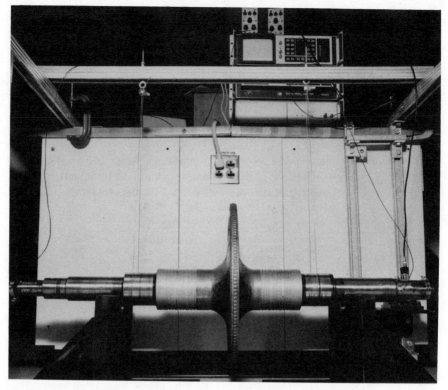

Figure 4.24. Single-stage steam turbine rotor. Photograph courtesy of Westhollow Research Center Graphics Dept., Shell Development Company, a division of Shell Oil Company.

It can be seen that the rotor is characterized by a turbine disk much larger in diameter than the shaft. Not surprisingly, it was found that the disk flexibility effect, described earlier as one of the computer program refinements, is significant in this rotor. Table 4-6 gives computed frequencies with and without this effect included, in addition to the measured frequencies. The percent error for each of the two cases is given in parenthesis. Note that without the disk flexibility effect, the third measured mode is missed completely.

The computed mode shapes (with the disk flexibility refinement) for this rotor are shown as solid lines in Fig. 4.25. The measured modal deflections are plotted as points on the same graph for comparison. It can be seen that the agreement of computer predictions with experiment is quite good, especially for the first two modes.

Note that both the second and third mode shapes have a node at the turbine disk and appear very similar. Detailed accelerometer measurements were made on and near the turbine disk to determine the significant differences between the two modes. It was found that the node of the second mode (602 Hz) is located very near the centerline of the disk, and the pitching motion of the disk is in phase with

TABLE 4-6.   Free–Free Natural Frequencies of a Single-Stage Steam Turbine Rotor

Measured (Hz)	Computed	
	No Flexible Disk [Hz (% Error)]	With Flexible Disk Hz (% Error)
373	376 (+0.8)	376 (+0.8)
602	622 (+3.3)	603 (+0.2)
1033	No ( )	907 (−12.2)
1262	1219 (−3.4)	1230 (−2.5)
1998	1642 (−17.8)	1795 (−10.2)
2509	2321 (−7.5)	2410 (−3.9)
	Overall RMS = 8.9%[a]	Overall RMS = 6.8%

[a]Not including third mode.

the oscillating slope of the shaft, at the disk location. The node of the third mode (1033 Hz) is located near the inside face of the turbine disk, and the pitching motion of the disk is out of phase with ("against") the oscillating slope of the shaft, at the disk location.

The latter mode, in which the disk pitches counter to the motion of the shaft,

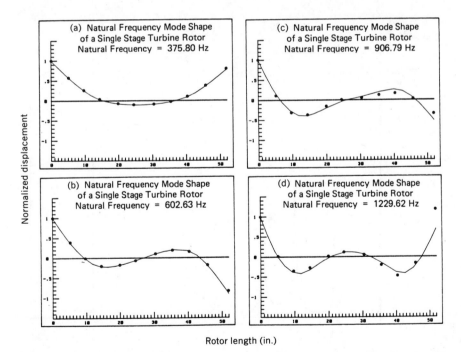

Figure 4.25.   A comparison of measured and calculated free–free mode shapes for the single-stage steam turbine rotor.

cannot be computed without the disk flexibility option in the program that adds an additional degree of freedom to the mathematical model.

Six-Stage Centrifugal Compressor Rotor

This rotor is a replacement spare for a compressor train at the Shell Deer Park Refinery. It was suspended in a horizontal position by ropes from a crane at the rotor storage facility (I. W. Hickham, Inc.) and excited in the horizontal plane by impact with a nylon hammer.

The measured frequency spectrum (see Chapter VIII for a description of frequency spectrum measurements) is shown in Fig. 4.26. Only the first three modes are clearly defined (i.e., is the fourth a true mode?).

Computed natural frequencies are compared to the measured frequencies for the first three modes in Table 4-7. Three different cases are presented:

1. From rotor station data as supplied by the manufacturer.
2. From the data in case 1, but as modified by actual dimensional measurements.
3. From the same data as in case 2, using the "flexible disk" option in the computer program.

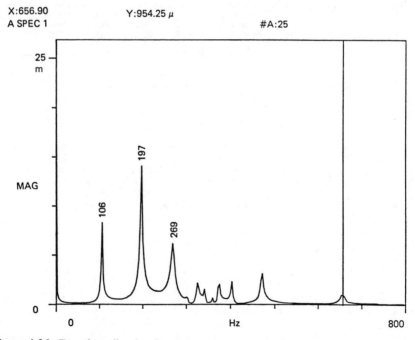

Figure 4.26. Free–free vibration frequency spectra measured on the six-stage compressor rotor.

TABLE 4-7.   **Free–Free Natural Frequencies of Six-Stage Compressor Rotor**

Measured (Hz)	Computed (% Error)		
	From Data as Received	With Disk I & Sleeve ODs[a]	With Flexible Disks
106	87 (−17%)	102 (−3.8%)	108 (+1.9%)
197	178 (−9.6%)	192 (−2.5%)	196 (−0.5%)
269	277 (−3.0%)	294 (+9.3%)	303 (+12.6%)
Overall error	RMS = 11.9%	RMS = 6.0%	RMs = 7.4%

[a]OD = outer diameter.

The percent error for each case is shown in parenthesis, and the overall RMS error for the three modes is the bottom line.

The larger errors associated with the first case are due mainly to deficiencies in the rotor data as supplied by the manufacturer. Specifically, both the impeller wheels and the spacer sleeves (between the impellers) are treated simply as added weights in the manufacturer's data.[3] Neither the mass moment of inertia of the impellers nor the added shaft stiffness due to the sleeves can be computed from the data supplied. Although these omissions partially cancel each other's effect on the computed frequency, the net results still average 12 percent in error for the three modes.

To improve the rotor model data, the impeller and sleeve diameters were measured in place on the rotor and used to modify the section diameters accordingly. As shown in second column of computed results in Table 4-7, these modifications reduce the error by a factor of almost 2.

This case emphasizes the importance of a well-defined rotor model when communicating design data. The definition of model parameters usually depends on the type of analysis being used by the supplier.

The RMS error in column 4 is seen to be misleading when the results for the first two modes alone (of greatest practical interest) are considered.

Eight-Stage Centrifugal Compressor Rotor #1

This rotor was obtained and measured in the same manner as the six-stage rotor described above.

Due to some practical constraints in the storage building, the support ropes and cables were so short in this test that they produced a "rigid-body" mode of horizontal swinging at 9 Hz. This frequency was high enough to indicate a measurable effect of the supports on the higher modes. Thus, unlike all of the other tests

---

[3]Apparently the compressor manufacturer's computer code was written to include the stiffening resulting from sleeves. Information needed for calculating this additional stiffness is not included with the data and thus was not used in our calculations.

reported here in which the frequency of the horizontal swing modes was less than 2 Hz, the results for this rotor were affected by support stiffness.

Table 4-8 gives the measured and computed natural frequencies for the first three modes, along with the errors in percent. Note that, in contrast to the six-stage rotor, this rotor has the smallest error associated with the third mode when the flexible disk effect is included. As in the case of the six-stage rotor, modes above the third were not clearly definable on the measured frequency spectrum, even when a log amplitude scale was used.

Eight-Stage Compressor Rotor #2

This rotor was obtained and measured in the same manner as the two preceding indiustrial rotors.

Figure 4.27 shows the measured spectrum, in which at least the first four modes are clearly definable in terms of their natural frequencies.

Table 4-9 shows a comparison of the computed frequencies with the measured ones for the first four modes. The three different cases for computed frequencies are defined in exactly the same way as described above for the six-stage rotor.

Thirteen-Stage Steam Turbine Rotor

This rotor is a spare steam turbine rotor, for use in the same compressor train as the previous three rotors, and was tested in the same way.

Figure 4.28 shows the measured frequency spectrum, in which at least the first four modes are readily definable.

The computed frequencies are compared to the measured ones in Table 4-10. There are only two cases under the "computed" heading for this rotor, either with a rigid or flexible attachment of the disk to the shaft. The rotor data as supplied by the manufacturer included all of the actual outside diameters of the wheels and shaft sections.

If only the first two modes are of practical interest, the flexible disk computer model is more attractive, since it produces the same error for the first mode but cuts the error in half for the second mode.

TABLE 4-8. Free–Free Natural Frequencies of Eight-Stage Compressor Rotor #1

Measured (Hz)	Computed (% Error)		
	From Data as Received	With Disk I & Sleve ODs	With Floppy Disks
60	39 (-35.0%)	58 (-3.3%)	58 ($+3.3\%$)
142	99 (-30.33%)	130 (-8.5%)	155 (-9.2%)
240	189 (-21.3%)	222 ($+7.5\%$)	243 ($+1.3\%$)
Overall error	RMS = 29.4%	RMS = 6.8%	RMS = 5.7%

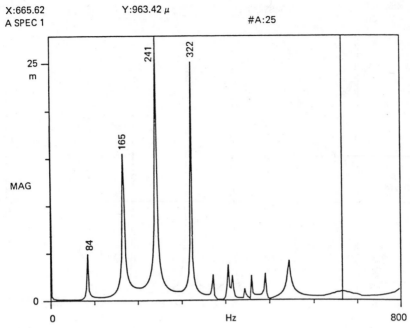

X:665.62   Y:963.42 μ
A SPEC 1   #A:25

Figure 4.27. Free–free vibration frequency spectra measured on the eight-stage compressor rotor #2.

Critical Speeds of a Three-Disk Laboratory Rotor–Bearing System

The measured free–free frequencies of the rotor alone were given above. The entire rotor was machined from one solid piece of steel and primarily consists of a 2.5-in diameter shaft 52.4 in long with three large identical disks each 10 in in diameter and 5 in long. The bearings are five-pad tilt-pad. Figure 4.29 is a photograph of the assembled test rig in place at the Shell Westhollow Research Center.

The critical speeds were obtained from a coast-down of the rotor during which the synchronous response to imbalance was measured and plotted.

TABLE 4-9. Free–Free Natural Frequencies of Eight-Stage Compressor Rotor #2

Measured (Hz)	Computed (% Error)		
	From Data as Received	With Disk I & Sleeve ODs	With Flexible Disks
84	69 (−17.9%)	88 (+4.8%)	88 (+4.8%)
165	131 (−20.6%)	173 (+4.8%)	172 (+4.2%)
241	178 (−26.1%)	267 (+10.8%)	267 (+10.8%)
322	264 (−18.0%)	382 (+18.6%)	404 (+20.3%)
Overall error	RMS = 20.9%	RMS = 11.3%	RMS = 11.9%

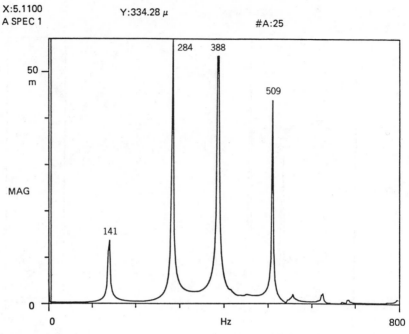

Figure 4.28. Free–free vibration frequency spectra of the 13-stage turbine rotor.

The mass–elastic computer model used for the rotor shaft is the same as used for the free–free measurements and is given in Table 4-11. The measured foundation parameters are also shown in this table as well as one set of bearing parameters. The bearing stiffness coefficients were measured with a special test apparatus [19]. Since no measurements were made of bearing damping, calculated values for damping were used from Ref. [20].

Table 4-12 shows the measured and calculated critical speeds, and Fig. 4.30 shows the mode shapes. In Fig. 4.30 the solid curves represent the calculated shaft modes and the letters "h" represent the calculated (horizontal) foundation displacements at each support. The measured shaft modes are shown as circles, and the measured motion of the test stand is represented as a straight line. Since

TABLE 4-10.   Free–Free Natural Frequencies of 13-Stage Steam Turbine Rotor

Measured (Hz)	Computed (% Error)	
	With Rigid Disks	With Flexible Disks
141	135 (−4.3%)	135 (−4.3%)
284	297 (+4.6%)	289 (+1.8%)
388	454 (+17.0%)	309 (−20.4%)
509	612 (+20.2%)	425 (−16.5%)
Overall error	RMS = 13.6%	RMS = 13.3%

Figure 4.29. Three-disk laboratory rotor installed in test rig. Photograph Courtesy of Westhollow Research Center Graphics Dept., Shell Development Company, a division of Shell Oil Company.

the third mode is vertical, the horizontal foundation plays no part here and thus is not shown. The computed fourth mode at 140 Hz has no corresponding measured mode. The coast-down vibration spectrum of the test rig (Fig. 4.31) shows no peak in the response near this frequency. The responses to hammer blows on the shaft and housing showed only a very slight response near a frequency of 140 Hz. The computed log dec (from the real part of the complex eigenvalue) for this mode is seen to be the highest of the five, and it may be that the response in this mode is being overshadowed by the two lesser damped modes immediately above and below it. If we compare the errors of the first four modes, the error in the computed frequency of the 155 Hz mode is seen to be the highest of the lot. This is probably due to the characteristics of the foundation. The foundation model used in the program simulates the fundamental mode of the foundation at 149 Hz. The foundation also has a second mode at 225 Hz not accounted for in the model. Since the fourth rotor mode is approaching this frequency, more error may be expected for this mode than the others.

Another comparison was made by using the natural frequencies measured from rap tests on the shaft and bearing housing. One such set of measurements was made at a shaft speed of 4000 rpm. The measured natural frequencies and their corresponding computed values are shown in Table 4-13.

Measured Damped Eigenvalues for a Single-Stage Steam Turbine

This is the steam turbine for which the rotor and its free–free modes are described above. The computer model used for this rotor is given in Table 4-14. It makes use of the flexible disk feature described earlier.

TABLE 4-11. Computer Shaft Model for Three-Disk Laboratory Rotor

Individual Station Data

	LEN (in)	DO (in)	DI (in)	AWT (lbf)	AI (in^{-4})	AP (lb-s^2-in)	AT (lb-s^2-in)
1	.753	4.000	0.000	1.304	12.566	.007	.004
2	1.752	2.501	0.000	2.591	1.921	.009	.006
3	1.752	2.501	0.000	2.436	1.921	.005	.004
4	.762	4.000	0.000	2.573	12.566	.009	.006
5	1.800	2.500	0.000	2.605	1.917	.010	.006
6	2.021	2.000	0.000	2.149	.785	.004	.004
7	3.000	2.000	0.000	2.232	.785	.003	.005
8	4.993	9.900	0.000	54.773	471.531	1.728	1.160
9	2.021	2.010	0.000	55.347	.801	1.727	1.158
10	3.000	2.010	0.000	2.254	.801	.003	.005
11	4.996	9.900	0.000	54.765	471.531	1.727	1.159
12	2.012	2.005	0.000	55.316	.793	1.727	1.157
13	3.000	2.005	0.000	2.239	.793	.003	.005
14	5.004	9.900	0.000	54.845	471.531	1.730	1.162
15	2.075	2.005	0.000	55.432	.793	1.729	1.160
16	3.000	2.005	0.000	2.267	.793	.003	.005
17	2.756	2.500	0.000	3.255	1.917	.006	.009
18	1.750	2.500	0.000	3.130	1.917	.006	.007
19	1.760	2.500	0.000	2.438	1.917	.005	.004
20	2.200	2.500	0.000	2.751	1.917	.006	.005
21	1.000	1.010	0.000	1.641	.051	.003	.003
22	1.001	1.010	0.000	.229	.051	0.000	0.000
23	0.000	1.000	0.000	.111	.049	0.000	0.000

Length = 52.413   Shaft Weight = 366.683
Internal Friction Coefficient: 0
Material Properties are the Same for all Stations
    Mass Density = 0.283 pci
Young's Modulus = 30.0E + 06 psi
    Shear Modulus = 12.0E + 06 psi

Stations with Bearings
            #3
            #19

Bearing Data

K_{XX} (lb/in)	K_{XY}	K_{YX}	K_{YY}	C_{XX} (lb-s/in)	C_{XY}	C_{YX}	C_{YY}
606,000	60,600	−49,700	497,000	500	0	0	2,100
606,600	60,600	−49,700	497,000	500	0	0	2,100

TABLE 4-11. (*Continued*)

Foundation Parameters

	Horizontal			Vertical		
Station	Stiffness lb/in	Damping lb-s/in	Weight lb	Stiffness	Damping	Weight
3	208,000.0	55.6000	92.0	0.0	0.000	0.0
19	208,000.0	55.6000	92.0	0.0	0.000	0.0

The fluid-film bearings supporting this rotor are five-pad tilt-pad bearings with the following properties:

load on pad
L = 1.25 in
D = 3 in
arch length = 60°
radial clearance = 3 mils
preload = 0
bearing load = 156 lb

Based on the tilt-pad bearing measurements, a value was calculated for the vertical bearing stiffness and the horizontal stiffness was then taken as twice the vertical. (This is the reverse of generally accepted analytical results to date, but the measurements were made carefully and repeated.) The damping coefficients were calculated using Ref. [20].

TABLE 4-12. Measured and Calculated Critical Speeds for the Three-Disk Laboratory Rotor

Measured		Computed	
Frequency (Hz)	Log Dec	Frequency (Hz)	Log Dec
30.5	0.11	32.5 (6.56%)	0.05
108.6	0.10	103.2 (−4.97%)	0.30
125.7	0.10	128.1 (1.91%)	0.25
Not measured		140.0	0.62
155.0	0.08	169.9 (9.61%)	0.40
Overall error: Avg. = 5.76%			

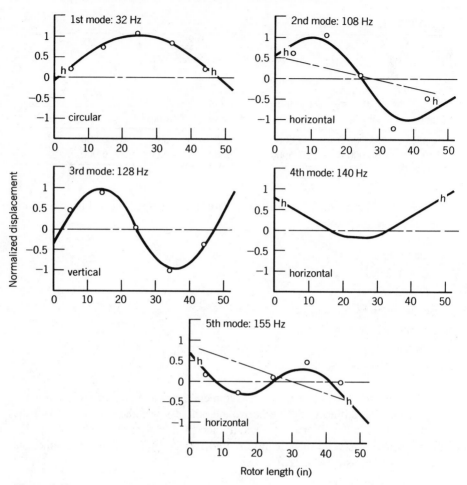

Figure 4.30. Comparison of measured and calculated mode shapes of the three-disk laboratory rotor.

When installed, the steam turbine drives the compressor through a gear coupling arrangement. The steam turbine therefore has a coupling hub mounted on one end, which in turn is connected to the power transmission shaft. This inertia must be carried by the steam turbine (all of the hub and part of the shaft) and therefore should be included in the model. Detailed inertia measurements were not available so 60 lb was used as the best estimate of the combined effect of the hub and shaft.

When this machine was purchased, the manufacturer predicted all critical speeds to be well above the top operating speed of 11,500 rpm. A rigid-support analysis performed with no coupling hub yields a first critical speed of 13,800 rpm. With

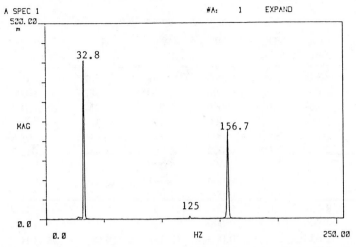

Figure 4.31. Instantaneous vibration spectrum taken during coast-down of the three-disk laboratory rotor.

the coupling and tilt-pad bearing parameters included, the computed critical speeds are in closer agreement to the measured eigenvalues shown in Figure 4.32 and Table 4-15. (A discussion of the measurement technique can be found in Chapter VIII.) The logarithmic decrements could not be obtained from the available field data.

The accuracy of the computed critical speeds is excellent in view of the fact that the coupling characteristics were approximated and the bearing parameters were obtained in an indirect manner from laboratory measurements on similar bearings. Since the dynamics of the machine's foundation are unknown and were not included in the computer model, they are apparently unimportant for this particular case.

TABLE 4-13.   Rap Tests vs. Calculated Natural Frequencies for Three-Disk Laboratory Rotor at 4000 rpm

Measured (Hz)	Computed (% Error): (Hz)
31.25	32.7 (4.64%)
110.0	105 (−4.55%)
125.0	128 (2.40%)
157.0	169 (7.64%)
	Overall error: Avg. = 4.81%

TABLE 4-14. Computer Shaft Model for Single-Stage Steam Turbine Rotor

Individual Station Data

	LEN (in)	DO (in)	DI (in)	AWT (lbf)	AL (in^{-4})	AP $(lb\text{-}s^2\text{-}in)$	AT $(lb\text{-}s^2\text{-}in)$
1	2.110	1.998	0.000	25.899	.782	.348	.194
2	2.000	1.998	0.000	1.860	.782	.002	.003
3	1.500	2.625	0.000	2.036	2.331	.004	.003
4	1.500	2.625	0.000	2.297	2.331	.005	.004
5	2.000	2.994	0.000	3.141	3.944	.008	.006
6	2.000	2.994	0.000	3.985	3.944	.012	.009
7	3.250	2.994	0.000	5.230	3.944	.015	.017
8	2.000	3.993	0.000	6.782	12.479	.028	.024
9	2.000	3.993	0.000	7.089	12.479	.037	.024
10	2.750	5.006	0.000	11.203	30.827	.080	.056
11	4.000	5.006	0.000	18.799	30.827	.152	.127
12	2.050	6.125	0.000	19.687	69.087	.194	.143
13	1.000	6.125	0.000	89.772	7269.087	10.698	5.358
14	2.050	6.125	0.000	12.716	69.087	.154	.086
15	2.700	5.006	0.000	16.067	30.827	.165	.102
16	3.000	5.006	0.000	15.875	30.827	.129	.092
17	4.050	3.990	0.000	15.521	12.441	.105	.094
18	3.100	2.995	0.000	10.256	3.950	.046	.055
19	2.300	2.995	0.000	5.383	3.950	.016	.017
20	2.300	2.995	0.000	4.586	3.950	.013	.012
21	2.550	2.500	0.000	4.064	1.917	.010	.010
22	2.000	2.500	0.000	3.160	1.917	.006	.007
23	.550	1.510	0.000	61.529	.255	.603	.303
24	.001	1.510	0.000	.140	.255	0.000	0.000
25	0.000	1.000	0.000	0.000	.049	0.000	0.000

Length = 52.761   Shaft Weight = 347.074

Internal Friction Coefficient: 0

Material Properties are the Same For all Stations

Mass Density = 0.283 pci

Young's Modulus = 30.0E + 06 psi

Shear Modulus = 12.0E + 06 psi

Bearing Data

K_{XX} (lb/in)	K_{XY}	K_{YX}	K_{YY}	C_{XX} (lb-s/in)	C_{XY}	C_{YX}	C_{YY}
700,000	0	0	350,000	1,200	0	0	1,500
700,000	0	0	350,000	1,200	0	0	1,500

Floppy Disk Parameters

Station #	Stiffness (in-lb/rad)
13	240,000,000.0

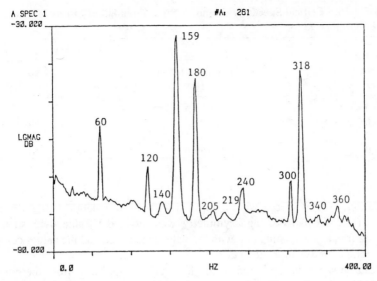

Figure 4.32. Vibration spectrum of the single-stage steam turbine, taken at constant speed = 158 Hz (9480 rpm).

The computed whirl directions are given for this case since the measured frequency values were taken from a spectrum generated by the turbine running at a constant 9500 rpm. All the measured frequencies except 158 Hz were nonsynchronous; they could have been either forward or backward whirl.

Importance of Foundation Dynamics and Bearing Asymmetry

Although foundation dynamics are apparently unimportant in the case just presented, the three-disk lab rotor is a counterexample. To show the effect of the

TABLE 4-15. Measured and Calculated Eigenvalues for Single-Stage Steam Turbine

Measured	Computed		
Frequency (Hz)	Frequency (Hz)	Log Dec	Whirl Direction
140	142 (1.42%)	0.18	Backward
158[a]	150	0.17	Forward
180[b]	185	0.55	Backward
205	201 (1.95%)	0.68	Forward
	Overall error: Avg. = 1.69%		

[a]Running speed: large amplitude.
[b]Harmonic of 60-Hz electrical noise.

TABLE 4-16. Critical Speeds With and Without Foundation Effects for the Three-Disk Laboratory Rotor

Measured Frequency (Hz)	Computed Frequency	
	With Foundation (Hz)	No Foundation (Hz)
30.5	32.5	32.8
108.6	103.2	Not predicted
125.7	128.1	139.8
155.0	169.9	Not predicted

foundation, critical speeds were computed with the foundation omitted from the model. (The tilt-pad bearing asymmetry was retained.) Table 4-16 shows the resulting frequencies, compared with the values measured and the values computed with foundation dynamics included.

It can be seen that omission of the foundation impedance from the computer model eliminates two of the four predicted critical speeds in the operating range.

Many critical speed codes developed in past years have a symmetric model incorporated for bearing stiffness ($K = K_{XX} = K_{YY}$). To show the effect of this simplification, critical speeds were calculated under the assumption of symmetric bearing stiffness for both the three-disk lab rotor (Table 4-17) and the single-stage steam turbine (Table 4-18). The symmetric bearing stiffnesses were taken to be the average of the vertical and horizontal values. The computer program was a damped eigenvalue code; and in the case of the lab rotor, foundation effects were retained in the model.

The tables show the new computed values of frequency compared to (1) the measured values and (2) the previously presented values computed with an asymmetric bearing model. It can be seen that, at least for the two machines studied here, a computer model with symmetric bearing stiffness can be used with no

TABLE 4-17. Critical Speeds With and Without Bearing Asymmetry for the Three-Disk Laboratory Rotor

Measured Frequency (Hz)	Computed Frequency	
	Asymmetric Bearings (Hz)	Symmetric Bearings (Hz)
30.5	32.5	32.6
108.6	103.2	102.4
125.7	128.1	128.1
Not measured	140.0	137.0
155.0	169.9	168.7

TABLE 4-18.   Eigenvalues With and Without Bearing Asymmetry for the Single-Stage Steam Turbine

Measured Frequency (Hz)	Computed Frequency	
	Asymmetric Bearings (Hz)	Symmetric Bearings (Hz)
140	142	141.7
158	150	150.6
180	185	184.8
205	201	199.6

sacrifice in critical speed accuracy when the stiffness is taken as the average of K_{XX} and K_{YY}.

Additional parameter studies on the computer showed that this insensitivity of the critical speeds to bearing asymmetry is valid only when the bearings have damping values that are at least as large as those predicted for tilt-pad bearings by Ref. [20]. With the computing power now available even in desktop computers, there is little reason not to include both damping and support asymmetry in eigenvalue codes, although an undamped symmetric critical speed code may still be found useful for quick preliminary estimates or for special cases.

REFERENCES

1. Yim, K. B., *Load-Induced Rotordynamic Instabilities in Turbomachinery*, Ph.D. Dissertation in Mechanical Engineering, Texas A&M University (December 1984), pp. 144–146.

2. Myklestad, N. O., "The Concept of Complex Damping," *Journal of Applied Mechanics*, pp. 284–287 (September 1952).

3. Timoshenko, S., and MacCullough, G. H., *Elements of Strength of Materials*, 3rd ed. Van Nostrand Reinhold, New York, 1949, Chap. VI.

4. Den Hartog, J. P., *Mechanical Vibrations*, 4th ed., McGraw-Hill, New York, 1956, p. 265.

5. Kleespies, H. S., *Calculation of Rotordynamic Unbalance Response Including Torque and Cross-Coupled Stiffness and Damping Effects*, M.S. Thesis in Mechanical Engineering, Texas A&M University (December 1986).

6. Yim [1], pp. 39–47.

7. Prohl, M. A., "A General Method for Calculating Critical Speeds of Flexible Rotors," *Journal of Applied Mechanics*, pp. A-142–A-148 (September 1945).

8. Myklestad, N. O., "A New Method of Calculating Natural Modes of Uncoupled Bending Vibration of Airplane Wings and Other Types of Beams," *Journal of the Aeronautical Sciences*, pp. 153–162 (April 1944).

9. Lund, J. W., "Stability and Damped Critical Speeds of a Flexible Rotor in Fluid-Film Bearings," *Journal of Engineering for Industry*, pp. 509–517 (May 1974).

10. Lund, J. W., and Orcutt, F. K., "Calculations and Experiments on the Unbalance Response of a Flexible Rotor," ASME *Journal of Engineering for Industry*, November 1967, pp. 785–796.

11. Vance, J. M., Murphy, B. T., and Tripp, H. A., "Critical Speeds of Turbomachinery: Computer Predictions vs. Experimental Measurements," *Proceedings of the Thirteenth Turbomachinery Symposium, Texas A&M University, November 13–15, 1984*, p. 105–130.

12. Vance, J. M., Murphy, B. T., and Tripp, H. A., "Critical Speeds of Turbomachinery: Computer Predictions vs. Experimental Measurements, Parts I and II," ASME Papers 85-DET-145 and 85-DET-146; also published in the *Journal of Vibration, Acoustics, Stress, and Reliability in Design*, **109**(1), 1–14 (1987).

13. Dopkin, J. A., and Shoup, T. E., "Rotor Resonant Speed Reduction Caused by Flexibility of Disks," *Journal of Engineering for Industry*, pp. 1328–1333, (November 1974).

14. R. J. Roark, *Formulas for Stress and Strain*, 4th ed. McGraw-Hill, New York, 1965, p. 242.

15. Nicholas, J. C., "Improving Critical Speed Calculations Using Flexible Support Modal Analysis Compliance Data," *Proceedings of the 15th Turbomachinery Symposium, Texas A&M University, November 10–13, 1986.*

16. Litang, Y., "Dynamic Analysis of Complex Composition Rotor Systems with Substructure Transfer Matrix Method," ASME Paper No. 85-GT-74, presented at the ASME Gas Turbine Conference, Houston, March 18–21, 1985.

17. Nelson, H. D., and McVaugh, J. M. "The Dynamics of Rotor–Bearing Systems Using Finite Elements," *Journal of Engineering for Industry*, pp. 594–600 (May 1976).

18. Rouch, K. E., *Finite Element Analysis of Rotor–Bearing Systems with Matrix Reduction*, Ph.D. Dissertation in Mechanical Engineering, Marquette University (December 1977).

19. Tripp, H., and Murphy, B. T., "Eccentricity Measurements on a 5-Pad Tilting-Pad Bearing," *ASLE Journal*, **28**(2), 217–224 (1984).

20. Nicholas, C. J., Gunter, E. J., Jr., and Allaire, P. E., "Stiffness and Damping Coefficients for the Five-Pad Tilting Pad Bearings," *ASLE Journal*, **22**(2), 113–124 (1979).

Chapter V

Rotor Balancing in Turbomachinery

The most common source of vibration in turbomachinery is rotor imbalance. If the center of mass of a rotating disk or shaft does not lie on the axis of rotation, it consequently orbits about the axis and generates a centrifugal force which must be reacted by the bearings and support structure. Since the force rotates at shaft speed, the vibrating frequency in the nonrotating structure is synchronous.

Even if the center of mass does lie on the axis of rotation, a misalignment of the principal axis of inertia will produce a rotating couple which will also excite synchronous vibration in the machine structure.

The rotating imbalance forces produce a whirling motion of the rotor and shaft known as synchronous whirl, or synchronous response to imbalance. (See Chapter I for an analysis of synchronous whirl in the Jeffcott rotor.)

As stated in Chapter I, there are three basic ways to reduce amplitudes of synchronous whirl: (1) balance the rotor, (2) change the speed (farther away from the closest critical speed), and (3) add damping. This chapter is concerned with method 1, which attacks the problem at its source.

In practice, real rotors can never be perfectly balanced because of measurement errors and because the rotating masses are not rigid, but high levels of synchronous vibration can almost always be reduced significantly by balancing.

CONCEPTS AND PRINCIPLES FOR RIGID-ROTOR BALANCING

Balancing would be much simpler if the shape of the rotor and shaft were invariable, even at high rotational speeds. Under some conditions,[1] the assumption of a

[1]Some examples are (1) low shaft speeds, (2) highly flexible bearing supports, compared to the shaft, and (3) narrow range of operating speeds.

171

rigid rotor is approximately true, justifying and allowing the use of rigid rotor balancing methods.

A rigid rotor is said to be perfectly balanced when a principal axis of inertia passing through the center of gravity (C.G.) is coincident with the designed axis of rotation (usually the geometric axis of revolution for the bearing journals). Some engineers use the following alternative definition for balancing in the field. A rotor is said to be perfectly balanced when the measured synchronous vibration in the machine is reduced to zero. It will be shown later in this chapter that this latter condition does not necessarily eliminate the vibratory bearing loads due to rotor imbalance.

A rigid rotor can be balanced by adding (or removing) correction weights in any two (separate) planes normal to the shaft axis. The correction weights are added to translate the C.G. until it lies on the rotational axis and to rotate the principal axis of inertia passing through the C.G. until it is parallel to (or coincident with) the rotational axis.

Also, the vibration measurements required to calculate the magnitude and location of the correction weights can be made in any two planes (not necessarily the same as the correction weight planes), normal to the shaft axis.

Figure 5.1(a) shows a rigid rotor of mass m, initially in perfect balance, with a coordinate system xyz fixed in the rotor and having its origin O at the center of mass. A small mass m_1 is added at radius R on the y axis to produce a *static imbalance*, expressed as

$$m_1 R = (m + m_1) u \quad \text{oz-in (g-cm)}, \tag{5-1}$$

or

$$u = \frac{m_1 R}{m + m_1} \quad \text{in (mm)}. \tag{5-2}$$

The latter quantity u is the static imbalance expressed as a *C.G. offset*—an offset of the center of gravity from the axis of rotation [see Fig. 5.1(b)].

If the unbalanced rotor is rested on knife-edge supports or mounted on very low friction bearings, gravity will rotate it until the y axis (and m_1) is down, thereby allowing the imbalance to be located. Thus, the basis of the ''static imbalance'' terminology is the idea that the direction of C.G. offset can be located without spinning up the rotor. Some balancing methods require rotor spinup to locate the C.G. but nevertheless accomplish only static balancing, since they do not locate and correct the principle axis angular misalignment as described below.

Figure 5.2 shows the same rotor, but with the imbalance mass m_1 moved axially some distance z_1 from the y axis. In both Figs. 5.1 and 5.2 the xyz axes are fixed in the rotor with z coincident with the axis of rotation.

The rotor in Fig. 5.2 still has the same static imbalance as in Fig. 5.1, but in addition it also has a *couple imbalance* due to the moment arm z_1. The couple imbalance is expressed as

$$P_{yz} = \int yz\, dm = m_1 z_1 R \quad \text{oz-in}^2 \text{ (g-cm}^2) \tag{5-3}$$

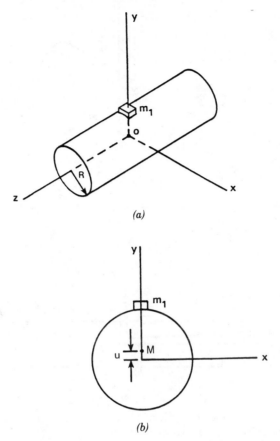

(a)

(b)

Figure 5.1. Static imbalance of a rigid rotor. (a) Location of m_1 on the y axis. (b) C.G. offset produced by m_1.

or

$$\alpha = \frac{P_{yz}}{I_y - I_z} \quad \text{mrad} \tag{5-4}$$

where I_y and I_z are the mass moments of inertia about the y and z axes, respectively.

The couple imbalance expressed as equation (5-3) is the product of inertia in the yz plane [1]. The alternative expression (5-4) is the angular misalignment between the principal axes z' of the unbalanced rotor and z of the initially balanced rotor (without m_1). The z axis is also the geometric axis of revolution of the bearing journals and can therefore be considered as the axis of rotation or bearing axis.

If the imbalance mass m_1 is placed in the xz plane (instead of the yz plane), it produces a product of inertia P_{xz} and the angle α lies in the xz plane instead of the yz plane. In general, there will be nonzero products of inertia in both planes.

Note that the unbalanced rotors in Figs. 5.1 and 5.2 can both be balanced by adding a single correction mass, equal to m_1, 180° around from m_1 at radius R in

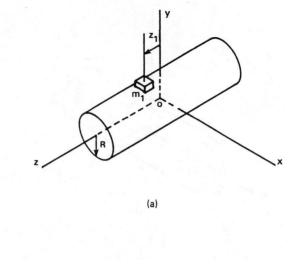

(a)

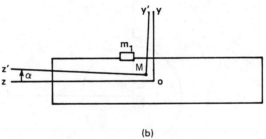

(b)

Figure 5.2. Couple and static imbalance of a rigid rotor. (a) Location of m_1 off the y axis. (b) Misalignment of the principal axes produced by m_1.

a (single) axial plane. In general, for a rigid rotor, two planes are required. For example, Fig. 5.3 shows the rotor from Fig. 5.2 with another imbalance mass, equal to m_1, added at coordinates $x = 0$, $y = -R$, $z = -z_1$. The resulting condition of rotor imbalance is purely couple (principal axis misalignment), since the C.G. now remains on the z (rotational) axis and cannot be corrected by any combination of balance masses added in a single axial plane.

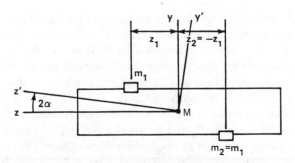

Figure 5.3. Purely couple imbalance.

For some very short rotors, couple imbalance may not be significant due to the small ratio of z_1/R [see Fig. 5.2(a)]. Such a rotor can be balanced by adding correction masses in a single plane.

In any specified axial plane normal to the rotational axis, any number of imbalance masses at specified circumferential locations can be resolved into x and y components, or into a single equivalent mass m_e at a resultant angle θ by vector addition, as shown in Fig. 5.4.

The x and y components of the resultant mass imbalance from m_1, m_2, and m_3 are

$$m_x = m_1 \cos \theta_1 - m_2 \sin \theta_2 + m_3 \cos \theta_3, \tag{5-5}$$

$$m_y = m_1 \sin \theta_1 + m_2 \cos \theta_2 - m_3 \sin \theta_3. \tag{5-6}$$

The single equivalent imbalance mass is given by

$$m_e = \sqrt{m_x^2 + m_y^2} \tag{5-7}$$

and is located circumferentially at

$$\theta_e = \tan^{-1} \left(\frac{m_y}{m_x} \right). \tag{5-8}$$

Similarly, in either the xz plane or the yz plane, any number of imbalance masses at specified axial locations can be resolved into a single equivalent mass m_x' or m_y' at a resultant axial location z_x or z_y.

For example, the three imbalance masses in Fig. 5.5 produce a product of inertia (couple imbalance) given by

$$P_{yz} = \left(m_{y_1} z_1 - m_{y_2} z_2 - m_{y_3} z_3 \right) R. \tag{5-9}$$

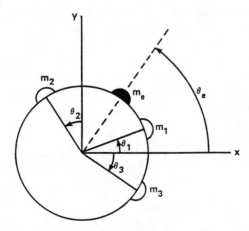

Figure 5.4. Resolution of three imbalance masses in an axial plane.

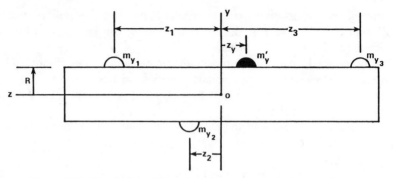

Figure 5.5. Resolution of three imbalance masses in the y–z plane.

The equivalent mass in the yz plane is given by

$$m_y' = m_{y_1} + m_{y_2} + m_{y_3}. \tag{5-10}$$

The resultant location is calculated from

$$z_y = \frac{m_{y_1} z_1 - m_{y_2} z_2 - m_{y_3} z_3}{m_y'}. \tag{5-11}$$

All of the above statements and equations with regard to the resolution of imbalance masses apply also to balance (correction) masses.

Notice that in Figs. 5.1–5.5 the imbalance masses are all attached at the same radius R. In practice, the axial location of the two balancing planes and the radius at which correction masses are attached are often fixed by the design of the machine. If the radius for each mass is different, then R cannot be factored out of equations (5-5)–(5-11) (as was done) and the masses m_i must be replaced by the product $m_i R_i$, where R_i is the attachment radius for the ith mass ($i = 1, 2, 3$).

To confirm that two separate balancing planes are required to balance a rigid rotor, consider a rotor of mass m with a general state of imbalance (called "dynamic imbalance"); that is, static imbalance described by a C.G. offset u at an angle θ_e from the x axis (see Fig. 5.4) and couple imbalance described by products of inertia P_{yz} and P_{xz}. Choose a balancing plane located at some axial distance z_1 from the C.G., with attachment radius R_1. The static imbalance can be corrected by a balance mass $m_1 = mu/R$, attached at $\theta_e + 180°$, which has x and y components

$$m_{x_1} = \left(\frac{mu}{R_1}\right) \cos\left(\theta_e + 180°\right) \tag{5-12}$$

and

$$m_{y_1} = \left(\frac{mu}{R_1}\right) \sin\left(\theta_e + 180°\right), \tag{5-13}$$

respectively. But the new products of inertia are not zero, as desired, but instead are given by

$$P'_{xz} = P_{xz} + m_{x_1} R_1 z_1 \tag{5-14}$$

and

$$P'_{yz} = P_{yz} + m_{y_1} R_1 z_1, \tag{5-15}$$

where the last term in equations (5-14) and (5-15) were predetermined to correct the static imbalance.

If a second balancing plane is now added at axial location z_2 ($\neq z_1$), the x and y components of the two balance masses (one in each plane) required to correct the static imbalance must satisfy the equations

$$m_{x_1} + m_{x_2} = \left(\frac{mu}{R_1} \right) \cos \left(\theta_e + 180° \right) \tag{5-16}$$

and

$$m_{y_1} + m_{y_2} = \left(\frac{mu}{R_1} \right) \sin \left(\theta_e + 180° \right). \tag{5-17}$$

Since there are now four balance mass components to be determined, the two products of inertia can also be balanced out by requiring

$$P_{xz} + m_{x_1} R_1 z_1 + m_{x_2} R_2 z_2 = 0 \tag{5-18}$$

and

$$P_{yz} + m_{y_1} R_1 z_1 + m_{y_2} R_2 z_2 = 0. \tag{5-19}$$

Equations (5-16)–(5-19) constitute a system of four algebraic equations in the four unknowns m_{x_1}, m_{x_2}, m_{y_1}, m_{y_2}. A solution of the equations allows the balance mass and its circumferential location in each plane to be determined as follows:

In plane 1, the balance mass is

$$m_1 = \sqrt{m_{x_1}^2 + m_{y_1}^2} \tag{5-20}$$

and should be located at an angle

$$\theta_1 = \tan^{-1} \left(\frac{m_{y_1}}{m_{x_1}} \right) \tag{5-21}$$

from the x axis.

In plane 2, the balance mass is

$$m_2 = \sqrt{m_{x_2}^2 + m_{y_2}^2} \tag{5-22}$$

and should be located at an angle

$$\theta_2 = \tan^{-1}\left(\frac{m_{y2}}{m_{x2}}\right) \tag{5-23}$$

from the x axis. Later in this chapter, the vectors m_1, θ_1 and m_2, θ_2 will be denoted as $\mathbf{M}_1$ and $\mathbf{M}_2$, respectively.

In practice the static imbalance vector (u, θ_e) usually cannot be measured separately from the couple imbalance (P_{xz}, P_{yz}), so equations (5-16)–(5-19) are not used directly in actual balancing procedure. Rather, a given state of rigid rotor imbalance is represented by the negatives of the balance mass vectors m_1, θ_1 [equations (5-20) and (5-21)] and m_2, θ_2 [equations (5-22) and (5-23)]. Individually, these vectors depend on the choice of balance planes 1 and 2, but they must always produce the same resultant static and couple imbalance.

EFFECT OF ROTOR FLEXIBILITY

Whenever conventional balancing techniques fail to reduce the level of synchronous vibration to acceptable levels throughout a desired range of shaft speed, it is usually because rotor flexibility has not been taken into account. If the operating speed range approaches or exceeds any of the critical speeds with mode shapes which contain a significant amount of shaft bending (see Chapters I and IV), the state of balance will vary with shaft speed. This is because elastic deformation of the rotor redistributes the mass about the rotational axis, which may result in a translation of the center of mass and/or a change in the angular orientation of the principal axes of inertia with respect to the rotational axis.

Figure 5.6 shows how the effect of balance masses on a flexible rotor changes with speed. The dark-shaded mass represents an initial unit imbalance on the uniform shaft. When we use balance correction planes A and B, rigid-rotor balancing theory requires the correction masses of one-half unit each, as shown in Fig. 5.6(a), to put the rotor in both static and couple balance.

At speeds close to the first critical speed (on stiff bearing supports), the mode shape will be a half sine wave, as shown in Fig. 5.6(b). Deflection of the rotor into this mode will be readily induced by the bending moments imposed by the centrifugal forces of the imbalance mass and correction masses. Furthermore, the imbalance mass will be displaced by the mode shape to a larger effective radius than the correction masses. The effect of the redistribution of mass will be to produce a static imbalance, with consequent rotating reactions at the bearing. The correction masses in planes A and B must be increased in size to move the center of mass of the deflected rotor back to the rotational axis, but this will compromise the state of balance at lower speeds.

Figure 5.6(c) shows the mode shape at speeds close to the second critical speed, which produces a product of inertia (couple imbalance), with consequent rotating forces at the bearings.

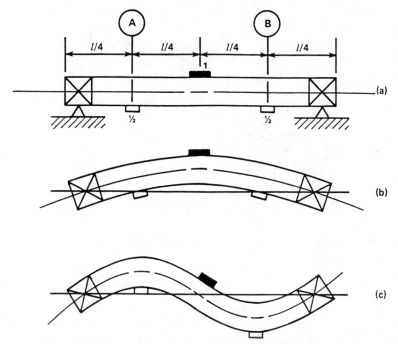

Figure 5.6. Balance mass distributions in a flexible rotor.

Intuitively, it seems that the rotor could be balanced for any one condition 5.6(a), 5.6(b), or 5.6(c) by adding different correction masses in planes A and B, but not for two (or three) of the conditions simultaneously. This has been demonstrated by experiments and analysis, so that a rule can be stated: *If only two correction planes are used, a flexible rotor can be balanced for only one speed.*

Furthermore, it can be seen that the principle of "indifferent dynamic equilibrium" at the critical speeds makes it very difficult to balance a rotor at all when it is operating close to a critical speed with a significant amount of bending in the associated mode shape.

Ideally, the number of balancing planes should be equal to the number of critical speeds traversed [2], and they should be carefully selected by considering the associated mode shapes. In contemporary turbomachinery, however, more than two correction mass planes are seldom conveniently accessible.

BALANCING MACHINES VS. IN-PLACE BALANCING

Most high speed turbomachinery rotors should be balanced on balancing machines at the time of manufacture (or overhaul) and again "in-place" after installation in the bearings.

The balancing machine is needed to balance each rotating component separately

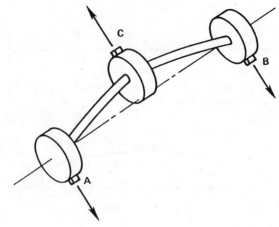

Figure 5.7. The effect of correction masses added at *A* and *B* to compensate for the central wheel imbalance at *C*.

and to balance each rotating subassembly after each new piece is added. Figure 5.7 shows a possible consequence of balancing only the final assembly of a rotor, when an individual component is unbalanced. In this case the originally unbalanced component is the central wheel, and the correction planes *A* and *B* for rigid-rotor balancing are on the end wheels. The assembly would only be properly balanced for low speed operation, with the shaft undeflected. High speed operation would bend the shaft, putting the rotor out of balance. Each of the wheels, as well as the shaft, should be balanced separately before assembly of the rotor to prevent this.

Even after balancing rotor components and the rotor assembly in a balancing machine, it is often necessary to "trim balance" the rotor "in-place" after installation in its bearings. One reason is that the rotor must be disassembled for installation in some turbomachines and the reassembly will never be exactly the same. Another reason is that the bearing stiffnesses have an influence on the whirling mode shapes and the balancing machine bearing stiffnesses will be different from those in actual operation. Rotor–bearing system damping will be different also, due to contributions from seals and aerodynamic forces.

The theory and operation of balancing machines is described in Ref. [3] and will not be addressed here. The rest of this chapter is devoted to a description of methods for "in-place" balancing of turbomachinery rotors.

INSTRUMENTATION AND MEASUREMENTS FOR BALANCING

Sophisticated and expensive electronic instrumentation is not required to balance rotating machinery, but it can greatly speed up the process by reducing the number of trials of different balance masses.

The minimum required instrumentation capability to balance a rotor is simply some way to measure the undesirable vibration due to imbalance. Several sizes of balance mass can be tried at different locations on the rotor until an acceptable level of vibration is attained. If a record is kept of the effect of each balance mass in each location, an orderly and "scientific" process can be evolved to intelligently select the magnitude and location of successive balance masses. A knowledge of the concepts already presented in this chapter can help make this process more efficient.

If both the vibration level and the speed can be measured fairly accurately (± 10 percent), then vibration measurements taken for four different balance mass locations can be used to *calculate* (by the "four-run" method described below) the required correction mass for single-plane balancing ("short" rotors).

If, in addition, the phase angle lag from a reference mark on the rotor to the rotating vibration vector ("high spot") can be measured, then more accurate single-plane balancing can be accomplished. If these measurements (vibration level, speed, and phase angle) can be made at two or more axial locations, then two-plane or multiplane "influence coefficient" balancing becomes possible. These methods allow the correction masses in two or more planes to be calculated from the measurements made with trial masses in each plane.

The most commonly used vibration transducers are the eddy-current proximity probe, which measures vibratory displacement of the shaft, and the accelerometer, which measures vibratory acceleration of a nonrotating part of the machine structure. Either can be used for balancing, and a velocity probe will also suffice. If the vibration signal is not predominantly synchronous, it must be passed through a narrow-band filter set to the frequency of shaft speed, since it is only the synchronous vibration which is produced by imbalance.

For phase angle measurements, a once-per-revolution voltage pulse should be generated, which is sometimes called a "keyphasor." The keyphasor signal indicates the angular position of the shaft at the time of the pulse and therefore locates an angular reference position from which angles can be measured in the rotor. Alternatively, a white mark on the rotor can be observed using a stroboscope light triggered by the vibration signal, and the phase angle measurement can be made visually.

The keyphasor signal or the stroboscope can also be used to actuate a tachometer for speed measurement.

A portable oscilloscope can be very helpful in balancing by the influence coefficient methods and is required when using the orbit method. A dual-trace model is the preferred choice.

There are several specialized "trim balance analyzers" available which take the signals from vibration transducers (and keyphasors) and electronically produce the required amplitudes, phase angles, and speeds. Examples are the Bently-Nevada DVF-2 and the Spectral Dynamics SD-119C. They make the balancing procedures described below more accurate and efficient, but are not required.

Figures 5.8(a) and 5.9(a) show how vibration and keyphasor signals can be fed into an oscilloscope to display the required information for two different balancing methods.

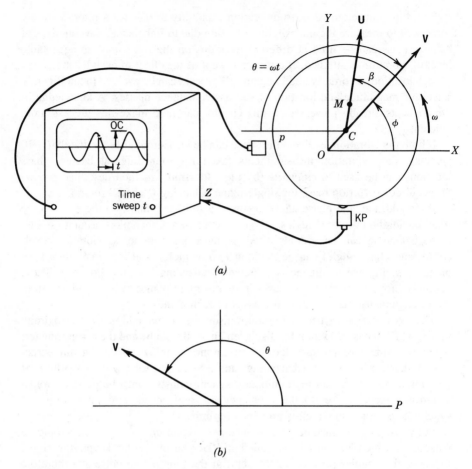

Figure 5.8. (a) Measurements for balancing by the influence coefficient method. (b) Phasor representation of the vibration vector.

SINGLE-PLANE BALANCING BY THE ORBIT METHOD

Rotors with most of the mass concentrated near a single plane (i.e., not axially distributed) can sometimes be balanced by using only one balance correction plane and one plane for vibration measurements (not necessarily the same). The inherent assumption is that the couple imbalance is negligible so that the imbalance is primarily static (C.G. offset).

Vibration displacement signals generated by transducers in two perpendicular directions (e.g., horizontal and vertical) on a shaft can be connected to an oscilloscope to display the whirl orbit as shown in Fig. 5.9. If the keyphasor signal (*KP*) is connected to the *z* axis blanking input at the back of the oscilloscope, the orbit trace will be blanked, or interrupted, at the time the keyphasor voltage pulse

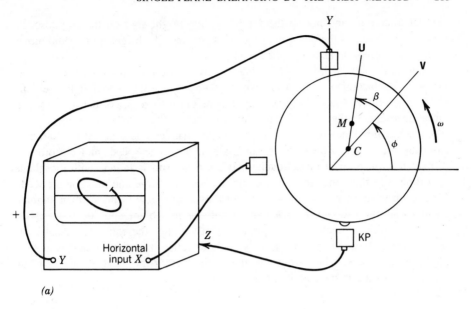

(a)

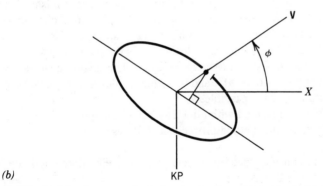

(b)

Figure 5.9. (a) Measurements for balancing by the orbit method. (b) Graphical determination of the location of the vibration vector by the orbit method.

is generated. This gives the angular orientation of the rotor and allows the vibration vector to be located with respect to a circumferential reference mark, such as the discontinuity which excites the keyphasor transducer.

For example, in Fig. 5.9 the displayed orbit and blank spot shows that the vibration vector **V** is located approximately 45° counterclockwise (CCW) from the *x* axis when the rotor is positioned with the keyphasor screw straight down. Care must be taken to set the correct polarity of the vibration transducers. This can be done by loading or bumping the rotor (when not rotating) in the horizontal and vertical directions and noting the motion of the bright spot on the oscilloscope screen (which marks the position of the shaft center). The bright spot should move in the same direction as the applied load.

In general, there will be a phase lag β between the rotor imbalance **U** and the vibration displacement vector **V**. However, at speeds less than about eighty percent of the first critical speed the phase lag will be close to zero, so that locating the vibration vector in the rotor is equivalent to locating the imbalance vector.

Once the imbalance is located, a trial balance mass can be attached 180° around from the imbalance. The rotor is run again at the same speed, and the orbit is displayed with the trial mass attached. If the new orbit is larger, with the blank spot flipped to the opposite side, the trial mass is too large and must be reduced. If the orbit is smaller, with the blank spot moved to a new location, then another smaller trial mass should be attached opposite the new blank spot location.

If the original imbalance was predominantly static, the rotor can usually be balanced in only a few iterations. This method, called the orbit method, was first published by Jackson [4]. Note that the method gives the location, but not the magnitude, of the imbalance in a single plane.

SINGLE-PLANE BALANCING BY THE FOUR-RUN METHOD

The four-run method allows single-plane balancing to be accomplished without phase measurement and without a keyphasor signal.[2] Three circumferential locations are marked on the rotor approximately 120° apart, at places where it is possible to attach trial balance masses. If location 1 is the reference, the trial mass locations are at $\Psi_1 = 0$, $\Psi_2 = 120°$, $\Psi_3 = 240°$. The angle between any two locations should not be less than 90°, and the actual angles Ψ_1, Ψ_2, and Ψ_3 should be determined within ±10° if possible.

After a transducer is installed with a meter or other readout device to measure the vibration amplitude, the method proceeds as follows:

1. A convenient constant speed is selected, and the vibration amplitude R_0 is measured and recorded at that speed.

2. A trial mass m' is attached at location 1 ($\Psi = 0$), the rotor is brought up to the same speed selected in step 1, and the new vibration amplitude R_1 is measured and recorded.

3. The trial mass m' is moved to location 2 ($\Psi = \Psi_2$), the rotor is again brought up to the selected speed, and the vibration amplitude R_2 is measured and recorded.

4. The trial mass m' is moved to location 3 ($\Psi = \Psi_3$), the rotor is brought up to the selected speed for the final measurement, and the vibration amplitude R_3 is measured and recorded.

---

[2]Provided that the rotor can be brought up to the same speed four times, sequentially. Some type of tachometer may be required to measure the speed.

The vibration amplitude measurements can be displacement, velocity, or acceleration and can be expressed in any convenient units, including volts. The same units must be used for all measurements. If the attachment radii for the trial mass are not the same at all three locations, then the vibration amplitudes must be modified by the ratio of the attachment radii to compensate (e.g., if the attachment radius is halved the measured amplitude must be doubled).

The vibration amplitude measurements R_0, R_1, R_2, and R_3 can be used to determine the balance correction weight from a graphical construction as shown in Fig. 5.10. A base circle of radius R_0 is drawn, and points 1, 2, and 3 are marked on the circle at angles $\Psi_1 = 0$, $\Psi = \Psi_2$, and $\Psi = \Psi_3$ measured from the x axis.

Using the same scale factor, an arc of radius R_1 is scribed from an origin at point 1, an arc of radius R_2 is scribed from an origin at point 2, and an arc of radius R_3 is scribed from an origin at point 3.

If the trial mass was well chosen, the arcs of radii R_1, R_2, and R_3 will come close to intersecting at a point as shown in Fig. 5.10. The radius $\overline{R}$ to the point of intersection is used to calculate the magnitude of the balance mass as

$$m = \frac{R_0}{\overline{R}} m'. \qquad (5\text{-}24)$$

The trial mass m' is then removed, and the balance mass m is attached to the

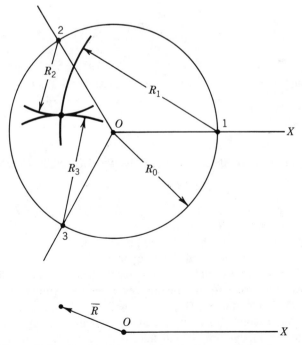

Figure 5.10. Graphical construction for four-run balancing.

rotor at the same angular location as $\overline{R}$. If the attachment radius for the balance mass m cannot be the same as the basis that was used for the trial mass m', then the magnitude of the balance mass must be modified accordingly.

This method was presented by Blake and Mitchell in their book on machinery vibration [5]. The balance mass and its location can also be determined by mathematical calculation, which is adaptable to a computer or pocket calculator, as follows [6]:

The balance mass magnitude is

$$m = \frac{m'}{\sqrt{R_x^2 + R_y^2}}, \tag{5-25}$$

where

$$R_x = \frac{2R_1^2 - R_2^2 - R_3^2}{4(1 - \cos \Psi_2) R_0^2}, \tag{5-26}$$

$$R_y = \frac{R_2^2 - R_3^2}{4R_0^2 \sin \Psi_2}. \tag{5-27}$$

The balance mass is located at an angle Ψ measured from the x axis (in the same direction as Ψ_2 and Ψ_3), where

$$\Psi = \tan^{-1} \left(\frac{R_y}{R_x}\right) + 180°. \tag{5-28}$$

An assumption used in the derivation of equations (5-25)–(5-28) is that $\Psi_2 = 360° - \Psi_3$, which is satisfied if $\Psi_2 = 120°$ and $\Psi_3 = 240°$ as suggested in the procedure.

SINGLE-PLANE BALANCING BY THE INFLUENCE COEFFICIENT METHOD

The method to be described here is widely known as the polar-plot method, using a graphical construction to determine the balance mass and its location from measurements of vibration amplitude and phase angle.

After giving the procedure and showing the graphical construction, the method will be explained in terms of influence coefficients. The method of influence coefficients is used as the basis for most computer codes for two-plane and multiplane balancing in the United States. This section will give the reader an opportunity to gain a basic understanding of what the influence coefficients are and how they are used in balance computations. It will thus also serve as an introduction to two-plane balancing, using influence coefficients.

For this method, one vibration transducer and one keyphasor transducer are required. Figure 5.8 shows how the required measurements can be made with an oscilloscope. The procedure is as follows:

1. With the machine at rest, turn the rotor until the keyphasor exciter (optical tape, keyway, etc.) triggers the keyphasor transducer; then make a mark P on the rotor under the vibration transducer.

2. With the machine running at a selected speed S (preferably not a critical speed), measure the synchronous vibration amplitude V and the phase lag θ from the keyphasor pulse to the next peak of the synchronous vibration signal. The measured phase lag θ is the angle from mark P around to the vibration vector, taken as positive in the direction opposite to rotor rotation. (The vibration vector rotates synchronously with the rotor and therefore may be considered as fixed in the rotor at any given speed.)

3. Stop the rotor and attach a trial mass m' at a measured angle Ψ' from mark P, using the same sign convention for angles as in step 2.

4. Run the rotor back up to the selected speed S and measure the new vibration amplitude V' and its phase lag angle θ'.

5. Use the measured data from steps 2–4 to determine the balance mass m and its location angle Ψ. The trial mass m' is removed when the balance mass is attached.

The graphical construction to determine the balance mass is shown in Fig. 5.11. Using a convenient scale factor, the amplitude V of the original vibration vector **V** is plotted as a phasor (polar plot) with its angle θ measured counterclockwise from the P axis (regardless of the direction of rotor rotation, but recalling that positive angles are opposite to the direction of rotor rotation).

The vibration vector **V'**, measured with the trial mass, is laid out by the same convention, with the same scale factor used for **V**.

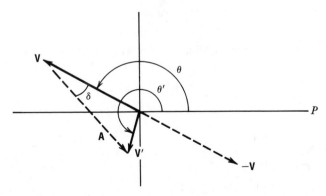

Figure 5.11. Graphical determination of the balance correction weight for single-plane balancing.

In Figure 5.11, the vector triangle can be interpreted as: the effect $\mathbf{V}$ of the original imbalance plus the effect $\mathbf{A}$ of the trial mass equals the effect $\mathbf{V}'$ of the trial mass together with the original imbalance. Therefore the vector $\mathbf{A}$ is the effect of the trial mass m'. If $\mathbf{A}$ had been equal to $-\mathbf{V}$, the $\mathbf{V}'$ would have been zero and the trial mass m' would be the correct balance mass m. It can be seen that the size of the trial mass should be modified by the V/A ratio and moved through an angle δ on the rotor (using the same sign convention as in the procedure and graphical construction).

This method is based on the assumption of a linear relationship between the imbalance and the vibration that results from it. That is, if the imbalance is doubled, the measured synchronous vibration amplitude will be doubled. If the imbalance is moved around from angle Ψ to $\Psi + \Delta\Psi$, then the resulting vibration vector will also be moved around through the same angle $\Delta\Psi$ to a new location $\theta + \Delta\Psi$ (measured from the P axis).

In terms of influence coefficients,

$$\mathbf{V} = \mathbf{CU}, \qquad (5\text{-}29)$$

where $\mathbf{U}$ is the imbalance vector[3] and $\mathbf{C}$ is an influence coefficient which determines the resulting vibration vector $\mathbf{V}$. The influence coefficient $\mathbf{C}$ consists of an amplitude C and an angle ξ, so $\mathbf{C} = Ce^{i\xi}$. The amplitude C is a measure of the sensitivity of a rotor–bearing system to imbalance, and the angle ξ is the angle by which the imbalance leads the vibration. In the Jeffcott rotor model of Chapter I, $\xi = \beta$. The influence coefficient $\mathbf{C}$ is determined by the proximity of rotor speed to a critical speed and will change as the rotor speed is varied.

Equations such as (5-29) can be easily manipulated by the rules of phasors in exponential form for multiplication and division [7]. Addition and subtraction operations are facilitated by using orthogonal components.

Referring to Fig. 5.11, the vibration $\mathbf{V}'$ (with the trial mass) equals the vector sum of the original vibration $\mathbf{V}$ and the effect $\mathbf{A}$ of the trial mass. That is,

$$\mathbf{V}' = \mathbf{V} + \mathbf{A}. \qquad (5\text{-}30)$$

In terms of the influence coefficient at the selected speed,

$$\mathbf{V}' = \mathbf{CU} + \mathbf{CM}', \qquad (5\text{-}31)$$

where

$$\begin{aligned} \mathbf{CM}' = \mathbf{A} &= Ae^{i\eta} \\ &= (Cm')\,e^{i(\xi + \Psi')}. \end{aligned} \qquad (5\text{-}32)$$

[3]Consisting of a mass m_1 at a location angle Ψ_1, so $\mathbf{U} = m_1 e^{i\Psi_1}$. If the effective radius R is not the same for all imbalance and balance terms, then m_1 must be replaced by the product $m_1 R_1$.

Note that the desired effect of the trial mass m' is $\mathbf{A} = -\mathbf{V}$, or

$$-\mathbf{V} = \mathbf{CM}, \tag{5-33}$$

where $\mathbf{M}$ is the correct balance mass vector.

Define a vector coefficient α such that

$$\mathbf{M} = \alpha\mathbf{M}' \tag{5-34}$$

and substitute equation (5-34) into equation (5-33) to give

$$-\mathbf{V} = \mathbf{C}\alpha\mathbf{M}' = \alpha\mathbf{A}, \tag{5-35}$$

so that

$$\alpha = -\frac{\mathbf{V}}{\mathbf{A}} \tag{5-36}$$

Equation (5-34) can now be used to compute the correct balance mass vector.

In terms of phasors in exponential form, the balance mass m and location Ψ are determined from

$$\mathbf{M} = me^{i\Psi} = \frac{V}{A}\,e^{i(\delta - \pi)}m'e^{i\Psi'} = \frac{Vm'}{A}\,e^{i(\Psi' + \delta)}, \tag{5-37}$$

where

$$\delta = (\theta + \pi) - \eta. \tag{5-38}$$

(See Fig. 5.11).

Every term on the right side of equation (5-37) is known from measured data, since $\mathbf{A} = \mathbf{V}' \rightarrow \mathbf{V} = Ae^{i\eta}$.

TWO-PLANE BALANCING WITH INFLUENCE COEFFICIENTS

In general, two planes are required to balance a rigid rotor, as described earlier in this chapter. However, if the method of the preceding section is applied sequentially to each of two chosen planes, say, planes A and B, then the computed balance mass added in plane B will destroy the balance achieved in plane A, and vice versa. This is because the rotor has cross-coupled influence coefficients, so that an imbalance mass added in one axial plane influences the vibration to a different extent in all planes of the rotor (i.e., the added mass influences both the static and the dynamic imbalance).

What is needed is an influence coefficient method modified to take the cross-coupled effects into account. Such a method was originally published by Thearle [8] in 1934. Although he did not employ the mathematical terminology of influence coefficients in describing the method, the concepts are verbally elucidated in the cited paper, which laid the foundation for the modern influence coefficient method.

Two vibration transducers (one in each of two separated planes) and a keyphasor transducer are required. The procedure is as follows:

1. Designate the vibration transducer nearest to balance plane A as transducer A, and the transducer nearest to balance plane B as transducer B. With the machine at rest, turn the rotor until the keyphasor exciter (optical tape, keyway, etc.) triggers the keyphasor transducer, then make a mark P_A on the rotor under transducer A and a mark P_B on the rotor under transducer B.

2. With the machine running at a selected speed S (preferably not a critical speed), measure the vibration vectors $\mathbf{V}_A$ and $\mathbf{V}_B$ with vibration transducers A and B, respectively.

Vibration vector $\mathbf{V}_A$ consists of an amplitude V_A and phase lag θ_A measured from mark P_A. Vibration vector $\mathbf{V}_B$ consists of an amplitude V_B and phase lag θ_B measured from mark P_B. Figure 5.8(a) shows how the phase lag angles are defined and measured from the respective marks P_A and P_B. Remember that positive angles are measured opposite to the direction of rotor rotation.

3. Stop the rotor and attach a trial mass m'_A in plane A at a measured angle Ψ'_A from mark P_A, using the same sign convention for angles as in step 2.

4. Run the rotor back up to the selected speed S, and measure the new vibration vectors $\mathbf{V}'_{AA}$ and $\mathbf{V}'_{BA}$.[4]

5. Stop the rotor and remove trial mass m'_A. Attach a trial mass m'_B in plane B at a measured angle Ψ'_B from mark P_B, using the same sign convention as in previous steps.

6. Run the rotor back up to the selected speed S and measure the new vibration vectors $\mathbf{V}'_{AB}$ and $\mathbf{V}'_{BB}$.

7. Use the measured data from steps 2–6 to calculate the balance mass vectors $\mathbf{M}_A$ and $\mathbf{M}_B$. A derivation of the calculation formulae now follows.

Using influence coefficients, the measured vibration vectors can be expressed in terms of the rotor imbalance vectors in each of the two planes A and B. Thus,

$$\mathbf{V}_A = \mathbf{C}_{AA}\mathbf{U}_A + \mathbf{C}_{AB}\mathbf{U}_B, \qquad (5\text{-}39)$$

$$\mathbf{V}_B = \mathbf{C}_{BA}\mathbf{U}_A + \mathbf{C}_{BB}\mathbf{U}_B. \qquad (5\text{-}40)$$

where $\mathbf{C}_{AB}$ is an influence coefficient that gives the effect of the imbalance in plane B on the vibration at transducer A.

---

[4]The first subscript denotes the measurement plane. The second subscript denotes the plane of the added trial mass.

Similarly, when the trial mass m'_A is added in plane A, the new vibration vectors are given by

$$\mathbf{V}'_{AA} = \mathbf{V}_A + \mathbf{A}_A, \tag{5-41}$$

where

$$\mathbf{A}_A = \mathbf{C}_{AA}\mathbf{M}'_A \tag{5-42}$$

is the effect of the trial mass vector $\mathbf{M}'_A$ on the vibration at transducer A, and

$$\mathbf{V}'_{BA} = \mathbf{V}_B + \mathbf{B}_A, \tag{5-43}$$

where

$$\mathbf{B}_A = \mathbf{C}_{BA}\mathbf{M}'_A \tag{5-44}$$

is the effect of the trial mass vector $\mathbf{M}'_A$ on the vibration at transducer B.

When the trial mass m'_B is added in plane B, the new vibration vectors are given by

$$\mathbf{V}'_{AB} = \mathbf{V}_A + \mathbf{A}_B, \tag{5-45}$$

where

$$\mathbf{A}_B = \mathbf{C}_{AB}\mathbf{M}'_B \tag{5-46}$$

is the effect of the trial mass vector $\mathbf{M}'_B$ on the vibration at transducer A, and

$$\mathbf{V}'_{BB} = \mathbf{V}_B + \mathbf{B}_B, \tag{5-47}$$

where

$$\mathbf{B}_B = \mathbf{C}_{BB}\mathbf{M}'_B \tag{5-48}$$

is the effect of the trial mass vector $\mathbf{M}'_B$ on the vibration at transducer B.

The desired balance mass vectors are the negatives of the original imbalance vectors, so that

$$\mathbf{M}_A = -\mathbf{U}_A, \tag{5-49}$$

$$\mathbf{M}_B = -\mathbf{U}_B. \tag{5-50}$$

Their desired effect is

$$-\mathbf{V}_A = \mathbf{C}_{AA}\mathbf{M}_A + \mathbf{C}_{AB}\mathbf{M}_B, \tag{5-51}$$

$$-\mathbf{V}_B = \mathbf{C}_{BA}\mathbf{M}_A + \mathbf{C}_{BB}\mathbf{M}_B. \tag{5-52}$$

Now define vector operators α and β as

$$\alpha = \frac{\mathbf{M}_A}{\mathbf{M}'_A}, \qquad \beta = \frac{\mathbf{M}_B}{\mathbf{M}'_B}, \qquad (5\text{-}53)$$

so that the balance mass vectors can be obtained from the trial mass vectors, once α and β are known.

Equations (5-51) and (5-52) can now be rewritten as

$$-\mathbf{V}_A = \mathbf{C}_{AA}\,\alpha\mathbf{M}'_A + \mathbf{C}_{AB}\,\beta\mathbf{M}'_B, \qquad (5\text{-}54)$$

$$-\mathbf{V}_B = \mathbf{C}_{BA}\,\alpha\mathbf{M}'_A + \mathbf{C}_{BB}\,\beta\mathbf{M}'_B. \qquad (5\text{-}55)$$

Substitution of equations (5-42), (5-44), (5-46), and (5-48) gives

$$-\mathbf{V}_A = \alpha\mathbf{A}_A + \beta\mathbf{A}_B,$$

$$-\mathbf{V}_B = \alpha\mathbf{B}_A + \beta\mathbf{B}_B, \qquad (5\text{-}56)$$

which can be inverted to give

$$\alpha = \frac{\mathbf{V}_B\mathbf{A}_B - \mathbf{V}_A\mathbf{B}_B}{\mathbf{A}_A\mathbf{B}_B - \mathbf{B}_A\mathbf{A}_B}, \qquad (5\text{-}57)$$

$$\beta = \frac{\mathbf{V}_A\mathbf{B}_A - \mathbf{V}_B\mathbf{A}_A}{\mathbf{A}_A\mathbf{B}_B - \mathbf{B}_A\mathbf{A}_B}. \qquad (5\text{-}58)$$

Each of the vector (phasor) quantities in equations (5-57) and (5-58) is known or can be computed directly from the measured data. For example, equation (5-45) gives

$$\mathbf{A}_B = \mathbf{V}'_{AB} - \mathbf{V}_A. \qquad (5\text{-}59)$$

Once α and β are computed, the balance masses are computed as

$$\mathbf{M}_A = \alpha\mathbf{M}'_A, \qquad (5\text{-}60)$$

$$\mathbf{M}_B = \beta\mathbf{M}'_B. \qquad (5\text{-}61)$$

In some cases it may be impractical to remove the trial mass m'_A as specified in step 5.[5] If trial mass m'_A is left in place while measuring $\overline{V}'_{AB}$ and $\overline{V}'_{BB}$, then equation (5-45) is modified to give

$$\mathbf{V}'_{AB} = \mathbf{V}'_{AA} + \mathbf{A}_B, \qquad (5\text{-}62)$$

---

[5]Especially when the trial masses are negative, which represents rotor material removed by drilling or grinding.

and equation (5-47) becomes

$$\mathbf{V}'_{BB} = \mathbf{V}'_{BA} + \mathbf{B}_B. \tag{5-63}$$

Equations (5-62) and (5-63) are then used to compute $\mathbf{A}_B$ and $\mathbf{B}_B$ for use in equations (5-57) and (5-58). The balance mass vectors computed from equations (5-60) and (5-61) now must include the trial mass vectors $\mathbf{M}'_A$ and $\mathbf{M}'_B$, since neither m'_A nor m'_B is to be removed. Therefore the final corrections to be made, designated $\mathbf{M}_{A-A}$ and $\mathbf{M}_{B-B}$, are

$$\mathbf{M}_{A-A} = \mathbf{M}_A - \mathbf{M}'_A, \tag{5-64}$$

$$\mathbf{M}_{B-B} = \mathbf{M}_B - \mathbf{M}'_B. \tag{5-65}$$

FLEXIBLE ROTOR BALANCING USING INFLUENCE COEFFICIENTS

A rigid rotor balanced by the method of the previous section will have minimum synchronous vibration over its entire range of operating speeds. If, after balancing in two planes, the machine exhibits low vibration at the speed used for balancing but high vibration at other speeds, then the rotor is probably not a rigid rotor. In such a case, repeated application of the two-plane method just presented may be fruitless, and a method employing more balance planes and/or more measurement transducers and/or more balance speeds will give improved results.

In theory, to achieve a perfect state of balance, the number of balance planes should equal the number of critical speeds traversed. Since a large number of balance planes is not always available, methods have been developed to achieve optimal results with a limited number of balance planes. The most popular of these methods is called "least-squares balancing," developed by Goodman at General Electric [9]. This method makes use of vibration measurements made at several different speeds. It is a special case of the analysis to follow, which is adapted from Ref. [10].

Given J balance planes, where balance trial and correction masses can be added, and the capability to make $I = K \times L$ vibration measurements, where K is the number of speed and/or load conditions and L is the number of vibration transducers, the objective is to find the set of J optimal balance mass vectors to minimize the amplitudes of the I vibration measurements.

There are two cases to be considered:

Case I. If $J \geq I$, the I amplitudes can theoretically be reduced to zero.

Case II. If $J \leq I$, there are not enough balance planes to reduce all I vibrations to zero. However, the sum of the squares of the vibration amplitudes can be minimized.

The rationale behind minimizing the sum of the squares is to prevent the canceling of large positive and negative values that can occur in ordinary averaging.

Let $\{ \mathbf{V}_i \}$ be the column vector[6] of I measured vibration phasors (amplitude and phase lag angle), $[\mathbf{C}_{ij}]$ be the matrix of influence coefficients, and $\{ \mathbf{U}_j \}$, $\{ \mathbf{M}_j' \}$, $\{ \mathbf{M}_j \}$ be the column vectors of J imbalance, trial, and balance mass phasors,[7] respectively.

The matrix equations to follow will be illustrated by a special case with $J = 2$ planes ($j = 1, 2$), $K = 2$ speeds, $L = 2$ transducers, so $I = 2 \times 2 = 4$ vibration measurements ($i = 1, 2, 3, 4$).

The original vibration produced by the uncorrected rotor imbalance is given by

$$\{ \mathbf{V}_i \} = [\mathbf{C}_{ij}] \{ \mathbf{U}_j \}. \tag{5-66}$$

For the special case, the component equations are

$$\mathbf{V}_1 = \mathbf{C}_{11}\mathbf{U}_1 + \mathbf{C}_{12}\mathbf{U}_2 \quad \text{(plane 1, speed 1)}, \tag{5-67}$$

$$\mathbf{V}_2 = \mathbf{C}_{21}\mathbf{U}_1 + \mathbf{C}_{22}\mathbf{U}_2 \quad \text{(plane 2, speed 1)}, \tag{5-68}$$

$$\mathbf{V}_3 = \mathbf{C}_{31}\mathbf{U}_1 + \mathbf{C}_{32}\mathbf{U}_2 \quad \text{(plane 1, speed 2)}, \tag{5-69}$$

$$\mathbf{V}_4 = \mathbf{C}_{41}\mathbf{U}_1 + \mathbf{C}_{42}\mathbf{U}_2 \quad \text{(plane 2, speed 2)}. \tag{5-70}$$

The vibration produced by a set of trial masses $\{ M_j' \}$ is

$$\{ \mathbf{V}_i' \} = \{ \mathbf{V}_i \} + [\mathbf{C}_{ij}] \{ \mathbf{M}_j' \}. \tag{5-71}$$

The component equations for the special case are

$$\mathbf{V}_1' = \mathbf{V}_1 + \mathbf{C}_{11}\mathbf{M}_1' + \mathbf{C}_{12}\mathbf{M}_2', \tag{5-72}$$

$$\mathbf{V}_2' = \mathbf{V}_2 + \mathbf{C}_{21}\mathbf{M}_1' + \mathbf{C}_{22}\mathbf{M}_2', \tag{5-73}$$

$$\mathbf{V}_3' = \mathbf{V}_3 + \mathbf{C}_{31}\mathbf{M}_1' + \mathbf{C}_{32}\mathbf{M}_2', \tag{5-74}$$

$$\mathbf{V}_4' = \mathbf{V}_4 + \mathbf{C}_{41}\mathbf{M}_1' + \mathbf{C}_{42}\mathbf{M}_2'. \tag{5-75}$$

If the $\mathbf{V}_i'$ are measured with the trial masses applied one at a time, the influence coefficients can be determined from the measurements. For example, equation (5-75) gives the vibration in plane 2 at speed 2 with trial mass $\mathbf{M}_1'$ (only) attached as

$$\mathbf{V}_{41}' = \mathbf{V}_4 + \mathbf{C}_{41}\mathbf{M}_1', \tag{5-76}$$

---

[6]As used in matrix algebra.

[7]To avoid confusion of the individual vectors with the "column vector," the term "phasor" is now used to denote an individual vector such as the imbalance $\mathbf{U}_j$ in plane j. It is useful to express the individual $\mathbf{V}_i$, $\mathbf{C}_{ij}$, $\mathbf{U}_j$, $\mathbf{M}_j'$, and $\mathbf{M}_j$ phasors as complex numbers and employ the rules of complex algebra.

which allows $\mathbf{C}_{41}$ to be calculated as

$$\mathbf{C}_{41} = \frac{\mathbf{V}'_{41} - \mathbf{V}_4}{\mathbf{M}'_1}. \qquad (5\text{-}77)$$

Every term on the right-side of equation (5-77) is known from measurements.

With the influence coefficients known, the optimum set of correction mass phasors $\{\mathbf{M}_j\}$ can be computed.

For the general case I, where there are at least as many balance planes as there are measurements, the matrix $[\mathbf{C}_{ij}]$ can be made square and therefore can be inverted.

The correction mass vector can then be computed as

$$\{\mathbf{M}_j\} = -[\mathbf{C}_{ij}]^{-1}\{\mathbf{V}_i\}, \qquad I = J. \qquad (5\text{-}78)$$

Substitution of this vector for $\mathbf{M}'_j$ in equation (5-71) shows that the new (corrected) vibration vector will be

$$\mathbf{V}_i - [\mathbf{C}_{ij}][\mathbf{C}_{ij}]^{-1}\{\mathbf{V}_i\} = \{0\}. \qquad (5\text{-}79)$$

For the general case II, where there are more measurements than balance planes, the objective is to minimize a function S, where

$$S = \sum_{i=1}^{I} |\mathbf{V}_i|^2 \qquad (5\text{-}80)$$

If the phasors $\mathbf{V}_i$ are represented by complex numbers, then S can be written as

$$S = \sum_{i=1}^{I} \mathbf{V}_i \cdot \hat{\mathbf{V}}_i, \qquad (5\text{-}81)$$

where $\hat{\mathbf{V}}_i$ is the conjugate of $\mathbf{V}_i$.

Since S is not analytic, it is minimized by requiring that

$$\frac{\partial S}{\partial \hat{\mathbf{M}}'_j} = 0, \qquad j = 1, J, \qquad (5\text{-}82)$$

where $\hat{\mathbf{M}}'_j$ is the conjugate of $\mathbf{M}'_j$.

The vector of correction mass phasors which satisfies equation (5-82) is given by (see Ref. [10])

$$\{\mathbf{M}_j\} = -[[\hat{\mathbf{C}}_{ij}]^{T}[\mathbf{C}_{ij}]]^{-1}[\hat{\mathbf{C}}_{ij}]^{T}\{\mathbf{V}_i\}, \qquad (5\text{-}83)$$

where $[\hat{\mathbf{C}}_{ij}]^{T}$ is the conjugate transpose of $[\mathbf{C}_{ij}]$.

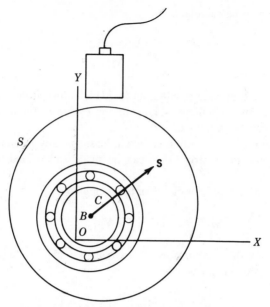

Figure 5.12. Shaft run-out.

EFFECT OF SHAFT RUNOUT

When the measured vibration signal is shaft displacement, taken from proximity probes, it may be necessary to compensate for shaft runout before using the measurements in balancing computations. Proximity probes generally measure the displacement of the geometric center of the circular surface to which they are proximate. If the objective of balancing is to minimize the displacement of this surface (e.g., to minimize a seal rub), then the probe signal should be used uncompensated. If the objective is to minimize dynamic bearing loads, then the signal should be compensated.

Figure 5.12 illustrates how the geometric center C of the transduced surface S may not be coincident with the center B of the bearing journal.[8] The dynamic deflection $\overline{OB}$ of the bearing journal produces the load on the bearing, but the proximity probe is measuring the deflection $\overline{OC}$, which is the vector sum of $\overline{OB}$ with the runout $\overline{BC}$. The point O is the location of the bearing center B with no dynamic load. The phasor of magnitude $\overline{BC}$ and angle σ is the shaft runout and will be denoted by $\overline{S}$.

Figure 5.13 shows how the runout can be measured with a dial indicator while the rotor is turned around very slowly. It can also be measured with the proximity probe, provided that the electronics of the measurement system will respond at the low frequency associated with a slow roll.

---

[8]Due to machining errors or imperfections.

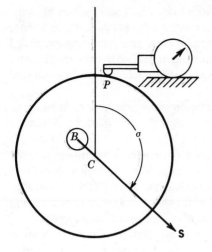

Figure 5.13. Measurement of shaft run-out.

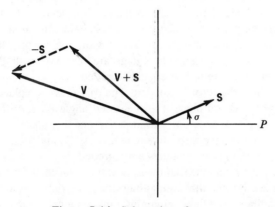

Figure 5.14. Subtraction of run-out.

Figure 5.14 shows how the runout can be subtracted vectorially from an uncompensated vibration phasor $\mathbf{V} + \mathbf{S}$, to produce the compensated vibration vector $\mathbf{V}$.

TWO-PLANE BALANCING WITHOUT PHASE MEASUREMENTS

The following method, which includes the effects of cross-coupling between the two planes, is due to Dr. Louis J. Everett of Texas A&M University [11].

Two planes are chosen for placing trial balance masses. A mark is made on each plane denoting the zero position. Designate the planes as balance planes A and B. Vibration transducers are placed in another two planes. Designate the transducer nearest balance plane A as transducer A and the remaining transducer as B.

When the unbalanced rotor is spun with a speed of S, vibration will be generated. The vibration at transducer A can be represented as phasor $\mathbf{V}_A$, consisting of magnitude V_A and phase lag θ_A measured from mark P_A. The vibration at transducer B can be represented as phasor $\mathbf{V}_B$, consisting of magnitude V_B and phase lag θ_B measured from mark P_B. The magnitudes of these phasors are measured and recorded. It will be convenient during the following derivation to express the vibration vectors as a magnitude times a unit phasor. For example, $\mathbf{V}_A$ can be expressed as $\mathbf{V}_A\boldsymbol{\theta}_A$ and $\mathbf{V}_B$ can be expressed as $\mathbf{V}_B\boldsymbol{\theta}_B$.

After recording the vibration magnitude measurements, the rotor is stopped and trial mass m is placed on plane A at a measured angle ψ_A^1 from mark P_A. When the rotor is again spun at speed S, vibration vectors $\mathbf{V}_{AA}^1$ and $\mathbf{V}_{BA}^1$ are developed. As in previous sections, the first subscript denotes the plane of the transducer, the second subscript denotes the plane of the trial mass. The vibration vectors have respective magnitudes V_{AA}^1 and V_{BA}^1. Their phase lags are respectively θ_{AA}^1 and θ_{BA}^1. As before, these vibration vectors will be expressed as magnitudes times unit phasors. Again only the magnitudes of vibration are measured and recorded.

After taking measurements, the rotor is again stopped. The first trial mass is removed and another identical trial mass m is placed on plane A at a measured angle ψ_A^2 from mark P_A. Notice that the superscript denotes the presence of the second trial mass on plane A. When the rotor is again spun at speed S, vibration vectors $\mathbf{V}_{AA}^2$ and $\mathbf{V}_{BA}^2$ are developed. These vibration vectors have respective magnitudes V_{AA}^2 and V_{BA}^2. Their phase lags are respectively θ_{AA}^2 and θ_{BA}^2. The magnitudes of vibration are measured and recorded.

For the third time, the rotor is stopped. The trial mass is replaced with another identical mass m placed on plane A at measured angle ψ_A^3 from mark P_A. When the rotor is again spun at speed S, vibration vectors $\mathbf{V}_{AA}^3$ and $\mathbf{V}_{BA}^3$ are developed. These vibration vectors have respective magnitudes V_{AA}^3 and V_{BA}^3. Their phase lags are θ_{AA}^3 and θ_{BA}^3. Vibration magnitudes are recorded.

After stopping the rotor, the trial mass must be replaced and rotor spun at speed S three more times. For the three remaining runs, the trial mass must be placed on plane B. For the first of the three remaining runs, an identical mass m is placed on plane B at a meaured angle ψ_B^1 from mark P_B. The second of the three runs is made with the mass located at ψ_B^2. The last run is made with the mass located at ψ_B^3. Each run produces two vibration vectors. These vectors, written as magnitudes times unit phasors are $V_{AB}^1\boldsymbol{\theta}_{AB}^1$, $V_{BB}^1\boldsymbol{\theta}_{BB}^1$ for the first of the three; $V_{AB}^2\boldsymbol{\theta}_{AB}^2$, $V_{BB}^2\boldsymbol{\theta}_{BB}^2$ for the second; and $V_{AB}^3\boldsymbol{\theta}_{AB}^3$, $V_{BB}^3\boldsymbol{\theta}_{BB}^3$ for the last run.

Assuming the response of the rotor to imbalance is described sufficiently with influence coefficient theory, it is possible to relate the vibration vectors to the imbalance in the rotor. The two vibration vectors caused by spinning the original unbalanced rotor can be expressed as

$$\mathbf{V}_A = \mathbf{C}_{AA}\mathbf{U}_A + \mathbf{C}_{AB}\mathbf{U}_B,$$

$$\mathbf{V}_B = \mathbf{C}_{BA}\mathbf{U}_A + \mathbf{C}_{BB}\mathbf{U}_B. \tag{5-84}$$

The vibration vectors generated by the three runs made with trial masses on balance plane A can be expressed as

$$\mathbf{V}_{AA}^{i} = \mathbf{C}_{AA}(\mathbf{U}_{A} + m\boldsymbol{\psi}_{A}^{i}) + \mathbf{C}_{AB}\mathbf{U}_{B},$$

$$\mathbf{V}_{BA}^{i} = \mathbf{C}_{BA}(\mathbf{U}_{A} + m\boldsymbol{\psi}_{A}^{i}) + \mathbf{C}_{BB}\mathbf{U}_{B}, \qquad \text{for } i = 1, 2, 3. \qquad (5\text{-}85)$$

The vibration vectors generated by the three runs made with trial masses on balance plane B can be expressed as

$$\mathbf{V}_{AB}^{i} = \mathbf{C}_{AA}\mathbf{U}_{A} + \mathbf{C}_{AB}(\mathbf{U}_{B} + m\boldsymbol{\psi}_{B}^{i}),$$

$$\mathbf{V}_{BB}^{i} = \mathbf{C}_{BA}\mathbf{U}_{A} + \mathbf{C}_{BB}(\mathbf{U}_{B} + m\boldsymbol{\psi}_{B}^{i}), \qquad \text{for } i = 1, 2, 3. \qquad (5\text{-}86)$$

The objective is to solve equations (5-84)–(5-86) for the imbalance vectors. The rotor is then balanced by placing correction masses at 180° from the imbalance.

To solve for the imbalance vectors, it is convenient to rewrite the equations. First, subtract equation (5-84) from (5-85) and (5-86) to provide

$$\mathbf{V}_{AA}^{i} - \mathbf{V}_{A} = \mathbf{C}_{AA}m\boldsymbol{\psi}_{A}^{i},$$

$$\mathbf{V}_{BA}^{i} - \mathbf{V}_{B} = \mathbf{C}_{BA}m\boldsymbol{\psi}_{A}^{i}, \qquad \text{for } i = 1, 2, 3; \qquad (5\text{-}87)$$

$$\mathbf{V}_{AB}^{i} - \mathbf{V}_{A} = \mathbf{C}_{AB}m\boldsymbol{\psi}_{B}^{i},$$

$$\mathbf{V}_{BB}^{i} - \mathbf{V}_{B} = \mathbf{C}_{BB}m\boldsymbol{\psi}_{B}^{i}, \qquad \text{for } i = 1, 2, 3. \qquad (5\text{-}88)$$

These equations are all of the same basic form, therefore only one will be used for the derivation. Consider i equal to 1 in equation set (5-87). First express the phasors as magnitudes times unit phasors. This gives

$$V_{AA}^{1}\boldsymbol{\theta}_{AA}^{1} - V_{A}\boldsymbol{\theta}_{A} = C_{AA}c_{AA}m\boldsymbol{\psi}_{A}^{1}. \qquad (5\text{-}89)$$

Rotate all phasors by $-\psi_{A}^{1}$ and $-\theta_{A}$. When this is done equation (5-89) becomes

$$\frac{V_{AA}^{1}\boldsymbol{\theta}_{AA}^{1}}{\boldsymbol{\psi}_{AA}^{1}\boldsymbol{\theta}_{A}} - \frac{V_{A}}{\boldsymbol{\psi}_{A}^{1}} = \frac{C_{AA}c_{AA}m}{\boldsymbol{\theta}_{A}}. \qquad (5\text{-}90)$$

By multiplying by negative one and reflecting about the real axis, equation (5-90) becomes:

$$C_{AA}m\left(\frac{-\boldsymbol{\theta}_{A}}{c_{AA}}\right) = V_{A}\boldsymbol{\psi}_{A}^{1} + V_{AA}^{1}\left(\frac{-\boldsymbol{\psi}_{A}^{1}\boldsymbol{\theta}_{A}}{\boldsymbol{\theta}_{AA}^{1}}\right). \qquad (5\text{-}91)$$

It is possible to place all equations (5-87) and (5-88) in the same form as (5-91). To do this one must perform the same operation on all the equations. When this is done, equations (5-87) become

$$C_{AA} m\left(\frac{-\theta_A}{\mathbf{c}_{AA}}\right) = V_A \psi_A^i + V_{AA}^i\left(\frac{-\psi_A^i \theta_A}{\theta_{AA}^i}\right), \qquad \text{for } i = 1, 2, 3; \quad (5\text{-}92)$$

$$C_{BA} m\left(\frac{-\theta_B}{\mathbf{c}_{BA}}\right) = V_B \psi_A^i + V_{BA}^i\left(\frac{-\psi_A^i \theta_B}{\theta_{BA}^i}\right), \qquad \text{for } i = 1, 2, 3. \quad (5\text{-}93)$$

Performing the same operations on the set of equations in (5-88) yields:

$$C_{AB} m\left(\frac{-\theta_A}{\mathbf{c}_{AB}}\right) = V_A \psi_B^i + V_{AB}^i\left(\frac{-\psi_B^i \theta_A}{\theta_{AB}^i}\right), \qquad \text{for } i = 1, 2, 3; \quad (5\text{-}94)$$

$$C_{BB} m\left(\frac{-\theta_B}{\mathbf{c}_{BB}}\right) = V_B \psi_B^i + V_{BB}^i\left(\frac{-\psi_B^i \theta_B}{\theta_{BB}^i}\right), \qquad \text{for } i = 1, 2, 3. \quad (5\text{-}95)$$

Equations (5-92)–(5-95) are used to determine a modification of the influence coefficient vectors. The method will be discussed by considering only equation (5-92) since the principles are the same for all four equation sets. The equation sets [i.e., equations in (5-92)] can be interpreted graphically. The right-hand side of equation (5-92) represents a vector of known magnitude and angle to which a vector of known magnitude is added. The loci of the right-hand side is drawn by first laying out vector $V_A \psi_A^1$, then drawing a circle of radius V_{AA}^1 centered at the tip of the vector. The vector $C_{AA} m(-\theta_A/\mathbf{c}_{AA})$ must lie somewhere on the circle. The loci can be pinpointed by constructing the two similar circles remaining in the equation set (i.e., $i = 2, 3$ in equation (5-92)). Since the three circles constructed represent the same vector, they must intersect. By dividing the distance from the origin to the circle intersection point by m and V_A the magnitude of $\mathbf{C}_{AA}/\mathbf{V}_A$ can be determined. Also, 180° minus the phase angle of the intersection is equal to the phase angle of $\mathbf{C}_{AA}/\mathbf{V}_A$. Likewise the other equation sets solve for modifications of the other influence coefficients.

To solve for the imbalance vectors, first rotate equation (5-84) by the initial vibration vectors to obtain

$$1 = \frac{\mathbf{C}_{AA}}{\mathbf{V}_A} \mathbf{U}_A + \frac{\mathbf{C}_{AB}}{\mathbf{V}_A} \mathbf{U}_B,$$

$$1 = \frac{\mathbf{C}_{BA}}{\mathbf{V}_B} \mathbf{U}_A + \frac{\mathbf{C}_{BB}}{\mathbf{V}_B} \mathbf{U}_B. \qquad (5\text{-}96)$$

Equations (5-96) can be inverted to obtain

$$\mathbf{U}_A = \frac{\mathbf{C}_{BB}/\mathbf{V}_B - \mathbf{C}_{AB}/\mathbf{V}_A}{\mathbf{C}_{AA}\mathbf{C}_{BB}/\mathbf{V}_A\mathbf{V}_B - \mathbf{C}_{AB}\mathbf{C}_{BA}/\mathbf{V}_A\mathbf{V}_B},$$

$$\mathbf{U}_B = \frac{\mathbf{C}_{AA}/\mathbf{V}_A - \mathbf{C}_{BA}/\mathbf{V}_B}{\mathbf{C}_{AA}\mathbf{C}_{BB}/\mathbf{V}_A\mathbf{V}_B - \mathbf{C}_{AB}\mathbf{C}_{BA}/\mathbf{V}_A\mathbf{V}_B}. \qquad (5\text{-}97)$$

The required balance masses are U_A and U_B, with their phase angles incremented by 180°.

REFERENCES

1. Beer, F. P., and Johnston, E. R., Jr., *Vector Mechanics for Engineers: Statics and Dynamics*, McGraw-Hill, New York, 1977, p. 397.

2. Kellenberger, W., "Should a Flexible Rotor Be Balanced in N or $(N + 2)$ Planes?" *Journal of Engineering for Industry*, Paper No. 71-Vibr-55, May 1972, pp. 548–560.

3. Wort, J. F. G., "Fundamentals of Balancing Machines," *B&K Technical Review*, **1** (1981).

4. Jackson, C., "Using the Orbit to Balance Rotating Equipment," *Mechanical Engineering*, **93**(2), 28–32, (1971).

5. Blake, M. P., and Mitchell, W. S., *Vibration and Acoustic Measurement Handbook*, Hayden (Spartan Books), Rochelle Park, NJ: 1972.

6. Lang, G. F., "How to Balance with Your Real Time Analyzer," Nicolet Scientific Corporation, Application Note 11 (October 1977).

7. Pipes, L. A., *Applied Mathematics for Engineers and Physicists*, McGraw-Hill, New York, 1958, pp. 37–40.

8. Thearle, E. L., "Dynamics Balancing of Rotating Machinery in the Field," *Journal of Applied Mechanics (Transactions of the ASME)*, **56,** 745–753 (1934).

9. Goodman, T. P., "A Least-Squares Method for Computing Balance Corrections," *Journal of Engineering for Industry (Transactions of the ASME)*, Ser. B, **86**(3), 273–279 (1964).

10. Lund, J. W., and Tonnesen, J., "Analysis and Experiments on Multi-Plane Balancing of a Flexible Rotor," *Journal of Engineering for Industry (Transactions of the ASME)*, **94**(1), 233, Appendix A (1972).

11. Everett, L. J. "Two-Plane Balancing of a Rotor System Without Phase Response Measurements," *Journal of Vibration, Acoustics, Stress and Reliability in Design*, April 1987, pp. 162–167.

Chapter VI

Bearings and Seals

The rotordynamic characteristics of a turbomachine are influenced strongly by the bearings on which the rotor runs. This is because the stiffness of the rotor–bearing system is mainly determined by the bearing support stiffness acting in series with shaft stiffness and the damping of the system is usually determined almost entirely by the bearing damping properties. In some machines, there are also significant fluid forces developed on the rotor by the rotating seals and at the centrifugal or axial-flow process wheels. The latter are traditionally outside the control of the machine designer.[1]

One of the most important design parameters for rotordynamics is the ratio of bearing support stiffness to shaft stiffness. As was shown in Chapters I and II, it is usually good design practice to keep this ratio as small as practical.

Another important parameter is the ratio of support damping to internal damping in the rotor. It was shown in Chapter I that this ratio should be kept as large as possible to insure rotordynamic stability.

In addition to direct stiffness and damping, some types of fluid bearings, as well as seals and process wheels, also produce cross-coupled stiffness and damping. As described in Chapter I, cross-coupling can have profound effects on rotordynamics, especially whirl stability.

In this chapter, bearings and seals will be classified, the mechanism of the forces that they exert on the rotor will be described, contemporary mathematical models for their design analysis will be explained, and some important examples of their effect on rotordynamics will be presented. A description of the fluid forces

[1]Analytical prediction and experimental measurement of these forces is the subject of current research, and it may soon be possible to modify rotordynamic characteristics by redesigning seals, impellers, and turbine stages.

developed by process wheels and their effect on rotordynamic stability will be reserved for Chapter VII.

CLASSIFICATION OF BEARINGS

For the purpose of turbomachinery design analysis, or troubleshooting, the most useful classification of bearings is in terms of their mechanism of rotor support, i.e., how the rotor support forces are produced. The two major types are

1. Rolling-element bearing.
2. Fluid-film bearing (sometimes called "sliding-element bearings").

In rolling-element bearings, the support forces are purely elastic, i.e., produced by elastic contact deformation of the balls or rollers, of the races, and of the local bearing housing structure. An appropriate model for this type of bearing is a linear or nonlinear bidirectional spring with little or no damping and no cross-coupling. Fairly accurate design predictions of the support stiffness can be made by contemporary computer codes based on the theory of elasticity. The reader is referred to Refs. [1] and [2].

Nearly all bearings in aircraft turbine engines are of the rolling-element type. This is because rolling-element bearings give an early warning of impending failure by high-frequency vibration and/or noise, whereas failure modes of fluid-film bearings can be quite sudden and catastrophic, thereby compromising flight safety.

For industrial and utility applications, fluid-film bearings are most often used because of their long life. In addition, if the design of a fluid-film bearing is optimized to minimize the running friction coefficient, the power losses can be considerably smaller than for a rolling-element bearing of the same load capacity.

For a sliding-element bearing to have high load capacity and low power loss, the journal must be kept separated from its bearing surface by a fluid film of finite thickness. This requires a fluid pressure in the film, which can be generated either hydrodynamically or hydrostatically by design. Thus we have a further subclassification of fluid-film bearings into (a) hydrostatic bearings and (b) hydrodynamic bearings.

HYDROSTATIC BEARINGS: MECHANISM OF LOAD SUPPORT

Figure 6.1 illustrates the configuration of a hydrostatic bearing. The pockets must be supplied with lubricating fluid at a regulated pressure, which is one reason why this type of bearing is used less commonly than the hydrodynamic type. As the fluid flows through each of the supply orifices, the pressure drop to the pocket increases with the rate of flow through the orifice.

Consider the effect of a displacement of the journal, off center, due to a load

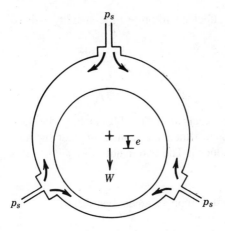

Figure 6.1. Hydrostatic bearing configuration.

applied to the rotor (e.g., the rotor weight) as shown in Fig. 6.1. The discharge flow from the pockets will now be unequal due to the variation of clearance around the bearing. Those pockets which have their discharge restricted by reduced local clearance will contain increased pressures, due to the smaller pressure drop across the associated supply orifice. The pockets with increased discharge clearances will contain reduced pressures. The unbalanced pressures on the journal are precisely in the direction required to react the applied load. Thus there is an equilibrium position of the journal, where the force exerted by the unbalanced fluid pressures equals the applied load. It can be seen that, from the standpoint of the applied load, the hydrostatic bearing acts like a *spring*. Forces are also produced by journal velocity, due to viscous and/or compressibility effects, so the bearing also acts as a *damper*.

HYDRODYNAMIC BEARINGS: MECHANISM OF LOAD SUPPORT

In hydrodynamic bearings, the fluid support pressure is generated entirely by motion of the journal and depends on the viscosity of the lubricating fluid. The fluid supply pressure need only be high enough to maintain a copious supply of lubricant in the load-supporting clearance around the journal. This is normally accomplished by introducing the fluid through one or more supply holes or grooves located in areas where the hydrodynamic pressure is low.

If there is an insufficient supply of fluid, or if any other factor prevents the generation of a high enough pressure in the fluid film to support the load, then the hydrodynamic film breaks down and the journal contacts the bearing surface. Bearings in which this continuously occurs are called "boundary-lubricated" bearings. Hydrodynamic bearings often act as boundary-lubricated bearings during the initial phase of a machine start-up, when the journal rotation speed is too slow to generate sufficient hydrodynamic pressure to support the load.

Boundary lubrication is characterized by higher friction and a much greater potential for overheating than is hydrodynamic lubrication. Figure 6.2 shows how the friction factor varies with viscosity μ, speed N, and load P, for the two different types of lubrication. Consider the temperature stability of the two cases, as follows:

With boundary lubrication, a rise in temperature reduces the viscosity, which raises the friction factor, which further raises the temperature. This cycle, repeated, tends to induce overheating. With hydrodynamic lubrication, Fig. 6.2 shows that a rise in temperature which reduces the viscosity will *reduce* the friction factor, thus reducing the heat generated. This cycle tends to produce a stable operating temperature.

Boundary-lubricated bearings are used satisfactorily in small mechanisms and appliances with light loads and light duty cycles, but they obviously are not desirable in industrial turbomachinery applications. Thus the maintenance of a load-supporting fluid film is of prime importance for reliability of all the various types of hydrodynamic bearings used in turbomachinery. As has already been mentioned, the supply pressure to the bearing is of limited importance in this regard.

Figure 6.3 illustrates how the hydrodynamic film pressure is generated in a journal bearing. The lubricating fluid is pulled by viscous shear into the converging wedge produced by off-center displacement of the journal. Note that it is rotation of the journal (or bearing) that produces the relative velocity along the film wall and induces the viscous shear. As the fluid is pulled into the converging wedge, its pressure is raised. Conversely, the fluid pressure decreases as the viscous shear pulls it out into the diverging wedge downstream from the point of minimum film thickness. The net effect of the distribution of hydrodynamic pressure around the journal is to produce a force which reacts the applied load.

It can be seen from Fig. 6.3 that the converging–diverging wedge effect becomes more pronounced as the off-center journal displacement e (referred to in the bearing

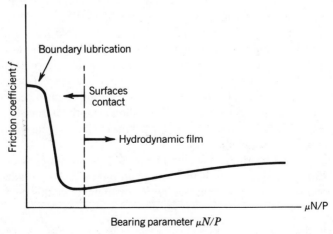

Figure 6.2. Variation of the friction coefficient with boundary lubrication and hydrodynamic lubrication.

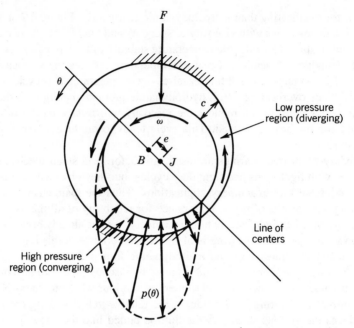

Figure 6.3. Hydrodynamic pressures generated by journal rotation in a cylindrical bearing.

literature as the "eccentricity") is increased. Thus, as in the case of the hydrostatic bearing, there is an equilibrium position where the support force developed by the fluid-film pressure equals the applied load. Furthermore, since a change in applied load produces a new equilibrium position with an equal change in support force, it can be seen that the hydrodynamic bearing also acts like a spring. Translational *velocity* of the journal also induces hydrodynamic pressure in the film (with a resulting force), so the bearing acts like a damper as well.

From the standpoint of rotordynamics, the most desirable feature of a hydrodynamic journal bearing is its damping, which is high compared to other types of bearings. This feature is also retained in the "squeeze film bearing damper," as illustrated in Fig. 6.4, which is simply a hydrodynamic journal bearing with a nonrotating journal provided by the outer race[2] of a rolling-element bearing (loosely pinned or keyed to prevent rotation while allowing orbital translation). Such dampers are commonly employed in aircraft turbine engines, where rolling-element bearings are used almost exclusively.

The plain journal bearing, as shown in Fig. 6.3, is the simplest of all hydrodynamic bearings. Its geometry is that of a plain right circular cylinder, and it is the least expensive bearing to manufacture. Rotors running on this type of bearing

[2]Usually the bearing outer race is mounted in a cylindrical housing which acts as the journal.

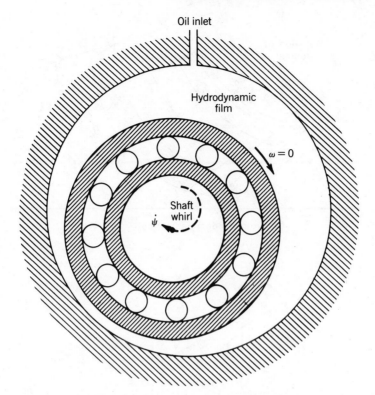

Figure 6.4. Squeeze film damper configuration.

are often speed limited by "oil whip," a form of rotordynamic instability to be described later in this chapter; therefore more complex types of hydrodynamic bearings have been developed to remove or reduce this limitation. Some examples are illustrated in Figs. 6.5–6.7.

Most industrial turbomachines are designed to use some type of hydrodynamic bearing, and nearly all aircraft turbine engines employ some type of hydrodynamic bearing damper (see Fig. 6.4), as described above. Consequently, the remainder of the material on bearings in this chapter will be devoted to the analysis of hydrodynamic bearings and their effect on rotordynamics.

HYDRODYNAMIC BEARING ANALYSIS

From the standpoint of rotordynamics, the most important questions about any hydrodynamic bearing are the following:

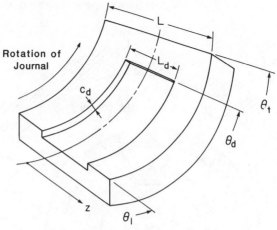

(a) Pressure dam bearing pad

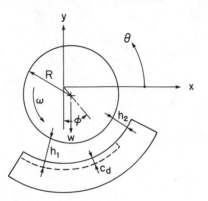

(b) Shaft and pad side view

Figure 6.5. Pressure-dam bearing. (Illustration courtesy of Prof. Paul Allaire, ROMAC, University of Virginia.)

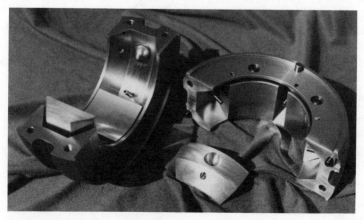

Figure 6.6. Tilting-pad bearing hardware, disassembled. (Photo courtesy of Centritech Corp.)

Figure 6.7. Tilting-pad bearing hardware, assembled and disassembled. (Photo courtesy of Centritech Corp.)

1. What are the magnitudes and directions of the forces exerted by the bearing on the rotor?
2. What is the static equilibrium position of the journal running in the bearing?
3. What are the magnitudes of the bearing stiffness and damping coefficients, as measured from a given static equilibrium position?
4. What is the dynamic equilibrium motion of the journal in the bearing (e.g., response to rotating imbalance)?
5. Are the equilibrium states (2 and 4, above) stable? If not, how can they be stabilized?

Since the fluid-film pressure forces (question 1) must be known before addressing questions 2–5, we will begin with an analysis of the film pressure distribution. This analysis, followed by a consideration of static equilibrium, will provide some additional physical insight into the mechanism of load support, as well as some useful definitions. Linearized stiffness and damping coefficients can then be defined, providing the simplest means of describing the dynamic forces exerted by the bearing on the rotor. At that point a discussion can be given of the effect of hydrodynamic bearings on rotordynamics, before going into the more detailed mathematical analyses to answer questions 4 and 5, above. The latter analyses require mathematical modeling of the rotor, in addition to the bearings.

FLUID-FILM PRESSURE DISTRIBUTION

It has been shown that the basic problem of hydrodynamic bearing analysis is a determination of the fluid-film pressure distribution $p(x, z)$ for a given bearing geometry (see Fig. 6.3). Bearing designers normally accomplish this by solving various forms of the Reynolds' equation for special cases.

The equation that bears his name was developed by Osborne Reynolds in 1886 to explain experimental measurements of the film pressure in railroad car bearings reported earlier by Beauchamp Tower [3]. The Reynolds' equation provides the basis of modern lubrication theory, and a number of its solutions for special cases of practical interest have been verified by experimental measurement. For any general thin film geometry, the equation is [4]

$$\frac{\partial}{\partial x}\left(h^3\frac{\partial p}{\partial x}\right) + \frac{\partial}{\partial z}\left(h^3\frac{\partial p}{\partial z}\right) = 6\mu\left\{\frac{\partial}{\partial x}\left[(U_0 + U_1)h\right] + 2\frac{\partial h}{\partial t}\right\}, \quad (6\text{-}1)$$

where $p = p(x, z)$ is the pressure distribution; x and z are coordinates that locate a point in the film; h is the local film thickness; μ is the viscosity of the fluid; U_0 and U_1 are tangential velocities at the two walls bounding the film; and $\partial h/\partial t$ is the time rate of change of the local film thickness. Reynolds employed a number of simplifying assumptions in deriving the equation, and the user must be aware of them so as to recognize the limitation of its solutions.

1. Viscous shear effects dominate, so the viscosity is the only important fluid parameter.
2. The fluid inertia forces are neglected (consistent with assumption 1).
3. The fluid is incompressible[3] (consistent with assumption 1).
4. The fluid film is very thin, so that the pressure does not vary across the thickness of the film and any curvature of the film (e.g., around a cylindrical journal) can be neglected.
5. The viscosity is constant throughout the film.
6. There is no slip at the wall (i.e., between the fluid–solid boundaries).

Most bearings employed for rotor support in turbomachinery have cylindrical geometry, or some variation of it, to match a cylindrical journal. In cylindrical coordinates, Reynolds' equation becomes

$$\frac{\partial}{\partial\theta}\left[(1 + \epsilon\cos\theta)^3\frac{\partial p}{\partial\theta}\right] + R^2\frac{\partial}{\partial z}\left[(1 + \epsilon\cos\theta)^3\frac{\partial p}{\partial z}\right]$$
$$= -6\mu\left(\frac{R}{C}\right)^2\left[(\omega - 2\dot{\psi})\epsilon\sin\theta - 2\dot{\epsilon}\cos\theta\right], \quad (6\text{-}2)$$

where

$$\partial/\partial\theta = R(\partial/\partial x)$$
$$\epsilon = e/c$$

---

[3]The equation must be modified to remove this assumption for the analysis of gas bearings or air bearings.

e = journal eccentricity

c = radial clearance

R = journal radius

Also $c(1 + \epsilon \cos \theta) = h$ = local film thickness; ω, $\dot{\psi}$ are the angular velocities of the journal and line of centers, respectively (the bearing surface is assumed to be stationary); and $\dot{\epsilon} = \dot{e}/c$ (dimensionless radial velocity).

For a plain journal bearing, open to the atmosphere at the ends $z = 0$ and $z = L$, and with an uncavitated fluid film, the boundary conditions are

$$p(\theta, 0) = p(\theta, L) = p_a \tag{6-3}$$

and

$$p(0, z) = p(2\pi, z) = p_0, \tag{6-4}$$

where p_a is atmospheric pressure and p_0 is determined by the supply pressure to the bearing.

Closed-form solutions to equation (6-2) in functional form, with realistic boundary conditions such as (6-3) and (6-4), have not been obtained to date, except for the special case of small eccentricities. There are also additional factors, such as film cavitation, which must be included for a realistic model in some cases and which make a functional solution even more difficult. To obtain useful solutions, two successful alternative approaches have been employed:

1. Simplification of the Reynolds' equation for special cases of practical interests, so that functional solutions can be obtained.
2. Reformulation of the equation into finite-difference or finite-element form, for numerical solution on a digital computer.

For journal bearings, the first approach has yielded solutions for two notable special cases: (a) the "long bearing" and (b) the "short bearing."

For the long bearing, the major simplifying assumption is that the second term in equation (6-2) is of negligible magnitude compared to the first; that is, the pressure distribution around the bearing is invariant along the length of the bearing. Thus the second term in the equation is omitted, which makes the equation integrable. Physically, this means that there is no flow in the axial (z) direction. The solution for the "long bearing" pressure distribution (uncavitated) is then

$$p(\theta) = p_0 + 6\mu \left(\frac{R}{C}\right)^2 \left\{ (\omega - 2\dot{\psi}) \frac{\epsilon(2 + \epsilon \cos \theta) \sin \theta}{(2 + \epsilon^2)(1 + \epsilon \cos \theta)^2} \right. \tag{6-5}$$

$$\left. + \dot{\epsilon} \frac{1}{\epsilon} \left[\frac{1}{(1 + \epsilon \cos \theta)^2} - \frac{1}{(1 + \epsilon)^2} \right] \right\},$$

where p_0 is a known pressure at $\theta = 0$ (location of maximum clearance), and the other variables are as previously defined and illustrated in Fig. 6.3.

For the short bearing, the first term of equation (6-2) is omitted on the basis that it has a negligible effect on the flow[4] compared to the other terms in the equation. Practically, this will occur when $L/D \leq 0.25$, where L is the axial length of the bearing and $D = 2R$ is the journal diameter. In this case the film pressure turns out to be a parabolic function of the axial coordinate z. The pressure distribution for a steady load ($\dot\epsilon = 0$) is given by [4]

$$p(\theta) = p_a + \frac{3\mu\omega}{c^2}\left(\frac{L^2}{4} - z^2\right)\frac{\epsilon \sin\theta}{(1 + \epsilon \cos\theta)^3}, \tag{6-6}$$

where p_a is the ambient pressure at the ends ($z = \pm L/2$) of the bearing.

Figure 6.8 is a plot of this pressure variation around the bearing midway between the ends ($z = 0$), for three different eccentricities, $\epsilon = 0.3, 0.4,$ and 0.5. The parameters used in equation (6-6) to produce the curves are representative of turbomachinery practice. Note that the peak pressure increases with eccentricity.

CAVITATION OF THE FLUID FILM

The lubricant is usually supplied to a hydrodynamic bearing through an inlet hole or groove in a region where low film pressures are predicted, such as the maximum clearance position $\theta = 0$. Typical supply pressures p_s are on the order of 20–80 psi, which will fix p_0 in equation (6-5) at about this value. Also, compared to the typical peak pressure in the film, the supply pressure is closer to p_a in equation (6-6), so the pressure p_a at $\theta = 0$ can also be taken as approximately the supply pressure.

Figure 6.8 shows that, for $p_a = 14.7$ psi, the film pressure in the diverging region $\pi < \theta < 2\pi$ is predicted to be highly negative. In practice, the lubricant cavitates and the actual pressure usually does not drop more than about 5 psi negative. Since the force exerted on the journal by the bearing is obtained by integrating the pressure over the surface area, it can be seen that cavitation has a major effect on this force.

The extent of cavitation is determined by the supply pressure and the eccentricity. A high enough supply pressure in a long bearing can eliminate cavitation and produce the full predicted pressure distribution, since the lowest pressure then does not go negative. Conversely, highly loaded bearings ($\epsilon > 0.7$) with low supply pressures may be cavitated over fully one half their circumference and are sometimes referred to as π-film bearings. For a static journal position ($\dot\epsilon = 0; \dot\psi$

---

[4]Each term in the Reynolds' equation represents fluid flow, either axially through the bearing or circumferentially around the bearing.

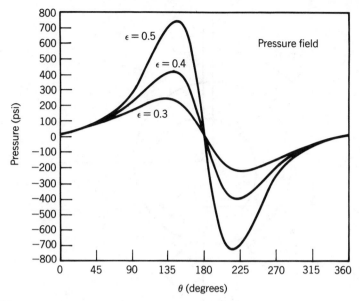

Figure 6.8. Variation of pressure around a short bearing, midway between the ends, for three different eccentricities.

= 0) the region of cavitation is $\pi < \theta < 2\pi$; when the journal is moving, the location of the cavitation is determined by the instantaneous values of $\dot\epsilon$ and $\dot\psi$. The cavitated π-film is considered superior to the 2π film (uncavitated) bearing in journal bearing applications because of its better whirl stability characteristics.

For moderate loads and supply pressures the extent of cavitation may be less than 180°. This will be referred to here as the partially cavitated case. For analysis of this case it is reasonable to set the predicted negative pressures equal to zero whenever they occur.

STATIC EQUILIBRIUM OF HYDRODYNAMIC BEARINGS

The static equilibrium position of the journal under a unidirectional load (such as rotor weight) can be calculated by setting components of the load equal to integrals of the fluid film pressure over the surface of the journal. It is usually convenient to take two components along and normal to the line of centers (i.e., in the direction of journal eccentricity and normal to it). As Fig. 6.9 indicates, the line of centers generally does not align itself with the applied load.

The two integrals represent the component reactions exerted by film pressure on the journal. They turn out to be functions of the journal eccentricity ϵ and the "attitude angle" ψ (see Fig. 6.9), and are given by

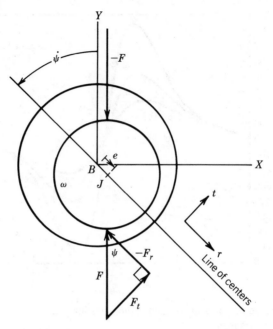

Figure 6.9. Components of the fluid-film force F, and the attitude angle ψ of the line of centers.

$$F_r = F \cos \psi = \int_0^l R \int^{2\pi} p(\theta, z) \cos \theta \, d\theta \, dz \qquad (6\text{-}7)$$

and

$$F_t = F \sin \psi = \int_0^l R \int^{2\pi} p(\theta, z) \sin \theta \, d\theta \, dz, \qquad (6\text{-}8)$$

where the film pressure distribution $p(\theta, z)$ generally is a nonlinear function of ϵ, as well as of θ and z. Equations (6-7) and (6-8) can be solved for ϵ and ψ giving the journal equilibrium position, assuming that the applied load F is known and that the functional form (or a numerical table) for $p(\theta, z)$ is known in terms of ϵ.

Figure 6.10 shows the locus of journal equilibrium positions of a fully cavitated (π-film) short journal bearing. The eccentricity $\epsilon \to 1$ and the attitude angle $\psi \to 0$ as the applied load F approaches the capacity of the bearing.

Bearing design and analysis is facilitated by expressing the load in terms of the dimensionless Sommerfeld number S, where

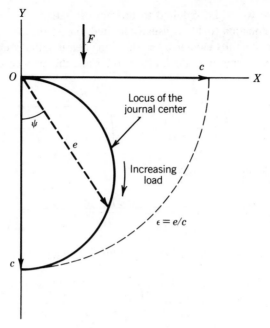

Figure 6.10. Locus of journal equilibrium positions for a cavitated short bearing.

$$S = \frac{\mu N}{P} \left(\frac{R}{C}\right)^2, \qquad N = \frac{\omega}{2\pi}, \tag{6-9}$$

and

$$P = \frac{F}{LD} \tag{6-10}$$

is the load per unit of projected area, sometimes called the "bearing pressure."

If the uncavitated long bearing pressure distribution (6-5) is substituted into equations (6-7) and (6-8) (with $\dot{\epsilon} = 0$), the Sommerfeld number is obtained as the following function of eccentricity ϵ:

$$S = \frac{(2 + \epsilon^2)\sqrt{1 - \epsilon^2}}{12\pi^2\epsilon}. \tag{6-11}$$

The attitude angle ψ for this case turns out to be $\pi/2$ for all loads, so the locus of journal equilibrium positions for a vertical load (as in Fig. 6.10) would be a horizontal straight line! It must be remembered, however, that an extremely high supply pressure would be required to prevent cavitation as $\epsilon \to 1$, so practical application of equation (6-11) is usually limited to $\epsilon < 0.3$.

For the case of a fully cavitated (π-film) long bearing, the integrals in equations (6-7) and (6-8) are zero over the limits π to 2π and the load–eccentricity relationship becomes

$$S = \frac{(2 + \epsilon^2)(1 - \epsilon^2)}{6\pi\epsilon} \sqrt{\frac{1}{\pi^2 + (4 - \pi^2)\epsilon^2}}, \tag{6-12}$$

with the attitude angle given by

$$\tan \psi = \frac{\pi \sqrt{1 - \epsilon^2}}{2\epsilon}. \tag{6-13}$$

The corresponding relationships for the fully cavitated short bearing are

$$S = \frac{(1 - \epsilon^2)}{\pi(L/D)^2\epsilon} \sqrt{\frac{1}{\pi^2 + (16 - \pi^2)\epsilon^2}} \tag{6-14}$$

and

$$\tan \psi = \frac{\pi \sqrt{1 - \epsilon^2}}{4\epsilon}. \tag{6-15}$$

DYNAMIC FORCE COMPONENTS

The load–eccentricity relationships (6-11)–(6-15) are valid only for static equilibrium conditions where $\dot{\epsilon} = 0$, and $\dot{\psi} = 0$, a rather hypothetical situation that would require a perfectly balanced and stable rotor. For rotordynamics analysis the equilibrium position is important as the starting point, but the dynamic forces exerted by the bearing must be evaluated for motions away from the equilibrium position.

Accordingly, the force components given by integration of equations (6-7) and (6-8) with $\dot{\epsilon}$, $\dot{\psi} \neq 0$ for the appropriate pressure distributions are now given for each of the special cases described in the previous section. Note that $\dot{\psi}$ is the angular velocity of journal whirling.

Uncavitated Long Bearing:

$$F_r = -12\mu RL \left(\frac{R}{C}\right)^2 \frac{\pi\dot{\epsilon}}{(1 - \epsilon^2)^{3/2}}, \tag{6-16}$$

$$F_t = 12\mu RL \left(\frac{R}{C}\right)^2 (\omega - 2\dot{\psi}) \frac{\pi\epsilon}{(2 + \epsilon^2)(1 - \epsilon^2)^{1/2}}. \tag{6-17}$$

Cavitated (π-Film) Long Bearing:

$$F_r = -6\mu RL \left(\frac{R}{C}\right)^2 \left[|\omega - 2\dot\psi| \frac{2\epsilon^2}{(2 + \epsilon^2)(1 - \epsilon^2)} + \frac{\pi\dot\epsilon}{(1 - \epsilon^2)^{3/2}} \right], \quad (6\text{-}18)$$

$$F_t = 6\mu RL \left(\frac{R}{C}\right)^2$$

$$\cdot \left[(\omega - 2\dot\psi) \frac{\pi\epsilon}{(2 + \epsilon^2)(1 - \epsilon^2)^{1/2}} + \frac{4\dot\epsilon}{(1 + \epsilon)(1 - \epsilon^2)} \right] + 2RLp_0.$$

$$(6\text{-}19)$$

Uncavitated Short Bearing:

$$F_r = -\mu RL \left(\frac{L}{C}\right)^2 \frac{\pi(1 + 2\epsilon^2)\dot\epsilon}{(1 - \epsilon^2)^{5/2}}, \quad (6\text{-}20)$$

$$F_t = \mu RL \left(\frac{L}{C}\right)^2 (\omega - 2\dot\psi) \frac{\epsilon}{2(1 - \epsilon^2)^{3/2}}. \quad (6\text{-}21)$$

Cavitated (π-Film) Short Bearing:

$$F_r = -\mu RL \left(\frac{L}{C}\right)^2 \left[|\omega - 2\dot\psi| \frac{\epsilon^2}{(1 - \epsilon^2)^2} + \frac{\pi}{2} \frac{(1 + 2\epsilon^2)\dot\epsilon}{(1 - \epsilon^2)^{5/2}} \right], \quad (6\text{-}22)$$

$$F_t = \mu RL \left(\frac{L}{C}\right)^2 \left[(\omega - 2\dot\psi) \frac{\pi\epsilon}{4(1 - \epsilon^2)^{3/2}} + \frac{2\epsilon\dot\epsilon}{(1 - \epsilon^2)^2} \right] + 2RLp_0. \quad (6\text{-}23)$$

LINEARIZED BEARING FORCE COEFFICIENTS

Figure 6.10 shows the locus of static operating points for a journal bearing supporting a balanced rotor. For very light loads the journal deflection is small and almost normal to the load ($\psi \rightarrow 90°$). For heavy loads, the deflection e approaches c in the direction of the load. The operating position is located by ϵ, ψ, or by the Cartesian coordinates X^*, Y^*. At the static operating position the oil-film force is in equilibrium with the load, as described earlier.

If the rotor–bearing system is unstable ("oil whip") or if imbalance is added to the rotor, the journal may execute orbits about the static operating points as shown in Fig. 6.11. To analyze the motion, the incremental variations of the oil-

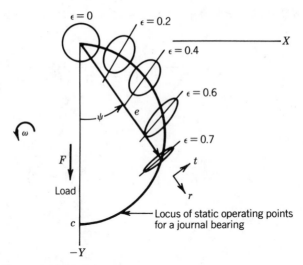

Figure 6.11. Whirl orbits about different static operating points.

film force on the journal must be expressed in terms of r–t components (along and normal to e) or in terms of X–Y components. These force components are functions of the journal displacement, measured from the static operating point, and of its translational velocity and acceleration. The incremental force functions are linear only if the orbit is small enough. For large displacements from the static operating point, the functions are highly nonlinear.

For example, Fig. 6.12 illustrates how the incremental force component ΔF_X might vary with the displacement X. The slope of this curve for small X measured from X^*, Y^* (the static operating point) is denoted as the direct stiffness K_{XX} of the journal bearing. Interestingly, the X displacement also generates an incremental force ΔF_Y, and the slope of this curve is the cross-coupled stiffness K_{YX}. Following the definition by Lund [5], the complete expressions for the X–Y force components can be written out in terms of linearized stiffness, damping, and inertia coefficients, both direct and cross-coupled, as follows:

$$\Delta F_X = -K_{XX}X - K_{XY}Y - C_{XX}\dot{X} - C_{XY}\dot{Y} - D_{XX}\ddot{X} - D_{XY}\ddot{Y},$$

$$\Delta F_Y = -K_{YX}X - K_{YY}Y - C_{YX}\dot{X} - C_{YY}\dot{Y} - D_{YX}\ddot{X} - D_{YY}\ddot{Y}. \quad (6\text{-}24)$$

Since both the damping coefficients C_{ij} and the inertia coefficients D_{ij} give forces proportional to motion, care must be taken to properly separate velocity from acceleration in cases where they are both nonzero.

Alternatively, the oil film incremental force can be resolved into r–t components as in the previous section and expressed as incremental components.

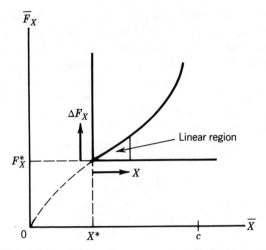

Figure 6.12. Oil film force versus journal displacement.

$$\Delta F_r = -K_{rr}r - K_{rt}t - C_{rr}v_r - C_{rt}v_t - D_{rr}a_r - D_{rt}a_t,$$

$$\Delta F_t = -K_{tr}r - K_{tt}t - C_{tr}v_r - C_{tt}v_t - D_{tr}a_r - D_{tt}a_t, \qquad (6\text{-}25)$$

where r is the incremental change of e, $t = e\,\Delta\psi$, $v_r = \dot{e}$, $v_t = e\dot{\psi}$, $a_r = \ddot{e} - \dot{\psi}^2 e$, and $a_t = e\ddot{\psi} + 2\dot{e}\dot{\psi}$. Here ΔF_r and ΔF_t are the incremental changes in force from their static equilibrium values. The advantage of form (6-25) is that most equations for oil-film forces found in the literature give r and t components, but form (6-24) is preferred for a Newtonian formulation of the equations of rotor dynamics. Therefore, a mathematical transformation of the bearing forces from r–t coordinates to X–Y coordinates is found to be useful.

Using the coordinate directions defined in Fig. 6.9, the transformation equations are given by

$$\begin{Bmatrix} F_X \\ F_Y \end{Bmatrix} = \begin{bmatrix} \sin\psi & \cos\psi \\ -\cos\psi & \sin\psi \end{bmatrix} \begin{Bmatrix} F_r \\ F_t \end{Bmatrix}. \qquad (6\text{-}26)$$

The linearized coefficients in equations (6-24) can be interpreted as partial derivatives. For example,

$$K_{XX} = -\frac{\partial F_X}{\partial X}, \qquad K_{YY} = -\frac{\partial F_Y}{\partial Y} \qquad (6\text{-}27)$$

are the direct stiffness coefficients, and X, Y are related to r, t by the same transformation matrix given in (6-26), above.

If F_r and F_t are known from hydrodynamic bearing theory, the linearized coefficients in X-Y coordinates can be obtained from transformation (6-26) and the chain rule of derivatives. This analysis has been carried out for several different types of bearings [5,6,9–11]. The results are usually presented in dimensionless form for universal application to all bearings of the type considered. Some representative curves are reproduced in Figs. 6.13–6.15. Two different parameter groups are used in rendering the stiffness and damping coefficients dimensionless. The dimensionless coefficients $\overline{K}_{ij}$ and $\overline{C}_{ij}$ given by Figs. 6.13 and 6.14, for example, are related to the coefficients K_{ij} and C_{ij} of equation (6-24) as follows:

$$K_{ij} = \frac{\mu \omega R}{2} \left(\frac{L}{C} \right)^3 \overline{K}_{ij},$$

$$C_{ij} = \frac{\mu R}{2} \left(\frac{L}{C} \right)^3 \overline{C}_{ij},$$

(6-28)

where $i, j = 1, 2$, and $1 \to X$, $2 \to Y$.

In Fig. 6.15, the dimensionless coefficients K'_{ij} and C'_{ij} give

$$K_{ij} = \frac{F}{C} K'_{ij},$$

$$C_{ij} = \frac{F}{C\omega} C'_{ij},$$

(6-29)

where F is the static load supported by the bearing.

For programming a computer, mathematical expressions for the coefficients are useful. The dimensionless stiffness and damping coefficients for the fully cavitated (π-film) short bearing, when perturbed from the static operating position ϵ, ψ, are given by [5, 6]

$$K'_{XX} = 4 \left\{ 2\pi^2 + (16 - \pi^2) \epsilon^2 \right\} Q(\epsilon),$$

(6-30)

$$K'_{XY} = \frac{\pi \left\{ -\pi^2 + 2\pi^2 \epsilon^2 + (16 - \pi^2) \epsilon^4 \right\} Q(\epsilon)}{\epsilon (1 - \epsilon^2)^{1/2}},$$

(6-31)

$$K'_{YX} = \frac{\pi \left\{ \pi^2 + (32 + \pi^2) \epsilon^2 + 2(16 - \pi^2) \epsilon^4 \right\} Q(\epsilon)}{\epsilon (1 - \epsilon^2)^{1/2}},$$

(6-32)

$$K'_{YY} = \frac{4 \left\{ \pi^2 + (32 + \pi^2) \epsilon^2 + 2(16 - \pi^2) \epsilon^4 \right\} Q(\epsilon)}{(1 - \epsilon^2)},$$

(6-33)

$$C'_{XX} = \frac{2\pi (1 - \epsilon^2)^{1/2} \left\{ \pi^2 + 2(\pi^2 - 8) \epsilon^2 \right\} Q(\epsilon)}{\epsilon},$$

(6-34)

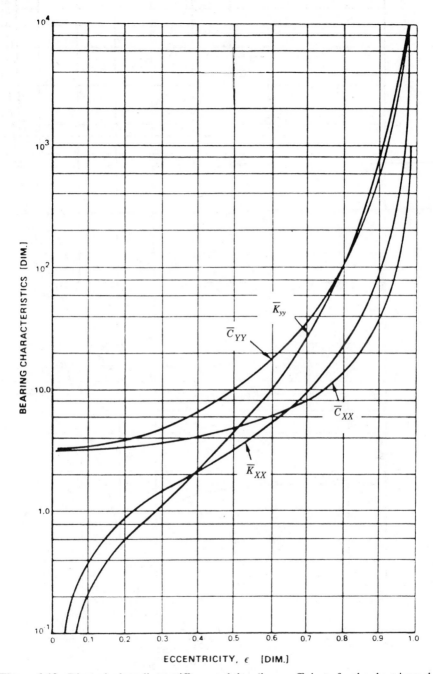

Figure 6.13. Dimensionless direct stiffness and damping coefficients for the short journal bearing. From Ref. [9].

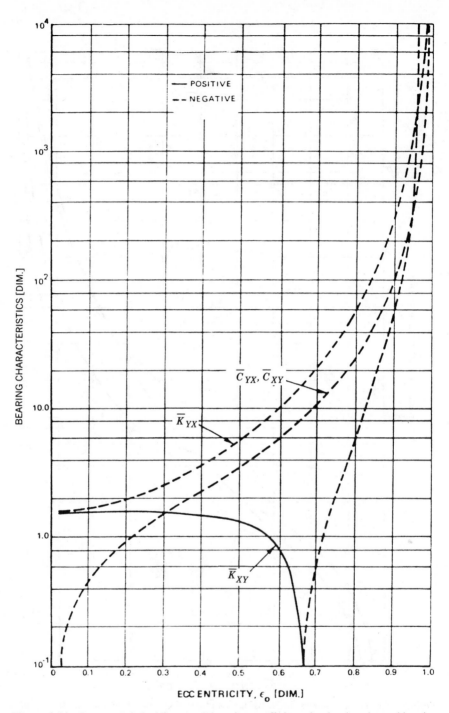

Figure 6.14. Cross-coupled stiffness and damping coefficients for the short journal bearing. From Ref. [9].

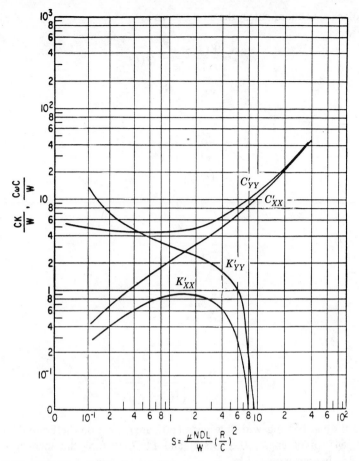

$\dfrac{CK}{W}, \dfrac{C\omega C}{W}$

$S = \dfrac{\mu NDL}{W}\left(\dfrac{R}{C}\right)^2$

Figure 6.15. Stiffness and damping coefficients for the 6 shoe tilting pad bearing, $L/D = 0.5$, load on pad. From Ref. [11].

$$C'_{XY} = 8\left\{\pi^2 + 2(\pi^2 - 8)\epsilon^2\right\} Q(\epsilon), \qquad (6\text{-}35)$$

$$C'_{YX} = C'_{XY} \qquad (6\text{-}36)$$

$$C'_{YY} = \frac{2\pi\left\{\pi^2 + 2(24 - \pi^2)\epsilon^2 + \pi^2\epsilon^4\right\}}{\epsilon(1 - \epsilon^2)^{1/2}}, \qquad (6\text{-}37)$$

where

$$Q(\epsilon) = \frac{1}{\left\{\pi^2 + (16 - \pi^2)\epsilon^2\right\}^{3/2}}. \qquad (6\text{-}38)$$

In polar $(r-t)$ coordinates, the expressions are

$$K'_{rr} = \frac{8(1 + \epsilon^2)}{(1 - \epsilon^2)} \overline{Q}(\epsilon), \qquad (6\text{-}39)$$

$$K'_{rt} = \frac{\pi(1 - \epsilon^2)^{1/2}}{\epsilon} \overline{Q}(\epsilon), \qquad (6\text{-}40)$$

$$K'_{tr} = \frac{\pi(1 + 2\epsilon^2)}{\epsilon(1 - \epsilon^2)^{1/2}} \overline{Q}(\epsilon), \qquad (6\text{-}41)$$

$$K'_{tt} = 4\overline{Q}(\epsilon), \qquad (6\text{-}42)$$

$$C'_{rt} = C'_{tr} = \frac{2}{\omega} k_{tt}, \qquad (6\text{-}43)$$

$$C'_{tt} = \frac{2}{\omega} k_{rt}, \qquad (6\text{-}44)$$

$$C'_{rr} = \frac{2}{\omega} k_{tr}, \qquad (6\text{-}45)$$

where

$$\overline{Q}(\epsilon) = \frac{1}{\left\{ \pi^2 + (16 - \pi^2)\epsilon^2 \right\}^{1/2}}. \qquad (6\text{-}46)$$

For a tilt-pad bearing with five pads (60° arc), load on the bottom pad, zero preload, 54° pivot angle, 0.5 offset, and $L/D < 0.5$, the load–eccentricity relationship is approximated by

$$S = \frac{6.125}{\pi(L/D)^2 e^{6.34\epsilon}}, \qquad (6\text{-}47)$$

where S is the Sommerfeld load number of equation (6-9). Approximate expressions for the stiffness and damping coefficients are

$$K'_{XX} = 4.01e^{-3.84\epsilon}, \qquad \epsilon > 0.3, \qquad (6\text{-}48a)$$

$$K'_{XX} = 2.01 - 2.24\epsilon, \qquad \epsilon \leq 0.3, \qquad (6\text{-}48b)$$

$$K'_{YY} = 1.67e^{2.47\epsilon}, \qquad (6\text{-}49)$$

$$C'_{XX} = 13.5(1 - \epsilon)^{2.11}, \qquad \epsilon > 0.3, \qquad (6\text{-}50a)$$

$$C'_{XX} = 35.0e^{-6.34\epsilon}, \qquad \epsilon \leq 0.3, \qquad (6\text{-}50b)$$

$$C'_{YY} = 30 - 83.2\epsilon + 76.4\epsilon^2. \qquad (6\text{-}51)$$

The conversion to coefficients with physical dimensions is given by equations (6-29) with i, j referring to X, Y or r, t as appropriate.

EFFECT OF HYDRODYNAMIC BEARINGS ON ROTORDYNAMICS

As mentioned earlier in this chapter, the feature of hydrodynamic bearings most beneficial with respect to rotordynamics is their relatively high damping. In addition, the stiffness of the oil film has a strong influence on critical speeds, and consequently the bearing design parameters can be used to raise or lower the critical speeds of a given rotor design.

Examination of the equations and curves just presented for hydrodynamic film forces shows that the film forces, and consequently the stiffness and damping, are highly sensitive to the radial clearance of the bearing. Thus, this parameter turns out to be one of the most important for controlling rotordynamics.

Rotors running on plain journal bearings have a threshold speed, above which the rotor–bearing system becomes unstable in "oil whip," characterized by subsynchronous whirling. The threshold speed of instability is generally about twice the first critical speed. It is the cross-coupled stiffness (K_{XY}, K_{YX}) of journal bearings which destabilizes the system. Although the damping of such bearings is high, it is not enough to suppress the oil whip at high rotor speeds.

Muszynska [7, 8] has explained the difference between "oil whip," just described above, and "oil whirl." When journal bearings are lightly loaded, the shaft is often observed to whirl at a frequency of one-half shaft speed, even at speeds below the threshold speed of oil-whip instability. This "oil whirl" is stable and its frequency "tracks" a changing shaft speed, always maintaining the one-half frequency ratio. Oil whirl cannot be predicted by linearized stability theory. A nonlinear model (accurate for large amplitudes) of the oil-film forces is required. The practical significance of oil whirl is that it may be a harbinger of unstable oil whip at a higher shaft speed.

Bearings designed to suppress oil whip, such as pressure dam bearings or tilt-pad bearings (see Figs. 6.5–6.7), generally have smaller cross-coupled stiffness at high speeds but may also have less damping at the critical speeds. The direct stiffness characteristics of the two types of bearings will in general be different. Thus, a bearing designed to suppress oil whip will have different critical speed characteristics when compared with a journal bearing. Furthermore, the actual critical speeds will be determined as much by the rotor design (rigid or flexible, light or massive) as by the bearings.

All these factors, as well as the highly nonlinear load characteristics of hydrodynamic bearings, make it difficult to generalize further as to their effect on rotordynamics. In the following section a modified Jeffcott rotor–bearing system will be analyzed to illustrate the effect of two types of hydrodynamic bearings on the response to imbalance and whirl stability as the shaft flexibility is varied.

RESPONSE TO IMBALANCE AND STABILITY OF A JEFFCOTT ROTOR ON HYDRODYNAMIC BEARINGS

Consider a Jeffcott rotor (see Chapter I) with a central disk of mass $2m$ and supported at each end by identical hydrodynamic bearings. The purely static imbalance of the disk is represented by the eccentricity u of its center of mass M from its geometric axis of revolution C. The elastic force in one half of the flexible shaft is applied to each bearing, and the reaction force of the shaft on the disk produces the acceleration of its center of mass according to Newton's second law. Figure 6.16 shows the geometry and coordinates of the rotor with its journal and shaft deflections. The rotor axis is horizontal, so the bearings must support the weight of the rotor, which determines the Sommerfield load number, and hence the static operating position, at any given speed.

Although this model is highly simplified by comparison with a multistage turbomachine, it is capable of approximating the response to imbalance (synchronous whirl amplitudes) of such a machine as its traverses any one of its critical speeds, and can also simulate whirl instability characteristics at the frequency of an associated eigenvalue. The shaft stiffness $2K_s$ and disk mass $2m$ may be regarded as modal values, associated with any chosen natural frequency of interest.

In Fig. 6.16, J is the journal center, C is the geometric center of the disk, and M is its center of gravity. Thus, e is the journal eccentricity in the bearing and δ is the shaft deflection. In terms of the X and Y coordinates of points C and J, the differential equations of motion for constant speed ω are

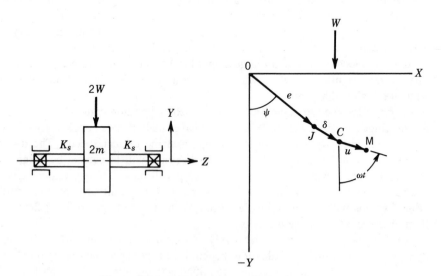

Figure 6.16. Jeffcott rotor on oil-film bearings.

$$m\ddot{X}_c + K_s(X_c - X_J) = m\omega^2 u \sin(\omega t),$$

$$m\ddot{Y}_c + K_s(Y_c - Y_J) = -m\omega^2 u \cos(\omega t) - W, \qquad (6\text{-}52)$$

where $W = mg$ is half the rotor weight, and $u = \overline{CM}$ is the static imbalance. The dynamic equilibrium of forces at each bearing is given by

$$F_{Bx} = -K_s(X_c - X_J),$$

$$F_{By} = -K_s(Y_c - Y_J), \qquad (6\text{-}53)$$

where F_{Bx} and F_{By} are horizontal and vertical components of the hydrodynamic film force exerted on the journal in each bearing.

With no imbalance ($u = 0$), the static equilibrium bearing force

$$F_{Bx}^* = -K_s(X_c^* - X_J^*) = 0, \qquad F_{By}^* = -K_s(Y_c^* - Y_J^*) = W, \qquad (6\text{-}54)$$

produces the static equilibrium positions of the disk and journal:

$$X_c^* = e \sin \psi, \qquad Y_c^* = -\delta_s - e \cos \psi,$$

$$X_J^* = e \sin \psi, \qquad Y_J^* = -e \cos \psi. \qquad (6\text{-}55)$$

where $\delta_s = W/K_s$.

If the rotor has a small imbalance, both the disk and the journal will execute orbits with the displacements given by

$$X_c = X_c^* + x_c, \qquad Y_c = Y_c^* + y_c,$$

$$X_J = X_J^* + x_J, \qquad Y_J = Y_J^* + y_J. \qquad (6\text{-}56)$$

The corresponding bearing forces will be

$$F_{Bx} = F_{Bx}^* - K_{XX}x_J - K_{XY}y_J - C_{XX}\dot{x}_J - C_{XY}\dot{y}_J,$$

$$F_{By} = F_{By}^* - K_{YY}y_J - K_{YX}x_J - C_{YX}\dot{x}_J - C_{YY}\dot{y}_J. \qquad (6\text{-}57)$$

The dynamic bearing force components given by (6-57) and (6-54) are substituted into equations (6-53), which in turn are substituted into the differential equations (6-52). For universal application to rotor–bearing systems of any size or speed, the differential equations are made dimensionless by substituting

$$\bar{x}_c = \frac{x_c}{c}, \qquad \bar{x}_J = \frac{x_J}{c},$$

$$\bar{y}_c = \frac{y_c}{c}, \qquad \bar{y}_J = \frac{y_J}{c},$$

$$\tau = \omega t, \qquad \bar{u} = \frac{u}{c}, \qquad (6\text{-}58)$$

$$\Omega^2 = \frac{\omega^2 m}{K_s}, \qquad \Delta_s = \frac{\delta_s}{c}.$$

In terms of the dimensionless variables, the differential equations of the small amplitude motion are

$$\Omega^2 \ddot{\bar{x}}_c + \bar{x}_c - \bar{x}_J = \Omega^2 \bar{u} \sin \tau,$$

$$\Omega^2 \ddot{\bar{y}}_c + \bar{y}_c - \bar{y}_J = -\Omega^2 \bar{u} \cos \tau, \qquad (6\text{-}59)$$

with

$$\bar{x}_c - \bar{x}_J = \Delta_s(K'_{XX}\bar{x}_J + K'_{XY}\bar{y}_J + C'_{XX}\dot{\bar{x}}_J + C'_{XY}\dot{\bar{y}}_J),$$

$$\bar{y}_c - \bar{y}_J = \Delta_s(K'_{YY}\bar{y}_J + K'_{YX}\bar{x}_J + C'_{YX}\dot{\bar{x}}_J + C'_{YY}\dot{\bar{y}}_J). \qquad (6\text{-}60)$$

In equations (6-59) and (6-60), the super-dot represents differentiation with respect to τ and the K'_{ij}, C'_{ij} terms are defined by equations (6-29).

The influence of various types of fluid-film bearings on the response to imbalance of a flexible rotor can be evaluated by substituting the appropriate bearing coefficients into equations (6-60) and computing the particular solution to equations (6-59). For evaluation of whirl stability characteristics, the eigenvalues of the homogeneous differential equations ($\bar{u} = 0$) are computed. Note that the Sommerfeld load number S, and therefore the static equilibrium eccentricity ϵ, changes with speed. Consequently the bearing stiffness and damping coefficients must be recomputed at each discrete speed ratio Ω. A computer program has been written to accomplish this and graph the results shown in the following example.

A cavitated short journal bearing is compared with a tilt-pad bearing in Figs. 6.17–6.21. The bearing coefficients are given by equations (6-30)–(6-38) and (6-48)–(6-51), respectively. It should be noted that the latter equations are for a tilt-pad bearing with zero preload. Preload is a geometric design parameter that has a strong effect on the tilt-pad stiffness and damping coefficients.

For this example, the shaft flexibility $\Delta_s = 0.1$ represents a fairly stiff rotor, and the Sommerfeld load number $S = 0.8$ at $\omega = \omega_n = \sqrt{K_s/m}$, the natural frequency of the rotor on rigid supports. The abscissa of the curves is $\Omega = \omega/\omega_n$, the ratio of shaft speed to the natural frequency on rigid supports.

Figure 6.17 shows the equilibrium eccentricity position versus speed for the two types of bearings. Figure 6.18 shows the attitude angle for the journal bearing. The attitude angle for the tilt-pad bearing is zero at all speeds.

Figure 6.19 shows the amplitude of the major axis of the whirl orbit versus speed for the journal bearing. The whirl orbits are skewed as shown in Fig. 6.11. Note that the rotor becomes unstable in "oil whip" when the speed reaches 90 percent of ω_n. This is where the real part of the eigenvalue becomes positive. The actual critical speed appears at about 33 percent of ω_n due to the relatively soft oil-film bearing support.

Figure 6.20 shows the modified response with the tilt-pad bearing. The critical speed has been raised to 80 percent of ω_n, and the whirl instability has been eliminated. (The real parts of the eigenvalues are now negative at all speeds.) The amplitude of the response to imbalance is cut in half at the critical speed but is higher at low speed.

Figure 6.21 shows the response of the minor axis of the whirl ellipse versus speed for both types of bearings. Only the journal bearing becomes unstable. Note that if vibration measurements could be made with a probe that measured only the minor axis, it would show two critical speeds (for the tilt-pad bearing only—the

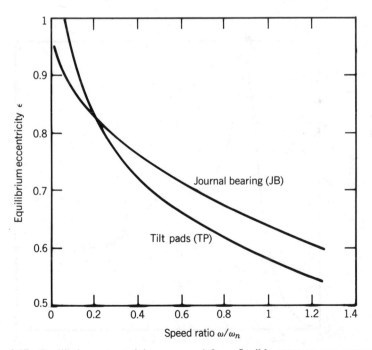

Figure 6.17. Equilibrium eccentricity vs. speed for a flexible rotor on two types of oil-film bearings.

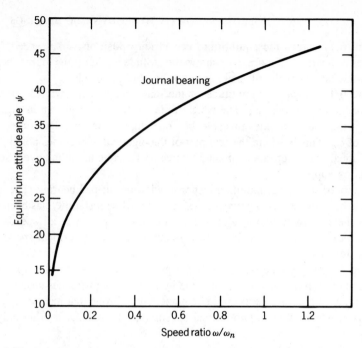

Figure 6.18. Attitude angle versus speed for a cavitated journal bearing supporting a flexible rotor.

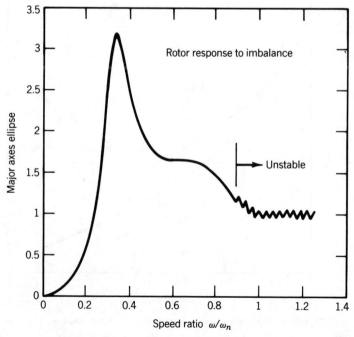

Figure 6.19. Dimensionless whirl amplitude (major axis) versus speed of a flexible rotor on cavitated journal bearings.

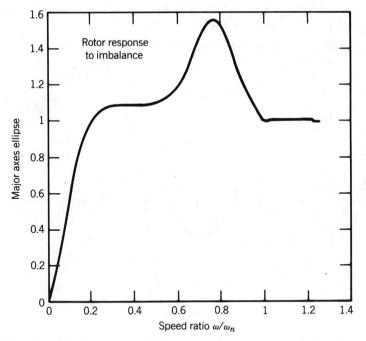

Figure 6.20. Dimensionless whirl amplitude (major axis) versus speed of a flexible rotor on tilt-pad bearings.

rotor with the journal bearing becomes unstable). Both the skew angle and the shape of the ellipse change with speed.

It is the dimensionless (ratio to the imbalance) whirl amplitude of the central disk which is plotted in Figs. 6.19–6.21. Rotor vibration measurements on turbomachines are usually made at the bearing journals wherer access is easier and temperatures are not too high, but it is the whirl amplitudes at the wheels that produce contact rubs of blades and seals. For a flexible rotor the latter amplitudes can be much larger than those measured at the bearings.

BEARING STABILITY CHARACTERISTICS

It is common practice in the rotordynamics literature [9, 12, 13] to describe the whirl stability characteristics of a hydrodynamic bearing with a stability map, as in Fig. 6.22, which gives the dimensionless threshold speed of instability ω_s versus the equilibrium eccentricity ϵ. The dimensionless speed $\omega_s = \omega \sqrt{c/g}$, where $\sqrt{g/c}$ is the natural frequency of a rigid rotor supported on a spring of static deflection c (the radial bearing clearance) and g is the acceleration of gravity. Typically, these curves are generated by computing the dimensionless speed at which the real part of the eigenvalue becomes positive, for various values of static eccentricity.

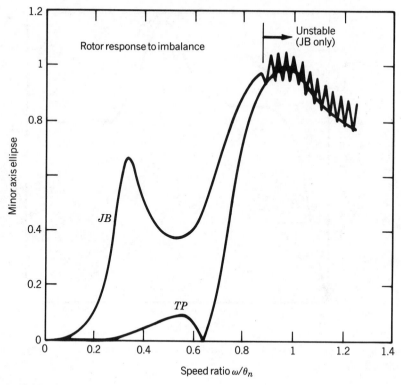

Figure 6.21. Dimensionless whirl amplitude (minor axis) versus speed of a flexible rotor on two types of oil-film bearings.

In Fig. 6.22, note that the fully cavitated journal bearing is stable for all speeds if it is very highly loaded ($\epsilon > 0.8$). Consequently, modified versions of the simple journal bearing (e.g., the pressure-dam bearing) are designed to supplement the actual supported load with hydrodynamic pressure in the oil film and thereby increase the operating eccentricity at high speed.

It should be realized that stability maps such as Fig. 6.22 are usually generated from a rigid rotor model, i.e., the special case $\Delta_s \to 0$ in equations (6-60), and therefore do not accurately represent the stability characteristics of the bearing on a flexible rotor. Also, the whirl stability characteristics of a turbomachine are determined by a number of factors in addition to the bearing coefficients. For example, the flexible rotor model of Figs. 6.17–6.21 is stable at all speeds on tilt-pad bearings since this type of bearing has no cross-coupled stiffness ($K_{XY} = K_{YX} = 0$). Yet there have been a number of cases in the field where turbomachines with tilt-pad bearings were observed to be unstable with violent subsynchronous rotor whirling. This is due to other (not in the bearing) sources of cross-coupled stiffness or negative damping, such as internal friction in the rotor assembly, seals,

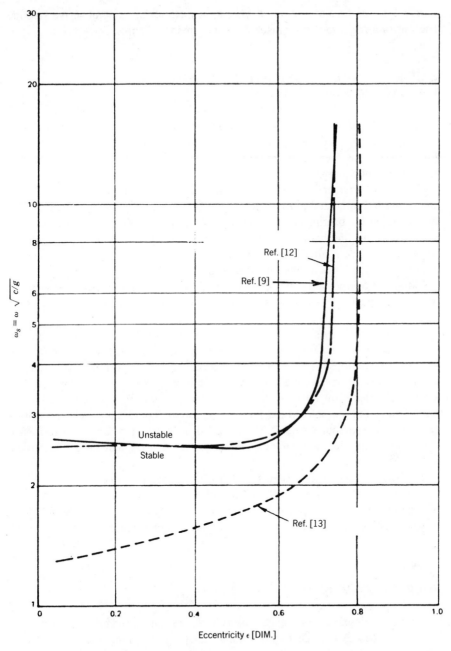

Figure 6.22. Stability maps for the cavitated journal bearing.

or aerodynamic forces on process wheels. A method for comprehensive stability analysis of a multistage turbomachine is described in Chapter VII.

EXPERIMENTAL VERIFICATION OF BEARING CHARACTERISTICS

Measurements of hydrodynamic bearing characteristics have been made on special test apparatus by engineering researchers. Agreement with the Reynolds theory just presented has been good, in most cases, or the discrepancies can be explained in terms of known errors.

Direct measurement of the bearing damping coefficients is especially difficult since both the force and journal motion are time-varying quantities. Parkins [14] measured journal bearing coefficients by shaking the test journal in horizontal and vertical straight line harmonic motions. Nordmann and Schöllhorn [15] used an impact impedance method, converting the vibratory response to frequency-dependent transfer functions to measure journal bearing coefficients.

Tripp and Murphy [16] measured the stiffness of tilt-pad bearings by applying a known static load at various shaft speeds and measuring the change in static eccentricity.

In a number of cases [17], the measured critical speeds and response to imbalance of turbomachines running on very large journal bearings, when compared to computer predictions based on theoretical bearing coefficients, have suggested that the Reynolds theory overpredicts the direct bearing coefficients. Nicholas and Barrett [18] have shown that the discrepancies may actually be due to the additional flexibility of the journal bearing housing or supporting structure, which is often neglected in a computer model. The effect of the housing flexibility is to reduce the effective stiffness and damping of the journal bearing and lower the critical speed.

For future researchers, Adams and Rashidi [19] have suggested the use of measured rotor–bearing instability threshold speeds to more accurately determine bearing properties. They propose to make use of the fact that the threshold speed of whirl instability is usually very sensitive to small changes in the bearing coefficients.

SQUEEZE FILM BEARING DAMPERS

As previously described, the squeeze film damper (SFD) is used primarily in aircraft turbine engines to provide hydrodynamic damping to rolling-element bearings, which otherwise would contribute no appreciable amount of damping to the rotor–bearing system. Figure 6.4 shows the basic configuration. Kinematically, the SFD is a journal bearing in which the journal can vibrate or whirl but not spin (that is, $\omega = 0$). The outer race of the rolling element bearing, or an attached sleeve, serves as the journal of the SFD. It is generally slotted or restrained loosely with a pin to prevent spin.

The SFD can be viewed as a journal bearing with zero spin, but the constraint on angular velocity has ramifications which may not be immediately obvious.

The most important effect of zero journal spin is that a rotor supported on SFDs does not execute oil whip, i.e., the SFD is rotordynamically stabilizing at all speeds. The nonlinearity of the SFD film force can produce nonsynchronous whirl orbits, but the motion is always stable (unless other destabilizing forces act on the rotor). This is a consequence of the fact that a SFD has no stiffness[5], either direct or cross-coupled, and its damping is always positive.

The purely translatory motion of a SFD also produces complete periodic flow reversals in the oil film that do not occur in journal bearings. Consequently, the assumption of Reynolds theory that fluid inertia in the film has no effect on the pressure distribution is invalid [20], at least for most aircraft turbine applications in which the squeeze film Reynolds number, Re, is typically larger than unity. The squeeze film Reynolds number is defined as

$$\text{Re} = \frac{\rho \omega c^2}{\mu},\tag{6-61}$$

where ω is the whirl orbit velocity, ρ is the fluid mass density, c is the radial clearance, and μ is the fluid viscosity.

San Andrés [21] has developed an experimentally verified analysis of squeeze film forces which includes the effects of fluid inertia (see also Refs. [22, 23]). The major effects are found to be an "added mass" (force directed radially outward from the center of the whirl orbit) and an increase in effective damping over the prediction of Reynolds bearing theory.

It should be remembered that there is an optimum amount of support damping for any rotor–bearing system (see Chapter I) and that too much damping can "lock up" the supports. Also, as will be seen below, the SFD produces frequency-dependent "dynamic stiffness" (actually cross-coupled damping) that affects critical speeds. Accurate prediction of the SFD film forces is therefore necessary for a good rotordynamic design analysis when SFD supports are employed.

The squeeze film damper forces can be expressed in the form (6-24) or (6-25), but the linearized interpretation is too restrictive for most SFD applications. Also, the static eccentricity ϵ_s is usually either zero or on the Y axis with $\psi = 0$ (see Fig. 6.10), since a stationary SFD journal generates no film pressure. Many SFD designs incorporate a mechanical centering spring, such as a squirrel cage roller bearing support, in parallel with the fluid film. The static eccentricity is then determined by the mechanical spring stiffness, with $\psi = 0$.

Most analyses of SFD forces have been done under the assumption of circular orbits around the bearing center, not necessarily of small amplitude. This is

---

[5]As defined by equation (6-24).

probably a reasonable assumption when a centering spring is used and the rotor has a moderate to large amount of imbalance. The forces are extremely nonlinear with orbit amplitude, and this has profound effects on the rotor–dynamic response.

Figure 6.23 shows the circular centered orbit (CCO) motion. If the orbit is small, equations (6-24) are entirely adequate and the linearized force coefficients give good results. In this case, the static operating force at $x^* = 0$ and $y^* = 0$ is zero, so ΔF_X and ΔF_Y represent components of the total oil film force for small oscillatory motions, since the X and Y coordinates are both harmonic functions of time:

$$X = e \sin(\dot{\psi}t),$$
$$Y = -e \cos(\dot{\psi}t). \tag{6-62}$$

Since displacement without velocity produces no film pressure, we have

$$K_{XX} = K_{XY} = K_{YY} = K_{YX} = 0,$$
$$K_{rr} = K_{rt} = K_{tt} = K_{tr} = 0. \tag{6-63}$$

The radial and tangential journal displacements and velocities for CCO are

$$r = e, \quad t = e\psi,$$
$$v_r = 0, \quad v_t = e\dot{\psi}. \tag{6-64}$$

If the orbit is not small, the form of equation (6-25) can be used, but the coefficients are no longer constants and therefore take on a different meaning. Figure

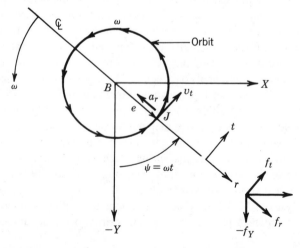

Figure 6.23. Coordinates and force sign convention for a squeeze film damper executing a circular centered whirl orbit at frequency ω.

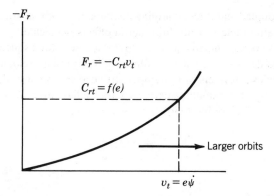

Figure 6.24. Nonlinear cross-coupled damping coefficient C_{rt} for $\dot{\psi}$ = constant.

6.24 shows, for example, the meaning of the cross-coupled damping coefficient C_{rt}. This coefficient gives the radial force produced by the tangential velocity, but it is not the slope of the force–velocity curve, except in an averaged sense. In fact C_{rt} is now a nonlinear function of e or the dimensionless eccentricity ϵ, and ΔF_r, ΔF_t are now the total film force components and so are written as F_r, F_t. The eccentricity ϵ now refers to the radius e of the whirl orbit divided by the radial damper clearance c, rather than a static operating point. The Reynolds short bearing theory, π-cavitated, predicts the radial and tangential force components as

$$F_r = -\frac{2\mu R L^3 \omega \epsilon^2}{c^2(1 - \epsilon^2)^2},\tag{6-65}$$

$$F_t = -\frac{\pi \mu R L^3 \omega \epsilon}{2c^2(1 - \epsilon^2)^{3/2}},\tag{6-66}$$

where ω is now the orbit whirl velocity $\dot{\psi}$.

Since all the K's are zero (no force without whirl velocity), and with fluid inertia neglected in Reynolds Theory, F_r and F_t can be expressed entirely in terms of the damping coefficients:

$$F_r = -C_{rt}v_t,\tag{6-67}$$

$$F_t = -C_{tt}v_t,\tag{6-68}$$

where $v_t = \omega e$, and

$$C_{rt} = \frac{2\mu R L^3 \epsilon}{c^3(1 - \epsilon^2)^2},\tag{6-69}$$

$$C_{tt} = \frac{\pi \mu R L^3}{2c^3(1 - \epsilon^2)^{3/2}}\tag{6-70}$$

are the cross-coupled and direct damping coefficients, both functions of the dimensionless whirl amplitude ϵ, with fluid inertia effects neglected.

The stiffness coefficients in equations (6-25) are zero for a squeeze film damper. This is reasonable, since a squeeze film requires journal velocity to generate a force. Nevertheless, references to damper stiffness are found in the literature.

Reference [24], for example, expresses the radial force (6-65) in terms of a direct stiffness coefficient by dividing F_r by e to give

$$F_r = -K_{rr}e, \tag{6-71}$$

where

$$K_{rr} = \frac{2\mu RL^3\omega\epsilon}{c^3(1 - \epsilon^2)^2} \tag{6-72}$$

is the direct stiffness, or "dynamic stiffness."

Note that the stiffness (6-72) not only depends on ϵ but is frequency dependent as well and goes to zero when the bearing is not whirling.

The rationale behind this approach is that a rigid rotor supported on a squeeze film (with no centering spring) has a critical speed and therefore should have a support stiffness. However, a rotordynamics analysis which models the damper forces as equations (6-67) and (6-68), with zero stiffness, yields the same critical speed. The cross-coupled damping C_{rt} plays the part of the stiffness and supports the rotor in its orbit.

It follows that the critical speeds of an aircraft turbine engine are significantly affected by the SFD even if no centering spring is used.

For a squeeze film damper with fluid inertia effects included, the force components are made up of both damping and inertia terms. That is,

$$F_r = -C_{rt}v_t - D_{rr}a_r$$
$$F_t = -C_{tt}v_t - D_{tr}a_r \tag{6-73}$$

are the nonzero terms from (6-25), where

$$a_r = -\omega^2 e$$
$$a_t = 0 \tag{6-74}$$

are respectively the radial and tangential accelerations of the "journal" for CCO.

The inertia coefficient D_{rr} is sometimes called the "added mass" because it gives a force proportional to the radial acceleration $\omega^2 e$, similar to a centrifugal force. When analyzing experimental damper force measurements, it is sometimes difficult to separate the inertia part from the damping part. One way is to compare the results with Reynolds (purely viscous) theory and assume the difference is due to fluid inertia. Another way is to show a dependence of the force on ω^2. The task is simplified when the film is uncavitated, since C_{rt} is then zero and the radial film force F_r is entirely an inertial effect.

In theory, the inertia terms are found as an additional force, proportional to the Reynolds number Re defined by (6-61).

Physically, the dimensionless group (6-61) represents the ratio of fluid inertia force to viscous force. Thus the radial and tangential components of the SFD film force can be expressed as

$$F_r = F_{r_0} + \text{Re} \, f_{r_1},$$
$$F_t = F_{t_0} + \text{Re} \, f_{t_1}, \tag{6-75}$$

where F_{r_0}, F_{t_0} represent the solution to the Reynolds equation (purely viscous forces), and the second term represents the added force due to the inclusion of inertia terms for the equations of fluid motion.

For short ($L/D < 0.5$) dampers with open ends (no seals) and a cavitated π-film, the damping coefficients in equations (6-73) turn out to be the same as equations (6-69) and (6-70). The inertia coefficients for Re $\gg$ 1 turn out to be

$$D_{rr} = -\frac{\pi \rho R L^3}{12c} \frac{\beta - 1}{\beta \epsilon^2} (2\beta - 1) \tag{6-76}$$

$$D_{tr} = \frac{-\rho R L^3}{c} \left(\frac{27}{70\epsilon}\right) \left[2 + \frac{1}{\epsilon} \ln\left(\frac{1 - \epsilon}{1 + \epsilon}\right)\right], \tag{6-77}$$

where $\beta = (1 - \epsilon^2)^{1/2}$.

The coefficients C_{rt} and D_{rr} can be combined to define an effective "dynamic stiffness" of the oil film for circular centered orbits at a given frequency ω. The result is

$$K_e = \frac{-F_r}{e} = C_{rt}\omega - D_{rr}\omega^2, \tag{6-78}$$

where it can be seen that the "added mass" or inertia coefficient D_{rr} reduces the stiffening effect of the cross-coupled damping. For small orbits, the inertia coefficient dominates so that the net radial force is positive (outward).

Similarly, the cross-coupled inertia coefficient D_{tr} adds to the tangential "drag" force of the direct damping coefficient, so the effective damping with fluid inertia effects included is

$$C_e = \frac{-F_t}{\omega e} = C_{tt} + D_{tr}\omega. \tag{6-79}$$

In applying equation (6-79), recall that the radial acceleration a_r in equation (6-73) is negative for a circular centered orbit, D_{tr} is also negative, so it increases the total damping force.

All of the equations just given are for a short π-cavitated damper with open ends. Some SFD designs have end seals that restrict axial flow through the damper. End seals have a large effect on the force coefficients. The analysis is quite complex, and the reader is referred to Ref. [21] for the complete treatment. The most important effect of end seals is to greatly increase the direct damping coefficient. Accurate prediction of the coefficients for an SFD with seals requires a precise knowledge of the end leakage condition.

EFFECT OF SQUEEZE FILM DAMPERS ON ROTORDYNAMICS

If the SFD were an ideal linear viscous damper with a constant direct damping coefficient, then its only effect on rotordynamics would be to attenuate synchronous response to imbalance and suppress rotordynamic instability (assuming a design to produce the optimum damping coefficient for the rotor–bearing system).

However, the nonlinear dependence of the force coefficients on whirl amplitude, the radial force produced by cross-coupled damping and fluid inertia, and the added effective damping produced by cross-coupled inertia combine to produce some rather complex effects on rotordynamics.

The most remarkable effect, under certain conditions, is a jump of the frequency response curve in which the response to imbalance is multivalued when the rotor imbalance exceeds a critical value. In this situation there are generally three possible whirl orbit amplitudes, two of which are stable motions, and only one of these is small enough to make the damper effective. The maximum critical value of static imbalance for a rigid rotor is 40 percent of the radial damper clearance, so the SFD to date has not been an effective device for controlling abnormally large rotor imbalances such as that produced by a turbine blade loss.

Experimental measurements of the effect of a squeeze film on rotordynamics were first published by Cooper [25] of Rolls-Royce Ltd. in 1963. In the same paper, Cooper showed analytically the possibility of two different orbit sizes at the same speed, both satisfying the equations of dynamic equilibrium. This analysis was later extended by other investigators [26, 27] to show the existence of three equilibrium orbits, with the intermediate one always unstable so that the whirling will jump out to the largest orbit (in which the damper amplifies the dynamic bearing load) or jump in to the smallest orbit (in which the bearing load is attenuated), depending on the operating conditions and damper design. These analyses were all for a rigid rotor, a special case emphasizing the effect of the damper.

In the design analysis of squeeze film dampers, a useful dimensionless group of variables is the "bearing parameter"

$$B = \frac{\mu R L^3}{m \omega C^3} \qquad (6\text{-}80)$$

or

$$B_k = \frac{\mu R L^3}{m \omega_n C^3},$$ (6-81)

where m is the rotor mass controlled by one damper, and ω_n is the natural frequency of the rotor on centering springs if they are used; B or B_K is a design parameter for the short damper, and $1/B$ can also be used as a dimensionless speed in response plots.

For circular centered orbits, and using the notation defined earlier in this chapter, Newton's second law in the radial and tangential directions gives

$$F_r - k_B e = -m\omega^2(e + u \cos \beta),$$ (6-82)

$$F_t = -m\omega^2 u \sin \beta,$$ (6-83)

where k_B is the centering spring stiffness. Figure 6.25 shows the imbalance phase angle β.

If we use damping and inertia coefficients defined in the previous section, the squeeze film force components are

$$F_r = -C_{rt}\omega e + D_{rr}\omega^2 e,$$ (6-84)

$$F_t = -C_{tt}\omega e + D_{tr}\omega^2 e.$$ (6-85)

For the case without fluid inertia (Re = 0; $D_{rr} = D_{tr} = 0$), these equations

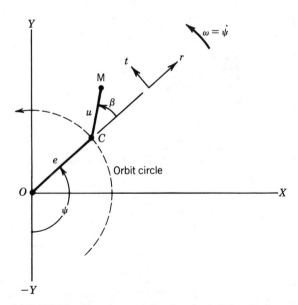

Figure 6.25. Whirl orbit geometry showing the imbalance phase angle β.

can be combined to eliminate β and written in terms of dimensionless groups to give

$$(B_k \overline{C}_{tt} \epsilon)^2 + \left(B_k \overline{C}_{rt} + \frac{1}{\Omega} - \Omega \right)^2 \epsilon^2 - U^2 \Omega^2 = 0 \qquad (6\text{-}86)$$

where

$$\overline{C}_{tt} = C_{tt}/C_s$$
$$\overline{C}_{rt} = C_{rt}/C_s$$
$$C_s = \mu RL^3/C^3$$
$$\Omega = \omega/\omega_n$$
$$U = u/C$$

For each dimensionless speed Ω and imbalance U equation (6-86) is a polynomial in ϵ, and its roots define the response curve.

For the case with no centering spring B_k is undefined and the dimensionless speed is $1/B$. The corresponding response equation is

$$(B\overline{C}_{tt}\epsilon)^2 + (B\overline{C}_{rt} - 1)^2 \epsilon^2 - U^2 = 0. \qquad (6\text{-}87)$$

The dashed line plotted in Figure 6.26 shows the predicted response of a symmetric rigid rotor of mass $2m$ on two short uncavitated ($C_{rt} = 0$) squeeze film

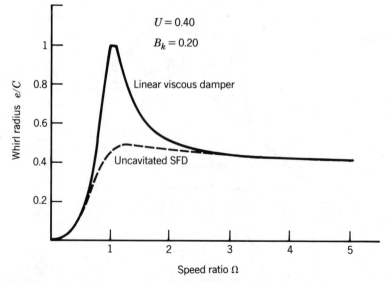

Figure 6.26. Imbalance response of a rigid rotor supported on uncavitated squeeze film dampers (without fluid inertia) and on linear viscous dampers.

dampers with fluid inertia neglected. Without cavitation there is no radial stiffening effect and the jump phenomenon does not appear, regardless of the amount of imbalance. For comparison, the solid line shows the response of the rotor on linear viscous (C_{tt} = constant) dampers. It can be seen that the nonlinear direct damping coefficient C_{tt} of the squeeze film damper is actually better than a linear coefficient in attenuating the response at the critical speed while still allowing the critical speed inversion to take place ($\beta \rightarrow 180°$ at high speed). The linear curve is based on a critical damping ratio $\xi = B_k$ for $\epsilon = 0.386$.

Figure 6.27 shows the response curve for the same amount of imbalance ($U = 0.4$) when the dampers are cavitated (π-film). Here the nonlinear stiffening effect of the C_{rt} coefficient produces a double-valued jump in the whirl amplitude. The higher whirl amplitudes produce a force transmissibility higher than unity and make this damper design undesirable. The jump can be eliminated by the following means:

1. Reducing the imbalance, i.e., make $U < 0.4$.
2. Raising the oil supply pressure high enough to eliminate cavitation (usually impractical).
3. Increasing the bearing parameter B_k. This might be done by increasing the lubricant viscosity μ or the damper bearing axial length L. Reducing the clearance C would not be desirable since this raises the dimensionless imbalance U. The result of increasing B_k to 0.5 is shown in Fig. 6.28.
4. Adding the effect of fluid inertia. In the physical system, this will occur naturally at high Reynolds numbers.

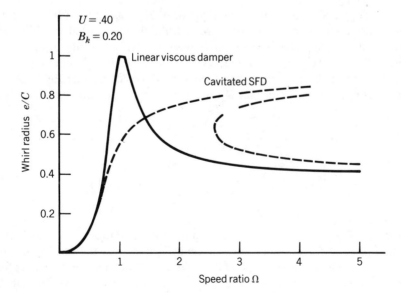

Figure 6.27. Imbalance response of a rigid rotor supported on cavitated squeeze film dampers (without fluid inertia) and on linear viscous dampers.

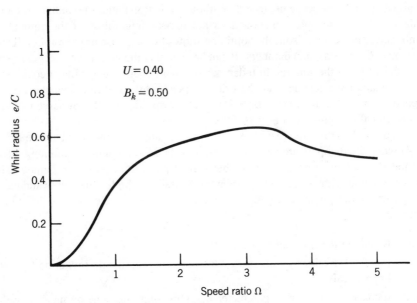

Figure 6.28. Imbalance response of a rigid rotor suppored on cavitated squeeze film dampers with the bearing parameter increased (without fluid inertia).

The inertia coefficient D_{rr} counteracts the radial stiffening effect of C_{rt}, and D_{tr} adds to the tangential damping force.

In practice, modern aircraft engine rotors with SFD operating at Re > 1 often show a very highly damped response through the first two (rigid mode) critical speeds, with the jump phenomenon rarely observed. This was difficult to explain theoretically until the fluid inertia effects were included in the prediction equations.

In terms of the bearing parameter B_k and the Reynolds number Re_n evaluated at the undamped natural frequency ω_n, the inertia coefficients are

$$D_{rr} = -m\left\{ B_k \,\mathrm{Re}_n \,\frac{\pi}{12}\left(\frac{\beta-1}{\beta\epsilon^2}\right)(2\beta-1)\right\}, \qquad (6\text{-}88)$$

$$D_{tr} = -m\left\{ B_k \,\mathrm{Re}_n \,\frac{27}{70\epsilon}\left[2 + \frac{1}{\epsilon}\ln\left(\frac{1-\epsilon}{1+\epsilon}\right)\right]\right\}, \qquad (6\text{-}89)$$

where $\beta = (1 - \epsilon^2)^{1/2}$.

Note that the { } bracketed quantity in (6-88) can be regarded as the fraction of rotor mass that produces an equivalent centrifugal force on the whirling motion.

Figure 6.29 shows the response of the rigid rotor with fluid inertia forces added to the conditions of Fig. 6.27, for $\mathrm{Re}_n = 10$ (a value typical of modern aircraft engine dampers).

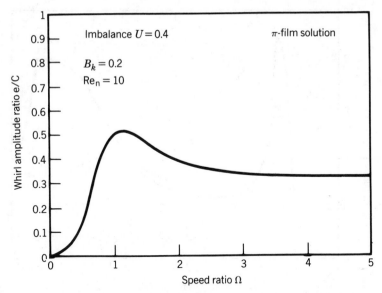

Figure 6.29. Imbalance response of a rigid rotor supported on cavitated squeeze film dampers with fluid inertia effects included.

For flexible rotors, the effect of a SFD on imbalance response is generally less pronounced, but it is more complex due to the multiplicity of critical speeds. Dampers located close to a node of the whirling mode shape have a small effect on that mode because of the relatively small amplitude of journal motion. However, the same type of nonlinear jump behavior is predicted for flexible rotors at small Reynolds numbers under certain conditions [28].

Figures 6.30–6.32 show the computed response of a flexible rotor on short SFD supports with fluid inertia effects included, for $\mathrm{Re}_n = 10$, from a nonlinear analysis by San Andrés [29]. The parameters used to generate the curves are representative of aircraft engine design conditions, and are shown in Table 6-1. If the fluid inertia coefficients are taken out, the predicted response becomes double-valued between the two critical speeds with a very high transmissibility associated with the larger amplitudes.

On each of the three figures, the curve with the highest peak value is for $B_k = 0.01$, a value typical of contemporary aircraft gas turbines. Larger values of B_k require a longer bearing length L and/or a higher oil viscosity μ, both of which may be impractical under other design constraints. However, note the very favorable effect of increasing the damper bearing parameter B_k to 0.1. Each rotor–bearing system must be analyzed to determine its optimum damper bearing parameters (e.g., changing to a damper with end seals will change the curves radically), but it can be seen from this illustration that rotordynamic response is sensitive to SFD design and its operating parameters.

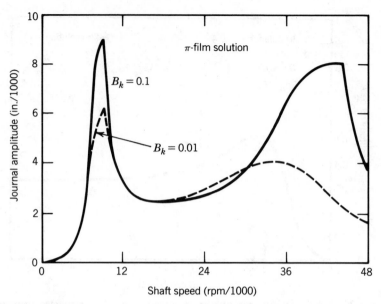

Figure 6.30. Imbalance response at the journal of a flexible rotor supported on cavitated squeeze film dampers with fluid inertia effects included, for two values of the bearing parameter (rotor–bearing data in Table 6-1).

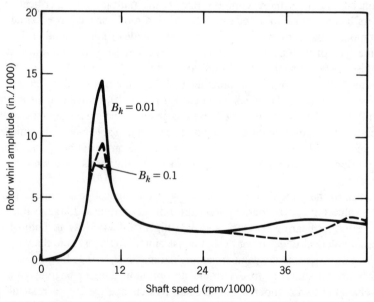

Figure 6.31. Imbalance response at midspan of a flexible rotor supported on cavitated squeeze film dampers with fluid inertia effects included, for two values of the bearing parameter (rotor–bearing data in Table 6-1).

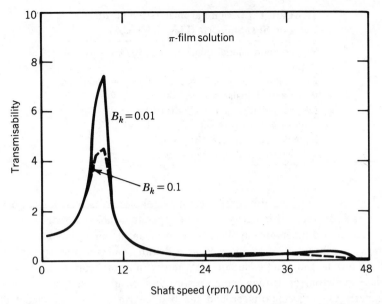

Figure 6.32. Force transmissibility of a flexible rotor supported on cavitated squeeze film dampers with fluid inertia effects included, for two values of the bearing parameter (rotor–bearing data in Table 6-1).

The flexible rotor model used for the illustration has a rigid support natural frequency of 16,000 rpm, and it can be seen on the response curves that this would be close to an optimum operating speed! Certainly this illustrates that the method of estimating the critical speeds from a "pin–pin" (rigid support) rotor model, sometimes practiced by engineers in the past, can yield useless results.[6]

The first critical speed at about 10,000 rpm might be referred to as a rigid rotor mode, whereas the second critical speed involves a large amount of rotor bend. The SFD damps the second critical speed better in this case because its mode shape has a large amplitude at the bearing and because the effective mass controlled by the damper in this mode is much smaller than for the first mode. As mentioned before, the reverse sometimes occurs, depending on the particular rotor–bearing configuration.

ELASTOMERIC BEARING DAMPERS

Elastomeric (synthetic rubber) bearing supports have been used for their damping properties in expendable aircraft drone engines and in rotor–bearing test rigs. Their

---

[6]There are some cases, notably flexible rotors supported on hydrodynamic journal bearings, where the method gives surprisingly accurate results.

TABLE 6-1.   Design Parameters for Flexible Rotor on SFD Supports (Figs. 6.30–6.32)

Rigid support natural frequency $\omega_s (30/\pi) = 16,000$ rpm

Rigid rotor natural frequency $\omega_r (30/\pi) = 9600$ rpm

Rotor static imbalance $u = 0.002$ in

Damper radial clearance $C = 0.010$ in

Damper bearing parameter $B_k = 0.01, 0.1$

Reynolds number at ω_s, $\mathrm{Re}_s = 10$

π-Cavitated short bearing damper

major limitation for long-life turbine engine applications is their loss of damping properties and structural integrity at high temperature (above 150°F).

For ambient temperature application in a clean air environment, elastomeric O'-rings have been successfully used as a flexible damped bearing support in a small air turbine [30]. In the author's laboratory, a rotor–bearing test apparatus to simulate industrial centrifugal compressors was designed and constructed with elastomeric O'-ring supports around the ball bearings. The equivalent critical damping ratio was found to be 0.12 at 80°F, compared to a maximum of about 0.06 without them. However, a considerable loss of damping was observed at temperatures in the range 140°–200°F.

The typical O'-ring support configuration for rolling-element bearings is illustrated in Figure 6.33. Darlow and Zorzi [31] have researched the damping properties of elastomers and suggest the use of cylindrical buttons arranged radially for bearing support, since their properties can be predicted with more accuracy and consistency than those of O'-rings. At the same laboratory, Smalley et al. measured the dynamic characteristics of O'-rings [32].

Elastomeric materials exhibit a hysteretic type of energy dissipation which can

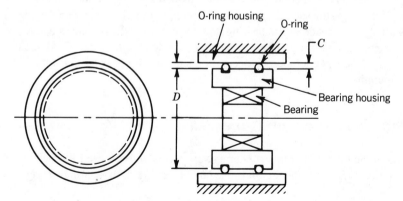

Figure 6.33. Elastomeric O'-ring bearing support. From Ref. [32].

best be modeled as "structural damping," which is a complex stiffness. The damping force F_d is the imaginary part of the complex dynamic force

$$F = k_e(1 + i\eta)X, \qquad (6\text{-}90)$$

or

$$F_d = k_e\eta X, \qquad (6\text{-}91)$$

where

k_e = stiffness of the elastomeric support

η = loss coefficient

X = displacement of the elastomer

Typical loss coefficients of currently available elastomers seldom exceed 0.4 and decrease with temperatures above 100°F. From equation (6-90) it can be seen that the combination of low support stiffness and high damping is difficult to achieve with these materials. Since a viscous damping force acting on a harmonic vibration has an amplitude of $c\omega X$, it can be seen that the equivalent viscous damping coefficient for an elastomeric damper is

$$C_{eq} = \frac{k_e\eta}{\omega}, \qquad (6\text{-}92)$$

which decreases with the frequency of vibration.

Since new synthetic materials are continually being developed and improved, it seems likely that some type of elastomeric damper will eventually compete with the squeeze film damper for many more turbomachinery applications.

FLUID SEALS

Various types of seal designs are used to reduce the leakage of working and lubricating fluids through the interface between rotors and stators in turbomachines. Some leakage is inevitable, and results in axial fluid velocities through the seal in the direction of pressure drop.

For example, Fig. 6.34 shows the working pressures and the designed fluid flow path through a two-stage centrifugal compressor or pump. Since $p_5 > p_4 > p_3 > p_2 > p_1 = p_e$, the working fluid will tend to leak backward through the impeller neck seals and the interstage seal, and forward past the balance piston.

The fluid velocity in a rotating seal has both axial and circumferential components, and the pressure within the seal generally varies circumferentially and with time due to rotor motion within the seal clearance.

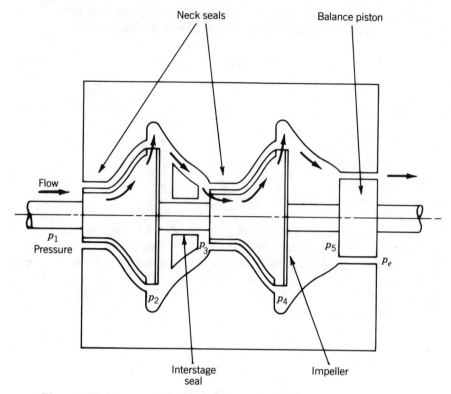

Figure 6.34. Flow path, fluid pressures, and seals in a two-stage compressor.

The major classifications of rotating seals used in turbomachinery are as follows:

1. *Plain Seals.* These can be straight, as in Fig. 6.34, or tapered or stepped as in Fig. 6.35. Plain seals are generally used in pumps. In geometry, they are similar to journal bearings but the clearance/radius ratio is usually two to ten times larger to avoid rotor/stator contact.

2. *Floating Ring Oil Seals.* See Fig. 6.36. In these seals lubricating oil is used to fill the clearance space and reduce leakage. The ring whirls or vibrates with the rotor, but does not spin ($\omega = 0$). They are used in high pressure multi-stage centrifugal compressors.

3. *Labyrinth Seals.* See Fig. 6.37. The circumferential blades in axial tandem produce a ''tortuous passage'' (Webster's dictionary) for the leaking fluid. The blades may be on the stator or on the rotor or on both. These seals are used in both centrifugal and axial compressors, and in turbines.

4. *Contact Seals.* In this case, no clearance is designed. Because of the rubbing, these seals are used mostly in low speed pumps, or where the working fluid can act as a coolant. Their effect on rotordynamics is unknown and probably minimal

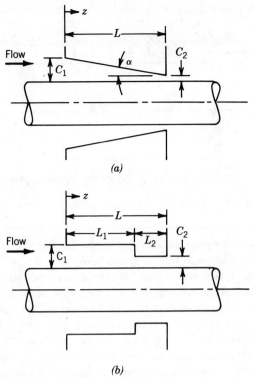

Figure 6.35. Tapered (a) and stepped (b) plain seals. From Ref. [40].

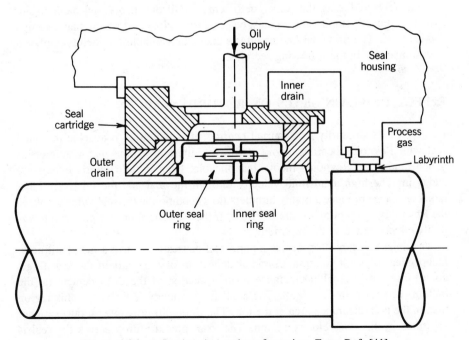

Figure 6.36. A floating ring seal configuration. From Ref. [41].

251

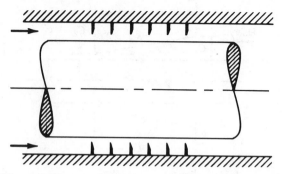

Figure 6.37. Labyrinth seal configuration, teeth on stator.

except for friction or hysteresis damping from the material in contact with the rotor. They are not discussed further in this book.

All of these seal types, with a possible exception of the last, produce forces on the rotor that affect rotordynamics. The seal forces are functions of rotor displacement, velocity, and acceleration at the seal location and so can be formulated in terms of stiffness, damping, and inertia coefficients, the same as for bearing forces. From the data available to date it appears that the linear interpretation of the force coefficients, as previously described for journal bearings, should be used for most seals under the assumption of small motions about the centered position. Seal forces apparently are not greatly sensitive to violations of the latter assumption. However, the analysis of seal forces is more difficult and generally has not been as successful in accurately predicting the force coefficients. Until the theory and modeling of seals is improved, empirical data must be primarily utilized for rotordynamic design and analysis. The principal known effects and data available for each type of seal are summarized in the following sections.

EFFECT OF PLAIN SEALS ON ROTORDYNAMICS

Because of its similarity to a journal bearing, it is tempting to analyze the plain seal with Reynolds bearing theory. However, there are some major differences which render the Reynolds theory inapplicable, even if fluid inertia effects are added in. The high axial fluid velocity through the seal and the relatively large radial clearance produce a highly turbulent flow condition in the seal, which violates the Reynolds assumption of laminar flow. If the working fluid is a gas, compressibility effects may also be important.

The axial pressure drop across a plain seal causes it to have a radial stiffness, independent of shaft rotation. A radial deflection of the shaft in the seal (e.g., down) produces a smaller clearance in the direction of the displacement (at the bottom) as shown in Fig. 6.38. The axial fluid velocity is slower in this region than in the high clearance area at the top. The Bernoulli principle produces higher pressures in the low velocity region. The axial pressure drop across the seal is

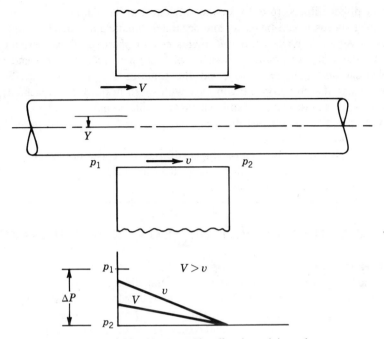

Figure 6.38. The Lomakin effect in a plain seal.

caused by an entrance loss and a friction loss along the length of the seal. As Fig.
6.38 shows, the entrance loss is greater in the high clearance region (at the top)
so the average pressure is lower there. The pressure difference produces a restoring
force $F_Y = -K_s Y$ upward, opposing the displacement Y. This radial stiffness K_s
of a plain seal, which occurs even for $\omega = 0$ provided the axial pressure difference
ΔP exists, is known as the Lomakin effect after its first investigator [33].

Black and his colleagues [34–36] and Childs [37, 38] (later) have formulated
and extended Lomakin's theory in terms applicable to the rotordynamic analysis
of centrifugal pumps. The Lomakin direct stiffness (K_{XX}) coefficient is given by

$$K_s = 4.7R\left(\frac{\Delta P}{\lambda}\right)\left(\frac{\sigma}{1.5 + 2\sigma}\right)^2 \qquad (6\text{-}93)$$

where

λ = friction coefficient for the axial flow

$\quad = 0.079/\mathrm{Re}^{1/4}$ for turbulent flow

$\sigma = \lambda L/C$

Here R, L, and C are the radius, axial length, and radial clearance of the seal,
respectively.

If the direct stiffness (6-93) were the only effect of the plain seal, then its effect on critical speeds would be easily and accurately predictable. Black, Childs, and others have shown, however, that K_s increases with shaft speed (at constant ΔP) and that the seal also produces cross-coupled stiffness (K_{XY}), direct and cross-coupled damping (C_{XX} and C_{XY}), and direct inertia (D_{XX}) coefficients [36,37]. Consider also that the drop ΔP will vary with speed in most turbomachines, and it can be seen that the rotordynamic effects are quite complex.

Experimentally measured stiffness and damping coefficients for plain seals have been published by Nelson et al. [39].

Fleming [40] has shown that the direct stiffness of stepped or tapered seals with converging clearance is 2–14 times higher than for a straight seal with clearance equal to the minimum of the converging seal.

EFFECT OF FLOATING RING OIL SEALS ON ROTORDYNAMICS

The floating rings in an oil seal are supposed to be free to move radially in a slot so that the only force exerted on the rotor is the sliding friction at the slot wall. This friction force increases with the axial pressure drop across the ring, which pushes the ring against the slot wall. If the friction force becomes excessive, the ring locks in place and acts as a journal bearing, Kirk and Miller [41] have analyzed the transient dynamics of the seal ring and its effect on the imbalance response and whirl stability of a multistage compressor.

When they are operating properly, oil ring seals can have a favorable effect in attenuating imbalance response at the critical speeds. If the rings become locked and operate as journal bearings, they can induce the rotordynamic instability known as oil whip. To suppress or eliminate the instability of the locked system, the effective axial hydrodynamic length of the ring can be reduced, but this tends also to reduce the damping effect on imbalance response at the critical speeds.

LABYRINTH SEALS

Labyrinth seals are known to produce a force on the rotor that can be described by direct stiffness, cross-coupled stiffness, and direct damping coefficients. It is the tangential (to the whirl orbit) component of the pressure force that has the greatest effect on rotordynamic response and stability, and this component can be divided into a part proportional to the radial displacement (cross-coupled stiffness K_{XY}) and a part proportional to tangential velocity (direct damping C_{XX}). Some designs apparently can produce a combination of cross-coupled stiffness and low (or negative) damping that destabilizes the rotor in subsynchronous whirl.

The author has tested labyrinth seal designs that are significantly stabilizing, in some cases doubling the logarithmic decrement of the rotor–bearing system and cutting the critical speed response amplitude in half.

The quantitative values of labyrinth seal force coefficients are the subject of current experimental research, and adequate theories for their accurate prediction are yet to be developed. Some of the results developed to date have been contradictory, probably because of the different test conditions employed in experiments and the diverse assumptions made by various investigators for simplified analyses.

The first published analysis of the labyrinth seal effect on rotordynamics was by Alford in 1965 [42]. Alford's analysis neglected swirl and viscous effects, and considered only the effect of compressibility of the gas. His results predicted that the pressure distribution around a converging seal (clearances decreasing in the direction of flow) would produce a negative (destabilizing) damping coefficient ($C_{XX} < 0$) and a small direct stiffness coefficient K_{XX}. A diverging seal was predicted to produce a positive damping. Alford's analysis was limited to a one-cavity (two-blade) seal, and employed an (erroneous) assumption that the gas flow would be choked at both the inlet blade and the exit blade.

In an analytical study published in 1980, Murphy and Vance [43] extended Alford's analysis to multibladed seals, considering choked flow to be possible only at the last (exit) blade. The signs of the computed coefficients (and hence the direction of the predicted forces) were the same as from Alford's analysis, but the magnitudes were modified. Earlier, in Germany in 1974, Spurk and Keiper [44] had published an analysis that predicted damping coefficients opposite in sign to Alford's.

From rap tests on a nonrotating rotor, with 60 psi air blowing down to an atmospheric exit across two 20-blade seals, Kurohashi et al. [45] showed that converging seals increased the damping coefficient, and diverging seals reduced it, for a natural frequency of 35.6 Hz (2136 rpm). From air pressure measurements around a rotating (but not whirling) rotor, Benckert and Wachter [46] showed that circumferential swirl of the gas through the seal produces a pressure distribution which gives a cross-coupled stiffness $K_{XY} > 0$ (drives forward whirl). The swirl was produced by shaft rotation and also by pre-inlet vanes. It was shown that the tangential force could be eliminated by inducing a reverse swirl at the inlet. Since the rotor axis was fixed in the test rig (no whirling), the seal damping coefficient could not be measured. Both of these papers also presented analyses that were said to predict the measured results just described.

In 1982, Iwatsubo et al. [47] published the analytical basis of a computer program to predict the seal coefficients. A program based on this analysis, written at Texas A&M by Scharrer and Childs, failed to agree well with experimental measurements when realistic values for the parameters were used as input.

Wright [48] presented experimental measurements of seal force from a unique test apparatus in which rotor whirl is driven (both forward and backward) by controllable electromagnetic forces. The paper contains a very complete set of measured data, so that comparison with other experiments or analysis is facilitated. However, Wright's test conditions limit the usefulness of his results for turbomachinery applications. The shaft speed was 1800 rpm, the whirl speed was 660 rpm, and the maximum pressure drop across the seal was 5 psi.

A test apparatus in the Turbomachinery Laboratories at Texas A&M is designed to test a variety of full size (6-in diameter) labyrinth seal types under conditions

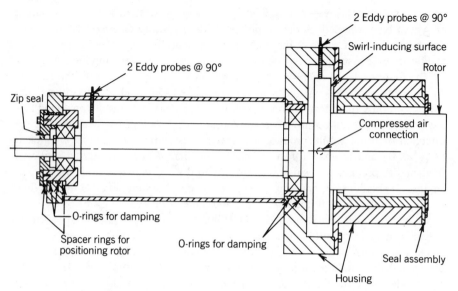

Figure 6.39. Labyrinth seal test rig, cross section.

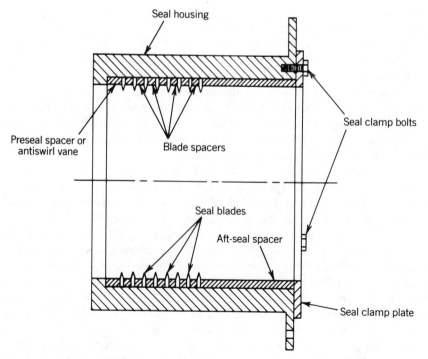

Figure 6.40. Labyrinth seal test section with replaceable blades.

Figure 6.41. Seal test section, photo.

representative of industrial and aircraft turbomachinery. Section drawings of the test rig and its seal test section are shown in Figs. 6.39 and 6.40. A photograph of the seal assembly is shown Fig. 6.41.

In Fig. 6.39, the nodal point corresponding to the first critical speed (3750 rpm) is located at the coupling end bearing (opposite end from the labyrinth seal). The second critical speed (approximately 10,000 rpm) has its nodal point located near the labyrinth seal test section. Thus, the effect of the labyrinth seal on rotordynamics is entirely and exclusively on the first mode.

This test rig was the source of the measurements described at the beginning of this section. Additional tests are in progress at this writing to determine the effect of various labyrinth seal designs on rotordynamics.

Another test rig in the same laboratory has been designed and constructed by Childs to directly measure the force exerted by the labyrinth seal, as the shaft executes a constrained vibratory motion (excited by an external shaker). Childs and Scharrer [49] have recently presented data from this rig which show that a seal with teeth on the stator is less destabilizing to forward whirl (smaller K_{XY}) than a seal with teeth on the rotor. The measured direct stiffness K_{XX} in both cases was negative (radial force outward), which would tend to lower natural frequencies of the rotor bearing system.

REFERENCES

1. Lewis, P., and Malanoski, S. B., *Rotor–Bearing Dynamics Design Technology*, Pt. IV: Ball-Bearing Design Data, Technical Report AFAPL-TR-65-45, Aeropropulsion Lab, Wright-Patterson Air Force Base, Ohio, May 1965.

2. Jones, A. B., "A General Theory for Elastically Constrained Ball and Radial Roller Bearings Under Arbitrary Load and Speed Conditions," *Journal of Basic Engineering*, pp. 309–320 (June 1960).

3. Tower, B., "First Report on Friction Experiments," *Proceedings of the Institution of Mechanical Engineers (London)*, pp. 632–666 (November 1883). Also Second, Third, and Fourth Reports, pp. 58–70 (1885), pp. 173–205 (1888), and pp. 111–140 (1891), respectively.

4. Trumpler, P. R., *Design of Film Bearings*, Macmillan, New York, 1966.

5. Lund, J. W., *Self-Excited, Stationary Whirl Orbits of a Journal in a Sleeve Bearing*, Ph.D. Dissertation in Engineering Mechanics, Rensselaer Polytechnic Institute (1966).

6. Lund, J. W., and Saibel, E., "Oil Whirl Orbits of a Rotor in Sleeve Bearings," *Journal of Engineering for Industry*, pp. 813–823 (November 1967).

7. Muszynska, A., "Improvements in Lightly Loaded Rotor/Bearing and Rotor/Seal Models," *Rotating Machinery Dynamics, Volume One*, Proceedings of the 1987 ASME Conference on Mechanical Vibrations and Noise, Boston, Mass., Sept. 27–30, 1987, pp. 91–98.

8. Muszynska, A., "Whirl and Whip-Rotor/Bearing Stability Problems," *Journal of Sound and Vibration*, **110** (3), London, U.K., 1986, pp. 443–462.

9. Kirk, R. G., and Gunter, E. J., "Stability and Transient Motion of a Plain Journal Mounted in Flexible Damped Supports," *Journal of Engineering for Industry*, May, pp. 576–592 (1976).

10. Holmes, R., "The Vibration of a Rigid Shaft on Short Sleeve Bearings," *Journal of Mechanical Engineering Science*, **2** (4) 337–341 (1960).

11. Lund, J. W., Arwas, E. B., Cheng, H. S., Ng, C. W., and Pan, C. H., *Rotor–Bearing Dynamics Design Technology*, Pt. III: *Design Handbook for Fluid Film Type Bearings*, Technical Report AFAPL-TR-65-45, Aero Propulsion Lab, Wright-Patterson Air Force Base, Ohio, May 1965.

12. Badgley, R. H., and Booker, J. F., "Turborotor Instability: Effect of Initial Transient on Plane Motion," *Journal of Lubrication Technology*, pp. 625–633 (October 1969).

13. Reddi, M. M., and Trumpler, P. R., "Stability of the High Speed Journal Bearing Under Steady Load-1: The Incompressible Film," *Journal of Engineering for Industry*, Ser. B., **84**, 351–358 (1962).

14. Parkins, D. W., "Theoretical and Experimental Determination of the Dynamic Characteristics of a Hydrodynamic Journal Bearing," *Journal of Lubrication Technology*, **101**, 129–139 (1979).

15. Nordmann, R., and Schöllhorn, K., "Identification of Stiffness and Damping Coefficients of Journal Bearings By Means of the Impact Method," Paper No. C285/80, *Proceedings of the 2nd International Conference on Vibrations in Rotating Machinery* (Institution of Mechanical Engineers), held at Churchill College, Cambridge University, September 2–4, 1980.

16. Tripp, H., and Murphy, B. T., "Eccentricity Measurements on a Tilting-Pad Bearing," *ASLE Transactions*, **28**(2), 217–224 (1984).

17. Morton, P. G., "Measurement of the Dynamic Characteristics of a Large Sleeve Bearing," *Journal of Lubrication Technology*, pp. 143–155 (Jan. 1971).

18. Nicholas, J. C., and Barrett, L. E., "The Effect of Bearing Support Flexibility on Critical Speed Prediction," ASLE Preprint No. 85-AM-2E-1 (1985).

19. Adams, M. L., and Rashidi, M., "On the Use of Rotor–Bearing Instability Thresholds to Accurately Measure Rotordynamic Properties," *Journal of Vibration, Acoustics, Stress, and Reliability in Design*, **107**(4), 404–409 (1985).

20. Tichy, J. A., "The Effect of Fluid Inertia in Squeeze Film Damper Bearings: A Heuristic and Physical Description," ASME Paper No. 83-GT-177 (1983).

21. San Andrés, L. A., *Effect of Fluid Inertia on Squeeze Film Damper Force Response*, Ph.D. Dissertation in Mechanical Engineering, Texas A&M University (December 1985).

22. San Andrés, L. A., and Vance, J. M., "Effects of Fluid Inertia and Turbulence on the Force Coefficients for Squeeze Film Dampers," ASME Paper No. 85-GT-191 (1985).

23. San Andrés, L. A., and Vance, J. M., "Experimental Measurement of Squeeze Film Bearing Force Coefficients for Circular Centered Orbits," ASLE Paper No. 86TC402, ASME/ASLE Tribology Conference, Pittsburgh (October 20–22, 1986).

24. Gunter, E. J., Barrett, L. E., and Allaire, P. E., "Design of Nonlinear Squeeze Film Dampers for Aircraft Engines," *Journal of Lubrication Technology*, **99**(1), 57–64 (1977).

25. Cooper, S., "Preliminary Investigation of Oil Films for the Control of Vibration," Paper No. 28, *Lubrication and Wear Convention, 1963* (Institution of Mechanical Engineers), pp. 305–315.

26. White, D. C., "The Dynamics of a Rigid Rotor Supported on Squeeze Film Bearings," A72-2212708.28, *Proceedings of the Conference on Vibrations in Rotating Systems, London, February 14–15, 1972* (Institution of Mechanical Engineers), pp. 213–229.

27. Mohan, S., and Hahn, E. J., "Design of Squeeze Film Damper Supports for Rigid Rotors," *Journal of Engineering for Industry*, pp. 976–982 (August 1974).

28. Rabinowitz, M. D., and Hahn, E. J., "Steady-State Performance of Squeeze Film Damper Supported Flexible Rotors," *Journal of Engineering for Power*, pp. 552–558 (October 1977).

29. San Andrés, L. A., and Vance, J. M., "Effect of Fluid Inertia on the Performance of Squeeze Film Damper Supported Rotors," *Journal of Engineering for Gas Turbines and Power*, **110**(1), 51–57, (1988).

30. Powell, J. W., and Tempest, M. C., "A Study of High Speed Machines with Rubber Stabilized Air Bearings," ASME Paper No. 68-LubS-9, *Journal of Lubrication Technology*, pp. 701–708 (1968).

31. Darlow, M., and Zorzi, E., *Mechanical Design Handbook for Elastomers*, Mechanical Technology Inc., Latham, NY (for NASA), January 1981.

32. Smalley, A. J., Darlow, M. S., and Mehta, R. K., "The Dynamic Characteristics of O'-Rings," ASME Paper No. 77-DET-27, *Journal of Mechanical Design*, **100**, 132–146 (Jan. 1978).

33. Lomakin, A. A., "Calculation of Critical Speed and Securing of Dynamic Stability of the Rotor of Hydraulic High Pressure Machines with Reference to Forces Arising in the Seal Gaps" [in Russian], *Energomashinostroenie*, **4**(4), 1–5 (April 1958).

34. Black, H. F., "Effects of Hydraulic Forces in Annular Pressure Seals on the Vibrations of Centrifugal Pump Rotors," *Journal of Mechanical Engineering Science*, **11**(2), 206–213 (1969).

35. Black, H. F., and Murray, J. L., "The Hydrostatic and Hybrid Bearing Properties of

Annular Pressure Seals in Centrifugal Pumps,'' The British Hydromechanics Research Association, Paper No. RR1026 (October 1969).

36. Black, H. F., and Jenssen, D. N., "Dynamic Hybrid Properties of Annular Pressure Seals," Paper 9, Advanced Class Boiler Feed Pumps, Fluid Plant and Machinery Group, Institution of Mechanical Engineers, September 1970; also *Proceedings of the Institution of Mechanical Engineers (London)*, **184**, 92–100 (1970).

37. Childs, D. W., "Dynamic Analysis of Turbulent Annular Seals Based on Hirs Lubrication Equations," *Journal of Lubrication Technology*, **105**, 429–436 (July 1983).

38. Childs, D. W., "Finite-Length Solutions for Rotordynamic Coefficients of Turbulent Annular Seals," *Journal of Lubrication Technology*, **105**, 437–444 (July 1983).

39. Nelson, C., Childs, D., Nicks, C., and Elrod, D., "Theory Versus Experiment for the Rotodynamic Coefficients of Annular Gas Seals: Part 2, Constant-Clearance and Convergent-Tapered Geometry," *Journal of Tribology*, pp. 433–438 (1985).

40. Fleming, D. P., "High Stiffness Seals for Rotor Critical Speed Control," ASME Paper No. 77-DET-10 (1977).

41. Kirk, R. G., and Miller, W. H., "The Influence of High Pressure Oil Seals on Turbo-Rotor Stability," *ASLE Transactions*, **22**(1), 14–24 (1979).

42. Alford, J. S., "Protecting Turbomachinery from Self-Excited Whirl," *Journal of Engineering for Power*, pp. 333–344 (October 1965).

43. Murphy, B. T., and Vance, J. M. "Labyrinth Seal Effects on Rotor Whirl Stability," Paper No. C306/80, *Proceedings of the 2nd International Conference on Vibrations in Rotating Machinery (Institution of Mechanical Engineers)*, held at Churchill College, Cambridge University, September 1–4, 1980, pp. 369–372.

44. Spurk, H. H., and Keiper, R., "Selbsterregte Schwingungen bei Turbomachinen infolge Labyrinths Tromung," *Ingenier-Archiv*, **43**, 127–135 (1974).

45. Kurohashi, T., Inoue, Y., Abe, T., and Fujikawa, T., "Spring and Damping Coefficients of the Labyrinth Seals," Paper No. C283/80, *Proceedings of the 2nd International Conference on Vibrations in Rotating Machinery (Institution of Mechanical Engineers)*, held at Churchill College, Cambridge University, September 1–4, 1980, pp. 215–222.

46. Benckert, H., and Wachter, J., "Flow Induced Spring Coefficients of Labyrinth Seals for Application in Rotor Dynamics," *Proceedings of the 1st Workshop on Rotordynamic Instability Problems in High Performance Turbomachinery, Texas A&M University, May 12–14, 1980*, NASA CP 2133, pp. 189–212.

47. Iwatsubo, T., Motooka, N., and Kawai, R., "Flow Induced Force of Labyrinth Seal," *Proceedings of the 2nd Workshop on Rotordynamic Instability Problems in High Performance Turbomachinery, Texas A&M University, May 10–12, 1982*, NASA CP 2250, pp. 205–222.

48. Wright, D. V., "Labyrinth Seal Forces on a Whirling Rotor," presented at the ASME Mechanics, Bioengineering, and Fluids Engineering Conference, University of Houston, June 20–22, 1983.

49. Childs, D. W., and Scharrer, J. K., "Experimental Rotordynamic Coefficient Results for Teeth-on-Rotor and Teeth-on-Stator Labyrinth Gas Seals," ASME *Journal of Engineering for Gas Turbines and Power*, **108**, 599–604 (Oct. 1986).

Chapter VII

Rotordynamic Instability
in Turbomachinery

Rotordynamic instability in turbomachinery is characterized by whirling of the rotor–bearing system at frequencies other than shaft speed. The cause of rotordynamic instability is never rotor imbalance but rather is usually associated with the variation of some fluid dynamic pressure around the circumference of a rotor component.

The word "instability" implies that the motion can tend to increase without limit, and indeed this sometimes occurs with destructive consequences to the machine.

On the other hand, nonsynchronous rotor whirling with bounded nondestructive amplitudes is often observed and tolerated in turbomachinery, and may persist for years of satisfactory operation. These cases need to be continually monitored, since a very small change in operating conditions or machine clearances can destabilize the system and produce a rapid growth of vibration amplitude.

TERMINOLOGY AND HISTORY

The most common source of vibration in rotating machinery is rotor or shaft imbalance. Unbalanced rotating parts produce vibration that is synchronous with rotor speed (i.e., at the same frequency as shaft revolution per minute). However, if the signal from a transducer measuring shaft vibration is displayed on an oscilloscope screen, it is seldom found to be a pure sine wave. Even in a relatively simple machine, the signal will be "complex," i.e., composed of several different frequencies, including the synchronous component due to imbalance. Figure 7.1 shows an example of a complex vibration waveform. The nonsynchronous

SIGNAL

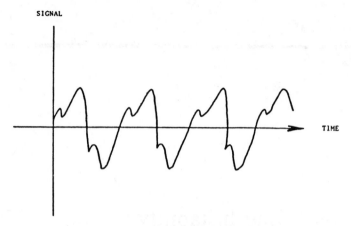

TIME

Figure 7.1. A complex vibration waveform.

frequencies in the waveform can be classified as either subsynchronous (frequency lower than shaft speed) or supersynchronous (frequency higher than shaft speed). An electronic microprocessor called an "RTA" (Real Time Analyzer) (see Chapter VIII) can break the complex signal into its various frequency components and display them on a screen for ready identification. Figure 7.2 shows how such a "spectrum" of the complex signal might appear in a case where the subsynchronous component is predominant.

Engineers who troubleshoot vibration problems in rotating machinery often refer to the *subsynchronous* frequencies as *instabilities*.[1]

This is not absolutely precise terminology, as will be shown below, but nevertheless it does have a sound basis for usage, both mathematically and in terms of practical experience. Large amplitudes of subsynchronous vibration do not occur frequently, but when they do occur they can be much more destructive and much more difficult to "cure" than imbalance problems. These "instabilities" can appear with little warning, under particular conditions of speed and load, and can build up in amplitude very rapidly to destructive levels.

Ever since the early 1920s, when the General Electric Company experienced instability problems with some of their (then) new turbocompressors developed for blast furnaces [1], the causes of rotordynamic instability have seemed somewhat mysterious. At the time, it took several years for the GE engineers assigned to the problem to determine that the cause of their instability was internal hysteresis, or friction in the rotor assembly. In the course of their investigation, they also identified oil film journal bearings as another source of instability ("oil whip") [2].[2]

---

[1]Especially when the subsynchronous frequencies are predominant in the signal.

[2]Incidentally, it is questionable how many manufacturers today would be as able and willing to publish research and development that identifies sources of problems encountered in their products, thus adding to the store of retrievable engineering knowledge. Recent product liability laws may be partially to blame.

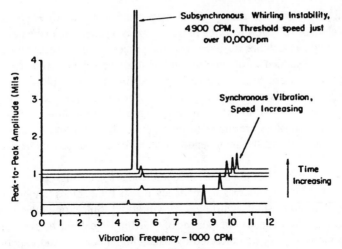

Figure 7.2. The frequency spectrum of a complex signal. From Ref. [16].

Subsequently, for several decades rotordynamic instability problems were an infrequent occurrence. When they did occur, they were almost always attributed to oil whip or internal friction.

More recently, as high speed turbomachinery has come into wider use (often replacing older reciprocating machines) and has been designed to ever higher performance specifications (resulting in higher shaft speeds and working pressures), instability problems have been occurring with increasing frequency. More importantly, many have resulted in failures that were exorbitantly expensive in terms of either lost production by the user or of delayed development programs and expensive redesigns by the manufacturer. The most nationally prominent of such problems was a rotordynamic instability encountered in the Space Shuttle main engine turbopumps [3]. A number of costly instability failures in more mundane machinery are described below. Among the different types of machinery involved are natural gas reinjection compressors, utility power-plant turbines, and aircraft turbojet and turboshaft engines.

The objectives of this chapter are to provide some useful definitions and insights so the reader can recognize instability in turbomachinery, to suggest design modifications and/or "fixes" that have been used to suppress instabilities, and to provide a foundation for understanding how rotordynamic instability can be simulated, analyzed, and predicted mathematically with the help of a computer.

PROBLEMS WITH CENTRIFUGAL COMPRESSORS

There are an increasing number of applications today in which turbomachinery is used to pump or compress fluids up to pressures of several hundred bars. Examples are found principally in the petroleum industry, involving hydrocarbon or petrochemical processing and natural gas reinjection. Somewhat lower but still significantly high (10–100 bars) working pressures are now common in the newest aircraft and industrial gas turbines, turbopumps, and process compressors. These machines are typically designed to operate through and above several critical speeds so as to maximize the work done by a given size machine. For example, a 17-in diameter impeller for an industrial centrifugal compressor can be designed for a work load well in excess of 2000 hp by running it at speeds approaching 9000 rpm. When eight or ten such disks are mounted on a single shaft, the resulting bearing span practically ensures that the shaft bending critical speed will be well below operating speed.

The result of this combination of supercritical speed, high pressure, and high work load has been an increasing tendency for such machines to exhibit problems of nonsynchronous shaft whirling and vibration. Since only the synchronous (frequency at shaft speed) component of whirling is caused by rotor imbalance, these problems cannot be corrected by balancing. Research and experience have shown that the driving forces have their sources in fluid-film bearings and seals, friction and hysteresis in rotating parts, and working fluid pressures on impellers and bladed disks. All these forces are "self-excited," that is, they are produced by the whirling motion itself, and often do not become large enough to sustain the motion until a "threshold" value of some critical parameter is exceeded (typically shaft speed).

DESTABILIZING FORCES IN TURBOMACHINERY

A number of destabilizing mechanisms have been identified or hypothesized to explain incidents of rotordynamic instability. The known or hypothesized sources of destabilizing forces in turbomachinery are listed below, along with a pertinent reference for each case (not included are instabilities caused by the time variation of parameters, to be described in a later section).

1. Hydrodynamic bearings ("oil whip") [2, 4].
2. Fluid ring seals (similar to oil whip) [5].
3. Internal friction in rotating parts [6].
4. Aerodynamic forces due to blade-tip clearance eccentricity (Alford's force) [7].
5. Trapped liquids inside a hollow shaft or rotor [8].
6. Dry friction whip (backward whirl driven by rubbing friction between rotor and stator) [9].
7. Labyrinth seals [7, 10].
8. Torquewhirl (the direct effect of torque on a disk misaligned by the mode shape of the rotor [11].

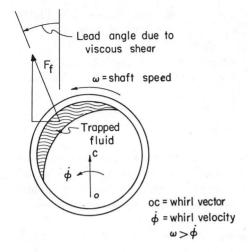

Lead angle due to
viscous shear

ω = shaft speed

F_f

Trapped
fluid

c

$\dot{\phi}$

o

oc = whirl vector
$\dot{\phi}$ = whirl velocity
$\omega > \dot{\phi}$

Figure 7.3. Whirl induced by trapped fluid.

9. Centrifugal impeller dynamics [12].
10. Propeller whirl flutter [13].

All but one (item 10) of the destabilizing forces listed above are forces that act tangentially to the shaft whirl orbit and therefore feed energy into the whirling motion.

An example is depicted in Fig. 7.3, which shows how trapped fluid inside a hollow rotor can produce a force component that is tangential to the whirl orbit. Viscous shear on the fluid produces a force in the same direction as the whirl, provided that the shaft speed is faster than the whirl speed. Since the latter condition is the definition of subsynchronous whirl, and since a rotor–bearing system tends to whirl at one of its natural frequencies, it can be seen that the instability will be subsynchronous and will be supported whenever the shaft speed exceeds a natural frequency (eigenvalue).

It will be shown in a later section how many of these destabilizing forces can be represented mathematically by cross-coupled stiffness coefficients.

WORK-LOAD DEPENDENCE

As the power and torque ratings of modern rotating machines have increased, serious nonsynchronous shaft whirling problems have appeared in some machines that are especially sensitive to the work load. In 1965, Alford [7] published a paper describing the phenomenon: ''Whirling occurs in the direction of rotation. . . . Large power inputs to the compressor rotor appear to increase the hazard of whirl. The vibration problem was encountered only at the full 100 percent power rating of the engine.'' Alford goes on to describe a theory of aerodynamic exciting forces due to variable blade-tip clearance in axial flow machines. According to Ehrich [14], the tip-clearance theory was first hypothesized in 1947 by A. H. Fiske, D. McClurkin, and R. O. Fehr of the General Electric Company.

Alford's work was a landmark paper, since it was the first published explanation of a torque-load-dependent destabilizing mechanism and is the basis for cross-coupled stiffness input to computerized stability analyses in current use by rotor dynamicists. However, it should be remembered that Alford developed the theory for axial flow machines, and it does not seem reasonable to apply it to centrifugal machines.

After listing eight mechanisms known to excite vibration instability in rotating machinery, Ehrich [14] states: "While this tabulation accounts for many observed incidents of instability in turbomachinery, there are occasions reported where other excitation mechanisms must be sought, particularly when the instability is induced by increases in mass flow rate."

In a paper published in 1975 [15], Wachel describes in detail the case histories of three different centrifugal compressors that exhibited load-dependent, nonsynchronous vibration. In his case No. 2: "The unit was running at 11,000 rpm with a small amplitude of instability at 4500 cpm. When the suction pressure was increased, the vibration trip-out occurred. . . . At a constant speed, the subsynchronous frequency was a function of the pressure ratio across the compressor. The design discharge pressure could not be reached."

In another paper presented at the same conference, Fowlie and Miles [16] describe similar problems with high pressure centrifugal compressors: "Destructive vibration of the reinjection centrifugal compressors delayed startup of Chevron's Kaybob Gas Plant. The main vibration was an unstable type whose frequency was substantially below the running speed. . . . Not even the leading rotor dynamics consultants could simulate these rotors. . . . We believe full-pressure, full-speed mechanical testing in the factory is essential for new machines near or beyond the boundaries of verified field experience. Research and development in this area is required by the manufacturers and the consultants to develop reliable prediction methods of the instability phenomena."

In 1975 and 1976, the author was involved in an engineering consulting job to identify and correct sources of nonsynchronous instability in a centrifugal compressor used for natural gas reinjection in an oilfield. Attempts to bring this machine up to the design pressure and load produced nonsynchronous whirling of an amplitude proportional to the load, so that vibration trip-outs were actuated long before full load was reached. This machine was fully stabilized by redesigning the shaft to a larger diameter (raising the critical eigenvalue), which necessarily reduced the rated design discharge pressure, due to the axial flow restriction by the larger shaft.

Most of the machines discussed above passed full speed, no-load shop tests before being installed for service, with no evidence of a vibration problem.

Field observations of some of the machines described above indicated that the nonsynchronous vibration was often acceleration dependent as well as load dependent. That is, a higher load and speed could be reached without vibration trip-out if it was approached slowly. Since the vibration in these cases was not synchronous, it must be a different phenomenon from the effect of acceleration on the critical speed response analyzed by several investigators in the past [17, 18].

Although the sensitivity of rotordynamic instabilities to the work load has been one of the most difficult aspects to predict or explain analytically and by computer simulation, recent progress has been made. Mathematical stability analysis has its limitations, but it offers the best chance for understanding these phenomena well enough so that intelligent design modifications can be suggested.

DESIGN MODIFICATIONS AND "FIXES"

Before going into the mathematical theory of dynamic instability, it is interesting to consider the modifications that have often been found to "cure" nonsyn-chronous vibrations and instabilities in turbomachinery. The reader should be warned that this is a field where experience, if applied without understanding, can be misleading. However, observations of the results from machine modifications can serve as guideposts in directing an analytical investigation.

The most commonly diagnosed source of rotordynamic instability is oil whip, caused by oil-film bearings, particularly plain cylindrical journal bearings (see Chapter VI). When oil whip is diagnosed, the typical modification is a change to a more stable type of bearing such as the tilt-pad type. This change usually, but not always, results in a more stable machine.

The author's observations have led him to believe that rotordynamic instability is seldom caused by one source acting alone. For example, a high pressure ratio centrifugal compressor with a built-up rotor supported on journal bearings certainly has destabilizing potential (cross-coupled stiffness) from the bearings, from internal friction in the rotor, and from fluid forces around the impellers. If the latter two sources of instability remain constant, the stability of the machine will depend on (a) the rotor–bearing system damping, which comes mainly from the bearings; (b) the magnitude of additional cross-coupled stiffness produced by the bearings; and (c) the ratio of critical speed to shaft speed, which is influenced by bearing stiffness.

Since tilt-pad bearings typically have no cross-coupling but may provide less damping than journal bearings, it is easy to see that a change to this type of bearing could either "cure" the instability or make it worse, depending on whether the cross-coupling was predominantly from the original bearings or from the rotor and impellers.

Fluid pressure seals are another source of both cross-coupled stiffness (destabilizing) and damping (stabilizing). Although their effect is similar in some respects to that of oil-film bearings, it is much more sensitive to changes in fluid pressure ratio and density, since the stiffness and damping coefficients are functions of the working fluid flow conditions across the seal.

In the past, most of the changes in seal configuration that have been made in an attempt to suppress instabilities have been of a trial-and-error nature. Some have been successful. Research in this area is currently underway at Texas A&M University [19].

Misalignment between the shafts of connected machines has been found in some cases to be an indirect cause of instability. As mentioned in Chapter I, misalignment can produce double-synchronous whirling, but this is usually not an unstable condition. Misalignment can also cause a subsynchronous whirling instability by taking the load off of a journal bearing. In Chapter VI, it is shown that the stability of a journal bearing is dependent on the load it supports and that a plain cylindrical journal bearing is unstable with no load.

Suppose, for example, that a turbogenerator set consists of a turbine with oil-film bearings driving a generator through a rigid coupling, with each machine on a separate foundation. If the turbine foundation sags at the coupling end, then part of the weight of the turbine rotor may be supported by the generator bearings. The load on the turbine bearing at the coupling end may go to zero or even negative (up), thereby causing the rotor–bearing system to become unstable in oil whip. In this case, alignment of the two shafts could "cure" the instability.

If a turbomachine has not been properly engineered to avoid instability, or if an old design has been uprated by increasing shaft speed without an engineering analysis to predict the threshold of stability, a major redesign of the machine may be necessary. Two major changes that are almost always in the right direction from the standpoint of stability are a reduction in bearing span and a stiffening of the rotor shaft (usually by increasing its diameter). Both of these changes will raise eigenvalues (and critical speeds) without a loss in effective damping from the bearings.

In some cases, the effective system damping can be increased by the addition of a squeeze film bearing damper. Figure 7.4 shows the configuration as applied to a rolling-element bearing in an aircraft turbojet engine. A clearance space (typically 0.005–0.010 in) is provided around the outer race and supplied with oil. The outer race is pinned or keyed to prevent rotation but is allowed to orbit and

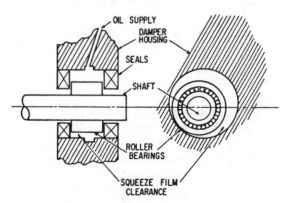

Figure 7.4. A squeeze film bearing damper.

"squeeze" the oil, thereby producing a damping force. Similar dampers have also been used on oil-film bearings [15].

It should be pointed out, however, that increasing damping at the bearings beyond the optimum value for a flexible rotor can actually decrease the effective damping for the rotor–bearing system and thereby be deleterious to rotordynamic stability [20].

This latter fact, somewhat surprising to some, emphasizes the importance of mathematical analysis and computer simulations to ensure that design modifications are in the right direction to achieve the desired results. Although there is

TABLE 7-1. Classification of Shaft Whirling Instabilities in Rotating Machinery

Diagnostic Information	Type of Vibration	
	Self-Excited Nonsynchronous Whirl	Parametrically Excited Whirl
Sources of excitation	Oil-film bearings ("oil whip"); internal friction in rotating parts; trapped fluid in rotor; tip clearance effects in axial flow bladed disks ("Alford's force"); labyrinth seals; ring seals; high gas pressure in centrifugal stages; rotor/stator rubbing friction (induces backward whirl); high torque loading on disks misaligned by the mode shape ("torquewhirl"); variable angle of attack on blades of axial stages ("propeller whirl flutter," can be forward or backward); dense or viscous fluid in impeller housings	Asymmetric shaft stiffness; asymmetric rotor inertia; pulsating torque, rotor/stator rubs, excessive ball bearing clearance
Whirl frequency ratio f/f_s	Almost always subsynchronous, typically 0.3–0.8	Can be supersynchronous, $f/f_s > 1.0$
Shaft speeds where encountered	Supercritical speeds, especially at $2w_{cr}$ and above	Subcritical and supercritical speeds
Effective solutions	Stiffen shafts or shorten bearing spans to raise bending critical speeds; asymmetric bearing supports; squeeze film bearing dampers; soften bearing supports to allow dampers to operate effectively	Squeeze film bearing dampers; remove asymmetries; isolate pulsating torque with torsionally soft coupling, align bearings to prevent rubs

much room for improvement in the accuracy of quantitative predictions, the state of the art today is such that trends can be predicted quite reliably.

Table 7-1 summarizes the known and hypothesized causes of rotordynamic instability, and the design modifications that have been used to suppress them.

MATHEMATICAL DEFINITIONS AND THEIR PRACTICAL IMPLICATIONS

In order to interpret mathematical results intelligently, it is instructive and useful to become familiar with certain definitions used in the development of rotordynamic instability theory and to compare these with the field engineers' definition of instability as "any subsynchronous vibration frequency."

First, it should be clear that a rotordynamic instability is different from a critical speed. Critical speed is defined as that speed at which synchronous response to imbalance is a maximum. An instability, on the other hand, is not related to imbalance response. Whereas a critical speed can usually be passed through without damage to the machine, a "threshold speed" of instability often cannot be exceeded without catastrophic effects.

In mathematical terms, a dynamic instability is defined as a solution to the linear differential equations of motion characterized by a complex eigenvalue with a positive real part. The real part of an eigenvalue gives the growth (or decay, if negative) factor of the solution; the imaginary part gives the frequency. Translated into the realm of rotordynamics, the "solution" is a function which describes the time-dependent amplitude of vibration. Figure 7.5 shows how an unstable solution is described by the eigenvalue. The motion associated with an instability becomes unbounded (infinitely large) with time, if linear stability theory holds. The vibration with a growing amplitude will cause seals or blades to rub, unless it ceases to grow due to *nonlinearities* in the system (not included in the linear differential

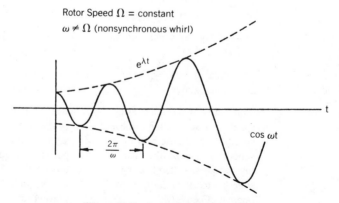

Figure 7.5. An unstable eigenvalue.

equations). An example of such a nonlinearity is a damping coefficient that increases with vibration amplitude.

Thus, linear stability theory can predict the *onset* and *frequency* of an instability quite accurately, even though it does not predict the final amplitude of the motion. Also, the *magnitude* of the positive real part of the eigenvalue can be a useful qualitative measure of the "strength" of an instability.

If a rotor–bearing system is modeled mathematically for both synchronous vibration and instabilities, an examination of the differential equations for each of the two phenomena shows only one essential difference: the type of force acting on the rotor. Synchronous response is caused by imbalance forces that rotate at shaft speed and *are independent of any vibratory motion.* Even if we could constrain the rotor not to whirl or vibrate, the imbalance force is still there. Its components in a plane normal to the shaft are harmonic functions of time only, with a frequency equal to shaft rotational speed and a magnitude determined by the amount of rotor imbalance.

On the other hand, the model for dynamic *instability* usually has forces that *depend on the rotor motion.* (The one class of exceptions is characterized by "parametric variations," described below). An example of such a force is viscous damping, which is proportional to rotor whirl velocity, but this is stabilizing unless it becomes negative. Most of the destabilizing forces in rotordynamics are "cross-coupled" in two directions, e.g., a radial deflection of the shaft away from its equilibrium position produces a tangential force that drives the shaft in an orbital

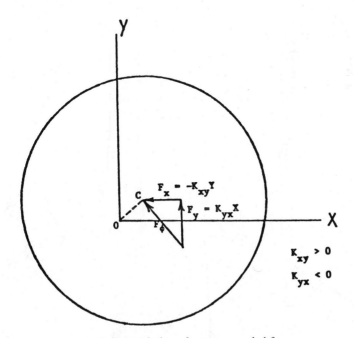

Figure 7.6. Resolution of a cross-coupled force.

motion. Figure 7.6 illustrates this type of force and shows how the tangential force can also be represented as the resultant of cross-coupled stiffness in Cartesian (X–Y) coordinates. Cross-coupled stiffness produces a tangential force proportional to shaft deflection, so that the force grows larger as the amplitude of whirl grows. Thus, it is clear why this type of instability is called "self-excited" motion.

It should also be apparent that the field engineers' definition of instability as "any subsynchronous vibration" is compatible in most cases with the mathematical definitions, so long as we recognize the limitations of linear theory.

DYNAMIC STABILITY THEORY: A SIMPLE EXAMPLE

The concepts of dynamic stability theory are best illustrated by a simplified example. Figure 7.7 shows a whirling pendulum of length l hinged to a vertical shaft. The assembly rotates with angular velocity ω.

The system of Fig. 7.7 has three different dynamic equilibrium states, each one a solution to the differential equation of motion in θ. The solutions are [21]

$$\theta_1 = 0 \quad \text{(pure spin)}, \tag{7-1}$$

$$\theta_2 = \cos^{-1} g/l\omega^2 \quad \text{(whirling)}, \tag{7-2}$$

$$\theta_3 = \pi \quad \text{(pure spin)}. \tag{7-3}$$

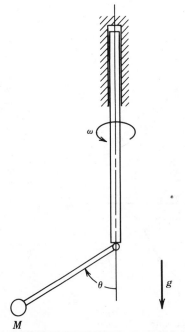

Figure 7.7. Model for stability analysis example.

Solution (7-3) implies that the shaft, hinge, and pendulum are constructed to allow $\theta = \pi$ without mechanical interference. Intuition suggests that equilibrium (7-3) is always unstable, and this can be verified by mathematical analysis. Intuition, however, is unreliable in predicting the stability of motions (7-1) and (7-2), and a linearized perturbation analysis is required to prove [21] that the required condition for stability of motion (7-1) is

$$\omega^2 < g/l \tag{7-4}$$

whereas

$$\omega^2 > g/l \tag{7-5}$$

is the required condition for stability of motion (7-2).

What this means physically is that, although the pure spin motion (7-1) can theoretically exist at any speed, any arbitrarily small perturbation of θ_1 will grow and become unbounded with time, whenever the shaft speed ω exceeds $\sqrt{g/l}$. On the other hand, the whirling motion (7-2) can only exist when its condition for stability (7-5) is satisfied. Then, any small perturbation on θ_2 will die out, just as a small perturbation on $\theta_1 = 0$ dies out whenever condition (7-4) is satisfied.

It is important to note that the stability condition for each one of the equilibrium states (7-1), (7-2), or (7-3) could have been obtained *without any knowledge of the other two equilibrium states*. This, in fact, is the usual situation when one is analyzing the dynamic stability of rotating machinery of realistic complexity. The differential equations are too complicated to solve for any of the dynamic equilibrium states except the pure spin solution. The objective of the analysis then becomes a determination of the conditions for stability of the pure spin motion.

Returning to the system of Fig. 7.7, consider now the meaning of dynamic instability as predicted by a perturbation analysis for equilibrium state (7-1) (pure spin), when $\omega^2 > g/l$. The required linearization procedure [21] makes the stability equation valid *only in the region of the equilibrium motion being analyzed* ($\theta_1 = 0$). Thus, the perturbation "growing unbounded with time" *ceases to be relevant to the real system* after θ has diverged appreciably away from $\theta_1 = 0$. All the perturbation analysis really tells us is that $\theta_1 = 0$ is unstable when $\omega^2 > g/l$. But the *complete* analysis [(7-1)–(7-5)]—not feasible for most real systems—tells us that, whenever $\omega^2 > g/l$, a perturbation of $\theta_1 = 0$ will cause the motion to diverge and approach the *stable* equilibrium θ_2 (whirling).

A ROTORDYNAMICS EXAMPLE

Figure 7.8 shows one type of destabilizing force that can occur in turbomachinery, known as "Alford's force" [7]. Alford's force is caused by the variation of blade-tip clearance around an unshrouded axial flow stage in a compressor, due to any deflection of the shaft away from the center of the housing. This is not a negative

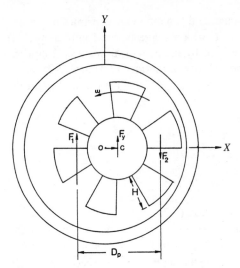

Figure 7.8. Alford's force.

damping force but a cross-coupled stiffness force (i.e., acting in a direction normal to the shaft deflection, and proportional to deflection instead of velocity) of the type shown in Fig. 7.6. It is the resultant of the variation in aerodynamic forces on the blades, produced by the instantaneous eccentricity of the bladed disk in its housing.

A simple rotordynamics model can be constructed by mounting the bladed disk (of mass m) on a shaft midway between two bearings with flexible supports (see Fig. 7.9). If the bearings and supports have completely symmetrical stiffness and damping properties (i.e., the same in all directions), then the motion of this rotor–bearing system is described by solutions to the following two coupled differential equations:

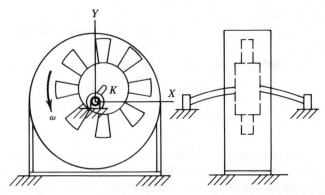

Figure 7.9. A simple rotordynamics model with Alford's force included.

$$\ddot{X} + \frac{c}{m}\dot{X} + \frac{k}{m}X + \frac{\mathcal{K}}{m}Y = 0, \qquad (7\text{-}6)$$

$$\ddot{Y} + \frac{c}{m}\dot{Y} + \frac{k}{m}Y - \frac{\mathcal{K}}{m}X = 0, \qquad (7\text{-}7)$$

where k and c are the combined effective stiffness and damping coefficients of the bearing supports and flexible shaft.

The last term in each equation contains Alford's force, which he hypothesized would be proportional to the eccentricity (X or Y), the stage torque (T), and the efficiency factor (β), and inversely proportional to the pitch diameter (D) and vane height (H). Thus, the cross-coupled stiffness coefficient in this case is

$$\mathcal{K} = K_{XY} = -K_{YX} = \frac{\beta T}{DH}. \qquad (7\text{-}8)$$

Application of Routh's method [22] shows that a necessary condition to prevent the real part of the complex eigenvalues of equations (7-6) and (7-7) from becoming positive (i.e., a necessary condition for stability) is

$$\mathcal{K} < c\sqrt{k/m}, \qquad (7\text{-}9)$$

which can also be written as

$$\frac{\beta T}{DH} < c\omega_n, \qquad (7\text{-}10)$$

where $\omega_n = \sqrt{k/m}$ is the undamped critical speed.

If the aerodynamic load torque T of the single stage increases with speed, then equation (7-10) shows that there may be a "threshold speed of instability" above which the inequality is no longer satisfied. Note that the threshold speed can be raised by stiffening the shaft (raising ω_n) or by increasing the effective damping coefficient c. It is known that increasing the bearing support stiffness usually *reduces* the effective damping coefficient and often has a very small effect on the critical speed.

Thus, in practice, the parameter most feasible to modify for improved stability is usually the bearing support damping. (Stiffening the shaft of an existing machine is an expensive proposition.) Sometimes the *effective* damping can be increased by reducing the bearing support stiffness [20]. An exception is found in turbo-pumps for cryogenic fluids, as used in liquid-fuel rocket engines, where extremely low temperatures make the achievement of any appreciable damping in the supports practically impossible. In this case, both the rotor and bearing supports must be stiffened to keep the whirling frequencies high.

PARAMETRIC EXCITATION

In differential equations (7-6) and (7-7), the coefficients are constant. A separate class of rotordynamic instabilities is characterized by conditions in the machine that are described by variable coefficients. Since the coefficients are made up of the parameters of the rotor–bearing system, this type of instability is referred to as "parametric excitation." In contrast to the self-excited instabilities described above, in which the whirling is always at a subsynchronous frequency equal to an eigenvalue of the system, parametric excitation produces whirling that can be synchronous [23], subsynchronous [24], or supersynchronous [25], depending on how the parameters vary in each particular case. Some of these "instabilities" are more like a critical speed or forced vibration than like a true instability, in both their mathematical form and their behavior. Indeed, some can even be "driven through" (i.e., a higher speed can be found where the rotor regains its stability). Sometimes, however, the "speed range of instability" is quite wide.

An example of this type of instability is caused by asymmetric shaft stiffness (i.e., a shaft with unequal stiffness in orthogonal directions). The stiffness asymmetry produces two distinct natural frequencies (for free vibration of the nonrotating shaft), which in turn produce two critical speeds. If damping is insufficient, the entire range of speeds between the two critical speeds is unstable in synchronous whirl. The same rotor can also exhibit a "gravity critical," sometimes called an instability, when the shaft speed is approximately one-half the first critical speed. In the latter case, the whirling is supersynchronous (double frequency).

Several different sources of parametric excitations that can occur in turbomachinery are as follows.

1. Shaft stiffness asymmetry [23].
2. Rotor mass inertia asymmetry [26].
3. Intermittent rotor/stator rub [24].
4. Pulsating torque [27].
5. Excessive ball bearing clearances [24].

STABILITY ANALYSIS BY DIGITAL COMPUTER (EIGENVALUE ANALYSIS)

Stability analysis of the differential equations, linearized by small perturbation theory, forms the basis of the most successful computer programs in current use for predicting nonsynchronous whirl in rotating machinery. Conditions for stability (or instability) and the onset frequency of whirling are both predictable, but the actual bounded whirling amplitudes are not part of the output, since the "unstable" solution theoretically is a perturbation growing unbounded with time. Usually, however, the actual (measured) *bounded* whirling motions occur at frequencies very close to the predicted frequencies for instability of the linearized system.

The bounded whirling motions that actually occur are, in reality, equilibrium solutions to the nonlinear equations of motion. If these solutions were available, amplitudes of nonsynchronous whirling could also be predicted. One such solution, for "torquewhirl" of a cantilevered disk, is described later in this chapter.

Lund [28] has extended the Myklestad–Prohl transfer matrix method (see Chapter IV) to rotor–bearing system models that include damping and destabilizing cross-coupled stiffness and damping coefficients. Whereas the Myklestad–Prohl method yields only the imaginary part of the eigenvalues (i.e., the natural frequencies),the Lund method yields complex eigenvalues (both the natural frequencies and the logarithmic decrement, which is a stability predictor).

The coordinates used in Lund's analysis are shown in Fig. 7.10. In general, the stiffness and damping coefficients form the matrices which define the forces and moments on each disk in the X, Y, α, and β directions.

For example, the force on a disk in the X direction, due to disk displacement and velocity, is

$$F_X = -K_{XX}X - K_{XY}Y - C_{XX}\dot{X} - C_{XY}\dot{Y} - K_{X\alpha}\alpha - K_{X\beta}\beta - C_{X\alpha}\dot{\alpha} - C_{X\beta}\dot{\beta}$$

$$(7\text{-}11)$$

and the matrix equation for the forces in all directions (on a single disk) is as shown in (7-12):

$$
\begin{Bmatrix} F_X \\ F_Y \\ M_Y \\ M_X \end{Bmatrix} = -
\begin{bmatrix}
K_{XX} & K_{XY} & K_{X\alpha} & K_{X\beta} \\
K_{YX} & K_{YY} & K_{Y\alpha} & K_{Y\beta} \\
K_{\alpha X} & K_{\alpha Y} & K_{\alpha\alpha} & K_{\alpha\beta} \\
K_{\beta X} & K_{\beta Y} & K_{\beta\alpha} & K_{\beta\beta}
\end{bmatrix}
\begin{Bmatrix} X \\ Y \\ \alpha \\ \beta \end{Bmatrix}
$$

$$
-
\begin{bmatrix}
C_{XX} & C_{XY} & C_{X\alpha} & C_{X\beta} \\
C_{YX} & C_{YY} & C_{Y\alpha} & C_{Y\beta} \\
C_{\alpha X} & C_{\alpha Y} & C_{\alpha\alpha} & C_{\alpha\beta} \\
C_{\beta X} & C_{\beta Y} & C_{\beta\alpha} & C_{\beta\beta}
\end{bmatrix}
\begin{Bmatrix} \dot{X} \\ \dot{Y} \\ \dot{\alpha} \\ \dot{\beta} \end{Bmatrix},
\qquad (7\text{-}12)
$$

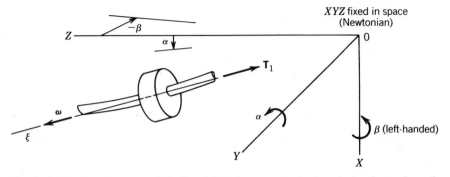

Figure 7.10. Coordinates used by Lund [28] for computerized analysis of rotordynamic instability.

where M_Y and M_X are *moments* on the disk. Equation (7-12) defines the stiffness (K) and damping (C) matrices, respectively.

It can be seen that there are a total of 32 stiffness and damping coefficients defining the forces and moments on each disk. The off-diagonal K elements are called the cross-coupled stiffness coefficients and are often associated with the destabilizing mechanisms. For example, Fig. 7.6 shows how a positive K_{XY} and negative $K_{YX} = -K_{XY}$ produces a tangential force driving forward whirl. It has been shown that oil-film (journal) bearings, internal rotor hysteresis, and Alford's force all produce cross-coupled stiffness coefficients [4, 6, 7]. The magnitude of most of these coefficients is not accurately known. In fact, the greatest limitation of Lund's method (as with all other stability analyses) is the lack of accurate information about the types and magnitudes of destabilizing excitations that exist in real machines and are used as input to the computer program.

A case in point is Alford's force, which had not actually been measured in a turbomachine or test rig until 1981, when Laudadio succeeded in making measurements of it for static eccentricities in a small high-speed axial blower at Texas A&M University [29]. Figure 7.11 shows the measured cross-coupled force for five different speeds, compared to the prediction of Alford's theory (in which speed is not a pertinent variable). Although it is doubtful that aerodynamic forces measured in a fractional horsepower blower can be scaled up to predict the forces

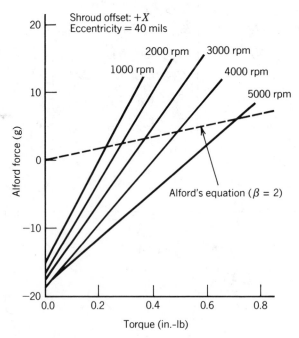

Figure 7.11. Measured Alford's force as a function of torque for different values of speed. From Ref. [29].

in large turbomachines, these measurements certainly show that existing theoretical predictions fall short of being quantitatively accurate.

The characteristic matrix method (see Chapter III) can also form the basis of a digital computer program for stability analysis. Each disk on the rotor has four degrees of freedom (X, Y, α, β), as in Lund's analysis (Fig. 7.10). However, in the characteristic matrix method the differential equations of motion are formulated simultaneously as a system of order $4N$, where N is the number of disk stations in the rotor model.

The forces on each disk are given by equation (7-12), provided the coefficients are known, plus the elastic forces from adjacent shafts. There is a second-order differential equation of motion for each of the coordinates.

As in the torsional vibration analysis of Chapter III, the eigenvalues are found by requiring the determinant of the characteristic matrix to be zero. The inclusion of cross-coupled coefficients and damping (for stability analysis) makes the eigenvalues complex, rather than purely imaginary.

The disadvantage of this method is that a rotor modeled with 10 disks will have 40 degrees of freedom and consequently the characteristic matrix will be of order 80. This requires a much larger computer memory than the Lund method, in which the largest working matrix is of order 4.

Another type of computer program computes a numerical "transient" solution to the full nonlinear equations, one time step after another, "marching" the solution out in time [30]. Although impressive simulations can be made by these numerical solutions when the input is accurate, usually the nonlinear parameters of the system (i.e., stiffness, damping, clearances) are not well enough known to warrant the large amounts of computer time required to generate only a few seconds of real simulation time. Also, the voluminous numerical output is often difficult to interpret. Certainly, however, it has been demonstrated that this type of analysis can give useful results when the parameters are accurately known, and the output becomes much more meaningful when presented in graphical form.

One of the most recent and powerful advances in computerized stability analysis involves the generation of the characteristic polynomial from the transfer matrices and the subsequent extraction of the eigenvalues from the polynomial [31]. This method retains the advantage of working with the 8×8 transfer matrices, with their smaller demands on computer memory, but overcomes most of the computational difficulties that have been experienced with the standard Lund program.

The difficulties referred to are a result of the iteration scheme used by Lund to converge on the system eigenvalues. It is a Newton–Raphson approach and involves taking the derivatives of all equations used in the program. When programmed for a digital computer, this technique works very nicely for many problems but runs into trouble on others, in that it fails to converge with sufficient accuracy on some eigenvalues and has been known to completely miss one or more of them in some applications.

By rearranging the calculations performed in a Lund-type program, one can calculate the coefficients of the characteristic polynomial for the rotor–bearing system. The natural frequencies, combined with their corresponding logarithmic decrements, are the complex roots of this polynomial. The logarithmic decrement provides the criteria for establishing system stability. With the polynomial known, the roots can be found and divided out in a more straightforward and efficient manner. Convergence can always be obtained, and no critical speeds will be missed.

The roots, or eigenvalues, are found in an iterative manner by both types of programs, but the polynomial program performs transfer calculations for the rotor–bearing system only one time in order to derive the characteristic polynomial. The iterations are then performed on the polynomial to find the roots. The number of iterations required to find a root is typically 5–10 (rarely more than 11), to achieve an accuracy of 6 digits. The Lund-type program also requires about the same number of iterations per root. But all the transfer calculations for the rotor–bearing system must be redone 4 times for each individual iteration. This difference accounts for a significant increase in computational efficiency for the polynomial program.

The rotor–bearing system is modeled in exactly the same manner as for the usual transfer matrix program (see Chapter IV). The rotor is represented by concentrated masses connected by massless shafts (see Fig. 7.12). Bearings are modeled as linearized forces acting on the masses at the appropriate axial locations. The masses have the inertia properties of rigid circular cylinders, and the shafts behave according to the Euler bending and Timoshenko shear formulas. A rotor "station" (see Fig. 7.13) is normally considered to consist of a mass element plus the shaft section immediately to its right. The notation on Figs. 7.12 and 7.13 is defined as in Chapter IV, except that the positive direction of β is reversed as in Fig. 7.10 to agree with beam deflection theory.

With the rotor–bearing system modeled as a connection of rotor stations, degrees of freedom (coordinates) are assigned at the junctions between the stations (i.e.,

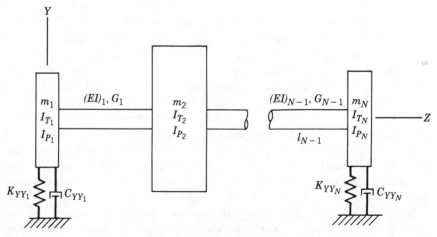

Figure 7.12. An N inertia rotor–bearing model.

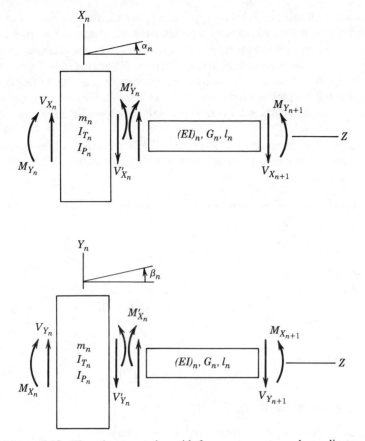

Figure 7.13. The nth rotor station with forces, moments, and coordinates.

at each concentrated mass). Linear differential equations are then written for each mass and put into the form of "transfer matrices." Since the equations are linear and homogeneous, the objective is to find the eigenvalues. When we use the homogeneous solution Xe^{st}, the matrices of differential equations are transformed into matrices of linear algebraic equations expressing the displacements and forces at the right end of the mass in terms of the displacements and forces at the left end [28]. The elements of these "inertia transfer matrices" (or coefficients of the equations) are not constant but are actually polynomials in the system eigenvalue.[3] For the shaft elements, strength of materials equations give the deflections and forces at the right end in terms of those at the left end.

By definition, a transfer matrix "transfers" displacements and forces from one end of the station to the other. If the transfer matrices for two adjoining rotor

---

[3] In the traditional transfer matrix program, the elements of the matrices are single numerical (complex) values, since a numerical estimate of the eigenvalue is made for the iteration calculations.

stations are multiplied together, using standard matrix multiplication, one obtains a single transfer matrix which fully represents the two stations. Since the elements of the two original matrices were polynomials, so will be the elements of the new matrix (and of correspondingly larger degree). Following this logic, one can multiply together all the transfer matrices for the system and obtain a single transfer matrix that fully represents the entire system. The determinant of a submatrix of the overall transfer matrix will be the characteristic polynomial of the rotor–bearing system.

The transfer matrices with X–Y orthotropy and a complex eigenvalue s are as follows. For the nth mass element, the transfer equations from left to right are as shown in (7-13):

$$
\left\{
\begin{array}{c}
X' \\
Y' \\
\alpha' \\
\beta' \\
V'_X \\
V'_Y \\
M'_Y \\
M'_X
\end{array}
\right\}_n
= [T_{I_n}]
\left\{
\begin{array}{c}
X \\
Y \\
\alpha \\
\beta \\
V_X \\
V_Y \\
M_Y \\
M_X
\end{array}
\right\}_n
\tag{7-13}
$$

where the m_{ij} elements of the $[T_{I_n}]$ are

$m_{ij} = 1$, for $i = j$ (all elements on the major diagonal)

$m_{51} = -s^2 M_n - C_{XX_n} s - K_{XX_n}$

$m_{52} = -C_{XY_n} s - K_{XY_n}$

$m_{53} = -C_{X\alpha_n} s - K_{X\alpha_n}$

$m_{54} = -C_{X\beta_n} s - K_{X\beta_n}$

$m_{61} = -C_{YX_n} s - K_{YX_n}$

$m_{62} = -s^2 M_n - C_{YY_n} s - K_{YY_n}$

$m_{63} = -C_{Y\alpha_n} s - K_{Y\alpha_n}$

$m_{64} = -C_{Y\beta_n} s - K_{Y\beta_n}$

$m_{71} = C_{\alpha X_n} s + K_{\alpha X_n}$

$m_{72} = C_{\alpha Y_n} s + K_{\alpha Y_n}$

$m_{73} = s^2 I_{T_n} + C_{\alpha\alpha_n} s + K_{\alpha\alpha_n}$

$m_{74} = (I_{P_n} \omega + C_{\alpha\beta}) s + K_{\alpha\beta}$

$m_{81} = C_{\beta X_n} s + K_{\beta X_n}$

$m_{82} = C_{\beta Y_n} + K_{\beta Y_n}$

$m_{83} = (-I_{P_n} \omega + C_{\beta\alpha_n}) s + K_{\beta\alpha_n}$

$m_{84} = s^2 I_{T_n} + C_{\beta\beta_n} s + K_{\beta\beta_n}$

All m_{ij} not defined above $= 0$. The transfer equations across the nth shaft element are given by (7-14):

$$
\begin{Bmatrix} X \\ Y \\ \alpha \\ \beta \\ V_X \\ V_Y \\ M_Y \\ M_X \end{Bmatrix}_{n+1}
= [T_{S_n}]
\begin{Bmatrix} X' \\ Y' \\ \alpha' \\ \beta' \\ V'_X \\ V'_Y \\ M'_Y \\ M'_X \end{Bmatrix}_{n} ,
\qquad (7\text{-}14)
$$

where the e_{ij} elements of $[T_{S_n}]$ are

$e_{ij} = 1$, for $i = j$ (all elements on the major diagonal)

$e_{13} = e_{24} = e_{75} = e_{86} = l_n$

$e_{15} = e_{26} = l_n^3/6EI_n - 1.33\, l_n/G_n A_n$

$e_{17} = e_{28} = e_{35} = e_{46} = l_n^2/2EI_n$

$e_{37} = e_{48} = l_n/EI_n$

All e_{ij} not defined above $= 0$.

The matrices in (7-13) and (7-14) will reduce to the ones in (4-37) and (4-38) for the special case of X–Y symmetry, no damping, infinite shear modulus G_n, and synchronous whirl ($s = i\omega$, where ω is shaft speed).

The elements of $[T_{S_n}]$ will be modified by high values of shaft torque. The modified elements are given later in this chapter in a section entitled "Torque-Dependent Transfer Matrices."

When we write a computer program to implement the polynomial method for the complex case, each matrix element should be stored as an array of its polynomial coefficients, as described in Chapter III for Holzer's method. As the transfer matrices are multiplied together, starting with station 1, the order of the polynomials will grow and the arrays of coefficients will become larger. Dynamic allocation of the arrays (a computer programming feature) can be used to advantage here. It should be realized that all of the displacements, shears, and moments are complex variables.

After we multiply out all of the station transfer matrices from $n = 1$ to $n = N$, the appropriate submatrix (i.e., the D matrix of Chapter IV) that determines the characteristic polynomial comes from the boundary conditions

$$
V_{X_1} = V_{Y_1} = M_{Y_1} = M_{X_1} = 0 \qquad (7\text{-}15)
$$

and

$$V'_{X_N} = V'_{Y_N} = M'_{Y_N} = M'_{X_N} = 0, \qquad (7\text{-}16)$$

which gives (7-16) as shown in (7-17):

$$\begin{Bmatrix} 0 \\ 0 \\ 0 \\ 0 \end{Bmatrix} = [D] \begin{Bmatrix} X \\ Y \\ \alpha \\ \beta \end{Bmatrix}_1 . \qquad (7\text{-}17)$$

The elements of the D matrix in (7-17) are now polynomials in s and the determinant of D gives the characteristic polynomial, the roots of which are the complex eigenvalues.

The degree of the characteristic polynomial depends on the complexity of the formulation (degrees of freedom per element) and the number of stations the shaft is divided into. The usual formulation has two displacements and two rotations per station (as above), and thus the degree of the polynomial is eight per station (i.e., 72-degree for a nine-station shaft model). When this method was conceived, it was first thought that finding the roots of such large-order polynomials might present serious numerical difficulties, in that the roots might be extremely sensitive to even the slightest computation errors in the coefficients. This was found not to be a problem, and it can be shown analytically [32] that this situation is confined to only the higher modes. This seems reasonable since if one were to change, say, one of the masses by even a very small amount, this could cause a large change in the hundredth critical speed, but certainly not in the first or second critical speeds. A scheme for numerical scaling of the coefficients, as well as other suggestions for programming the method, can be found in Chapter III.

A complete comparison has been made of the polynomial-type program with a Lund-type program written directly from Ref. [28], using the space shuttle hydrogen turbopump as a model [33]. The model used, Fig. 7.14, consists of nine stations and contains asymmetric bearing stiffnesses, destabilizing cross-coupling at three axial locations due to the pump interstage seals, and the aerodynamics of the turbines. With the cross-coupling coefficients used, the first forward mode will be unstable.

All roots were found to an accuracy of six digits in both programs, and a direct comparison shows all corresponding roots to be identical, within the limits of this accuracy. The comparison illustrates a considerable improvement by the polynomial program in efficiency of execution time. To find eight modes, the Lund-type program required 27.6 sec of execution time on a Prime 750 digital computer. Performing the same tasks and giving the same results, the polynomial program took just 4.4 sec. Also, since the model has nine stations, it has 72 possible roots (36 conjugate pairs). The Lund-type program did not find all these roots, since

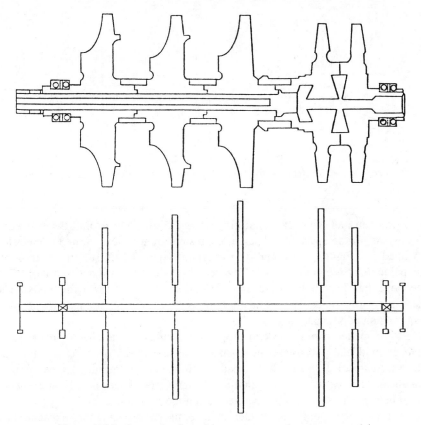

Figure 7.14. Space shuttle turbopump rotor and computer model.

convergence breaks down for the higher modes. The polynomial program found all roots, requiring one additional second of execution time.

The Space Shuttle model of Fig. 7.14 can be considered to be of minimum complexity while still giving reasonable accurate results. The system is divided into just nine stations: one station for each disk (pump and turbine stages); one station for each bearing and for each portion of overhung shaft. This "minimum number of stations" modeling philosophy is used by some engineers because of its inherent simplicity. Another modeling philosophy is to divide the rotor into many small stations, in order to improve accuracy. To test the performance of the polynomial program on this type of model, another comparison was made using an eight-stage centrifugal compressor typical of the petrochemical industry. The rotor weighs 1400 lb, is supported on two tilt-pad bearings with a bearing span of 80.7 in, and is 103 in in overall length. There are also a coupling and thrust collar overhung at the shaft ends. The rotor is modeled (see Fig. 7.15) as 35 stations, supported in two flexible and damped bearings at stations 4 and 32. Both the Lund-

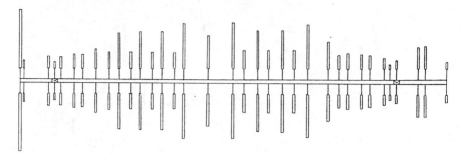

Figure 7.15. A 35-station rotor model for the eight-stage centrifugal compressor.

type program and the polynomial program were executed to find the first eight modes. As for the Space Shuttle case, the results are identical within the precision obtained (six digits). The Lund-type program required 72.5 sec to execute, and the polynomial program required 22.8 sec (14.3 to derive the polynomial). Since the model has 35 stations, there are 280 possible roots (140 conjugate pairs). The Lund-type program could not find all these roots, but the polynomial program did, requiring an additional 30 sec.

As for mode shapes, the two programs calculate mode shapes in the same manner, with the polynomial program coming out ahead on execution time by about a factor of 2.75 (roughly constant). However, this result may be of small significance, since the time required to find the mode shapes is only in the range of 10 percent of the total time required to run the program.

As discussed earlier in this chapter, many of the destabilizing forces that produce rotordynamic instability have not been quantitatively measured and cannot be accurately predicted at present. Therefore, the most productive method of analysis for troubleshooting often is to vary the undetermined coefficients in repeated runs of the computer program until the field-observed critical speeds, the stability threshold speeds, and the instability frequencies are reasonably well simulated by the program. Subsequently, the program can be used to investigate the effect of changes in the design parameters which are known and which can be practically modified (e.g., bearing or seal stiffness and damping).

As an example, consider the eight-stage centrifugal compressor described above. This type of machine (multistage, high discharge pressure, high speed, with centrifugal impellers) has a history of costly rotordynamic instability problems, difficult to diagnose and cure. There is a considerable body of experimental evidence [12, 15, 16, 34] which suggests that large destabilizing forces are produced by the dynamics of the working fluid around the impellers.

The forces on each impeller can be modeled in the program as linearized stiffness and damping, including cross-coupled terms, originally formulated by Lund and described earlier. Remember that destabilizing effects from the working fluid often can be modeled using the cross-coupled stiffness coefficients $K_{XY} = -K_{YX}$. Figure 7.16 shows how the logarithmic decrement (stability indicator) for the 3090

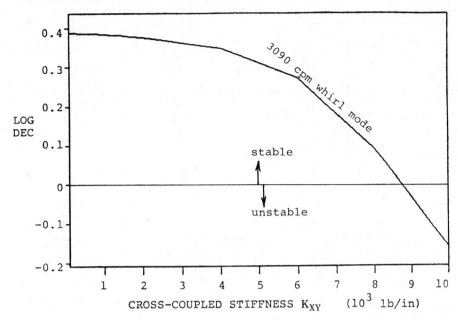

Figure 7.16. Stability curve for the first forward whirl mode when shaft speed equals operating speed. From Ref. [31].

cpm mode of the compressor varies with the coefficient K_{XY} over a range from 100 to 10,000 lb/in. The 3090 cpm whirl mode is predicted to be unstable for all K_{XY} > 8750 lb/in. Since this compressor actually exhibited an instability near this frequency in a field installation, it can be surmised that the cross-coupled stiffness in this machine exceeded 8750 lb/in (assumed equal on each impeller wheel). Until verified theories are available to predict the destabilizing coefficients, this value may be the best estimate available for this type of machine. At least one effort has been made to develop empirical equations to fit this type of data [34].

Changing the design parameters in a compressor sometimes results in a shift of the instability from one mode to another [15]. In such a case, the greatest advantage of the polynomial program is that the "new" unstable modes and frequencies would not be missed by a failure to converge.

Such a program can also be used to investigate the effect of damping added at specific locations in a machine (usually the bearings). For example, squeeze film bearing dampers are used in many aircraft turbine engines, since their rolling-element bearings have very little inherent damping. Although the primary purpose of the dampers is to reduce synchronous whirl amplitudes (caused by rotor imbalance), they also serve to suppress potential rotordynamic instability. Figure 7.17 shows a rotor-bearing model of the high-pressure spool from this type of engine, with the effect of a damper at bearing A to be investigated.

Figures 7.18 and 7.19 show the logarithmic decrement for three whirl modes with weak and strong cross-coupling (K_{XY}), as a function of damping at bearing

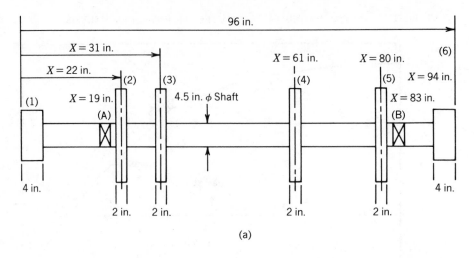

(a)

	Mass Properties			Bearing Properties			Cross-Coupling	
	W (lb)	I_T (lb-in^2)	I_P (lb-in^2)	$K_{XX} = K_{YY}$ (lb/in)	$C_{XX} = C_{YY}$ (lb-sec/in)		$K_{XY} = -K_{YX}$	(lb/in)
							I	II
(1)	80	720	360					
(2)	100	3600	1800	(A) 50,000	variable (0–1000)	(2)	300	8750
(3)	100	3600	1800			(3)	300	8750
(4)	100	3600	1800	(B) 20,000	0	(4)	300	8750
(5)	100	3600	1800			(5)	300	8750
(6)	80	720	360					

Total rotor weight = 920 lb

Figure 7.17. Rotordynamic model of a simulated high pressure (or HP) rotor in an aircraft engine.

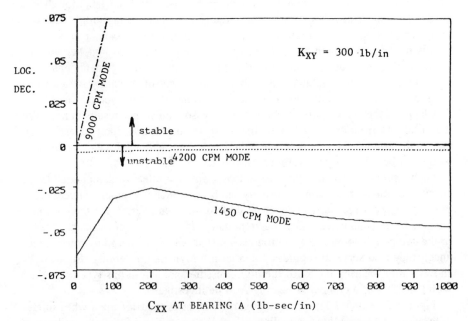

Figure 7.18. Effect of damper at bearing A on stability of simulated high pressure rotor at 15,000 rpm with weak cross-coupling.

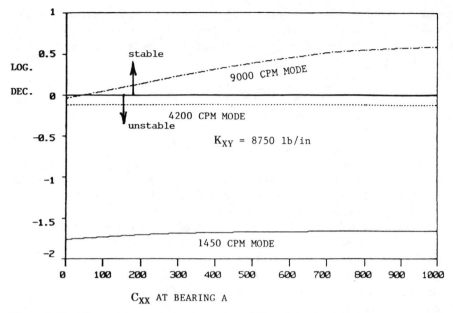

Figure 7.19. Effect of damper at bearing A on stability of simulated high pressure rotor at 15,000 rpm with strong cross-coupling.

A. Note that only the 9000-cpm whirl mode responds favorably to the added damping. This is because bearing A is too close to a node of the whirl mode shape for both the 1450-cpm mode and the 4200-cpm mode. The damper simply cannot act effectively unless it is subjected to a significant amplitude of motion.

Note also that, under some conditions, added damping at bearing A could actually worsen an instability at the 1450-cpm frequency.

With the exception of the numerical marching method, which was only briefly mentioned above, the computer stability analyses just described are based on linearized differential equations. It is instructive to recall the significance of the linearization, since most real systems are actually nonlinear when large motions occur.

SIGNIFICANCE OF THE LINEARIZED EQUATIONS

The linearized equations of motions have realistic meaning only for small motions about the equilibrium, since the forces are all described by "constants" that do not remain constant for large motions. The (unstable) solution that grows indefinitely with time soon grows into a new regime in which the *nonlinear* equations govern. This is the regime which may have bounded solutions, even though the linearized analysis predicts instability.

For example, Ref. [35] shows analytically that a rotor on journal bearings can execute bounded whirl orbits over a significant range of speeds above the threshold speed of instability in oil whip. This is shown to be due to the nonlinear behavior of the oil-film forces. For a rigid rotor of mass m on journal bearings, the differential equations of motion are given by

$$m\ddot{X} = F_X(X, Y, \dot{X}, \dot{Y}) + W,$$

$$m\ddot{Y} = F_Y(X, Y, \dot{X}, \dot{Y}),$$

(7-18)

where F_X and F_Y are the nonlinear components of oil-film force (see Chapter VI) in the X and Y directions, respectively, and W is the rotor weight on the bearing. For small perturbations away from an equilibrium position, the change in forces F_X and F_Y can be represented by equations such as (7-11). (The angular coefficients are usually neglected in bearing analysis.)

Figure 7.20 shows these loads and the locus of static equilibrium points for the journal. (The vertical X axis is a convention used by Lund [35]). The latter points represent the solution $\ddot{X} = \ddot{Y} = 0$ to the equations of motion (7-18). At speeds below the threshold speed of instability, a perturbation of the static equilibrium will die out. For a range of speeds above the threshold speed, a perturbation will create a whirl orbit of a constant finite size. The shape and orientation of this "limit cycle orbit" depends on the original equilibrium position (i.e., the bearing load) as shown in Fig. 7.21. Figure 7.22 shows the corresponding typical $X(t)$ [or $Y(t)$] motion which might be measured by a unidirectional vibration probe.

This analysis is supported by observations in the field. There have been reported cases [34] where a rotor has operated with stable subsynchronous whirl orbits at speeds well above the threshold speed of "instability."

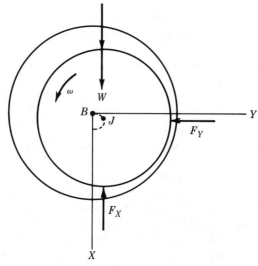

Figure 7.20. Oil-film force components and static operating points of a journal bearing.

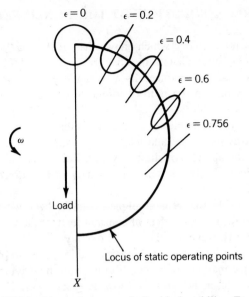

$\epsilon = 0$ $\epsilon = 0.2$

$\epsilon = 0.4$

$\epsilon = 0.6$

$\epsilon = 0.756$

ω

Load

Locus of static operating points

X

Figure 7.21. Whirl orbits at the onset of oil whip instability. From Ref. [35].

The analysis in the section to follow gives a bounded "limit-cycle" solution for a simple rotor model, where load torque drives the whirling motion (torque-whirl). Ref. [37] shows that the "pure spin" solution to the linearized equations of the torquewhirl model is unstable in the sense of Lyapunov. Thus, we see that the field engineer's definition of "instability" in rotating machines (i.e., any nonsynchronous vibration) is actually compatible with mathematical predictions, once the nonlinear regime is considered with its bounded solutions.

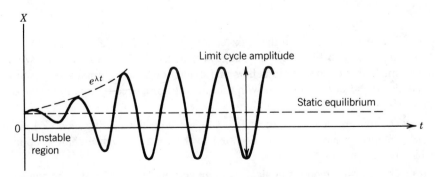

X

Limit cycle amplitude

$e^{\lambda t}$

Static equilibrium

0

Unstable region

t

Figure 7.22. Typical $X(t)$ motion for an unstable equilibrium point growing to a limit cycle motion.

TORQUEWHIRL: A SOLUTION TO THE NONLINEAR EQUATIONS

This section presents a theory which shows that, if the load torque on a coning disk tends to remain aligned with the disk axis, as in a fluid impeller, nonsynchronous whirling can be produced as a dynamic equilibrium motion, with an amplitude that depends on the ratio of torque to damping. The theory has been experimentally verified [38].

A review of the rotordynamics literature shows that it has been common practice to eliminate the driving torque and load torque from the equations of motion in order to simplify the analysis of shaft whirling. This procedure is mathematically correct, but obviously it is not suited to a study of the effect of torque on the motion.

For rotors with disks that remain aligned with the bearing axis (e.g., centered on the bearing span), the main effect of constant load torque is simply to lower the critical speed [27]. However, if the disk is overhung, or located near one end of a bearing span so that it can execute a coning motion, Bousso [39] has shown that the disk load torque may not be in equilibrium with the driving torque. Bousso's analysis is incomplete, as he does not show necessary or sufficient conditions for torque-driven whirl to occur, but his vector diagrams show revealingly how a component of the driving torque can act on the precession coordinate of a coning disk to feed energy into the whirling motion.

This latter effect, which does not require a time-varying torque, should be properly differentiated from the effects of torque on rotor response described in Ref. [27] by Eshleman and Eubanks. Their experimental study showed that a pulsating torque of small magnitude, superimposed on a constant torque, produces unstable whirling over a range of speed that becomes wider with increasing torque. The instability disappears when the pulsating component of the torque is removed. Coning motion of the disk apparently is not a significant factor in their model.

The analysis presented below shows that constant torque at constant speed can produce nonsynchronous whirling if the disk motion is conical.

Figure 7.23 shows the model analyzed. This is the simplest possible model which has all of the characteristics necessary to produce the phenomenon under study:

1. The load torque T_L on the disk remains parallel to the disk axis (z'). The vanes in the disk in Fig. 7.23 are suggestive of the type of machine that would approximate this condition. Impellers and bladed disks are normally designed to maximize the torque (associated with useful work) produced by rotation about the disk axis.

2. The driving or shaft torque T_s is aligned with the bearing axis (Z). In a machine, this torque is often transmitted to the rotor through a shaft coupling, which is idealized by the joint at O'.

3. The whirling mode is conical, with an amplitude described by the angle θ.

4. Whirling of the disk produces a damping (drag) force on the disk, not shown

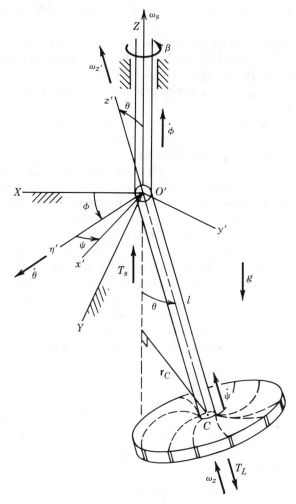

Figure 7.23. Cantilevered rotor with a flexible joint and rigid shaft.

in the figure. This force F_d is tangential to the path of the disk center C and produces the moments required for dynamic equilibrium under steady-state conditions. Note that T_L and T_s cannot be in equilibrium without F_d, unless $\theta = 0$ [21].

To avoid unnecessary analytical complexity, the shaft is assumed rigid (except at O') and all of the mass (M) of the rotor is concentrated in the disk. Imbalance is not included in the present analysis, so $u = 0$.

The mass properties of the rotor are completely represented by M and by mass moments of inertia $I_{x'} = I_{y'}$ and $I_{z'}$ about principal axes x', y', z' through the point O'. These axes, and parallel axes x, y, z through C, (note $z = z'$), are fixed in—

and rotate with—the rotor. Axes X, Y, Z are fixed in space, with Z along the bearing axis.

The rotor stiffness K_θ is assumed to produce a restoring moment proportional to θ, $M_{n'} = -K_\theta \theta$. This could be due to shaft bending stiffness or coupling stiffness at O', or a flexible bearing support outboard of the disk. In addition, the rotor is assumed to be vertical, so that gravity produces an additional restoring moment[4] of magnitude $Mgl \sin \theta$.

Both viscous and aerodynamic models are used to describe the velocity dependence of the load torque and the damping force. The corresponding expressions for generalized forces are given below.

The motion and instantaneous position of the rotor can be completely described by specifying the three coordinates ϕ, θ, and ψ as functions of time. These may be recognized as the three Euler angles defined by Goldstein [40]. For conical motion, with circular orbits of the disk center C, the precession or whirling velocity is $d\phi/dt = \dot{\phi}$, the amplitude of whirling is $\theta = $ constant, and the shaft speed $\omega_s = \omega_z = \dot{\phi} + \dot{\psi}$. The rotational speed of the disk is $\omega_z = \dot{\phi} \cos \theta + \dot{\psi}$. The inequality of the latter two angular velocities is central to an understanding of how torquewhirl is produced. Note that the rotational velocity ω_z of the disk becomes zero when $\theta = 90°$, $\dot{\psi} = 0$, $\omega_s = \dot{\phi}$. Under this (improbable) condition, all of the shaft work would have to be dissipated by damping, and $\overline{T}_L = 0$.

In terms of the angular velocities about body-fixed principal axes x', y', z through O' (see Fig. 7.23) the kinetic energy of the rotor is

$$T = \tfrac{1}{2} I_{x'} \omega_{x'}^2 + \tfrac{1}{2} I_{y'} \omega_{y'}^2 + \tfrac{1}{2} I_z \omega_z^2. \tag{7-19}$$

The kinematic relationships required to express T in terms of the generalized coordinates are

$$\omega_{x'} = \dot{\theta} \cos \psi + \dot{\phi} \sin \psi \sin \theta,$$
$$\omega_{y'} = -\dot{\theta} \sin \psi + \dot{\phi} \cos \psi \sin \theta, \tag{7-20}$$
$$\omega_z = \dot{\phi} \cos \theta + \dot{\psi}.$$

The potential energy is

$$V = \tfrac{1}{2} K_\theta \theta^2 + Mgl(1 - \cos \theta) \tag{7-21}$$

---

[4]In most real machines, this term is insignificant when compared to the restoring moment of the shaft or coupling.

Substitution of (7-20) into (7-19), with $I_{x'} = I_{y'}$, gives the Lagrangian as

$$L = T - V = \tfrac{1}{2}[I_x + Ml^2](\dot{\theta}^2 + \dot{\phi}^2 \sin^2 \theta)$$
$$+ \tfrac{1}{2}I_z(\dot{\psi}^2 + \dot{\phi}^2 \cos^2 \theta + 2\dot{\phi}\dot{\psi} \cos \theta) \qquad (7\text{-}22)$$
$$- \tfrac{1}{2}K_\theta \theta^2 - Mgl(1 - \cos \theta).$$

References [41] and [42] show how first-order differential equations of motion can be obtained by direct application of Hamilton's principle, using only the first-order partial derivatives of the Lagrangian.

Using the methods of Refs. [41] and [42], the six first-order equations of motion are obtained from (7-22) in canonical form as

$$\dot{\theta} = \frac{1}{I_{x'}} p_\theta, \qquad (7\text{-}23)$$

$$\dot{\phi} = \frac{1}{I_{x'} \sin^2 \theta} p_\phi - \frac{\cos \theta}{I_{x'} \sin^2 \theta} p_\psi, \qquad (7\text{-}24)$$

$$\dot{\psi} = \frac{\cos \theta}{I_{x'} \sin^2 \theta} p_\phi + \left[\frac{1}{I_z} + \frac{\cot^2 \theta}{I_{x'}} \right] p_\psi, \qquad (7\text{-}25)$$

$$\dot{p}_\theta = \frac{\cos \theta}{I_{x'} \sin^3 \theta} p_\phi^2 - \left[\frac{1}{I_{x'} \sin \theta} + \frac{2 \cos^2 \theta}{I_{x'} \sin^3 \theta} \right] p_\phi p_\psi$$
$$+ \left[\frac{\cot \theta}{I_{x'}} + \frac{\cot^3 \theta}{I_{x'}} \right] p_\psi^2 - K_\theta \theta - Mgl \sin \theta + Q_\theta, \qquad (7\text{-}26)$$

$$\dot{p}_\phi = Q_\phi, \qquad (7\text{-}27)$$

$$\dot{p}_\psi = Q_\psi, \qquad (7\text{-}28)$$

where the momenta p_θ, p_ϕ, and p_ψ are defined as

$$p_\theta = \frac{\partial L}{\partial \dot{\theta}} = I_x \dot{\theta}, \qquad (7\text{-}29)$$

$$p_\phi = \frac{\partial L}{\partial \dot{\phi}} = I_{x'} \dot{\phi} \sin^2 \theta + I_z(\dot{\psi} + \dot{\phi} \cos \theta) \cos \theta, \qquad (7\text{-}30)$$

$$p_\psi = \frac{\partial L}{\partial \dot{\psi}} = I_z(\dot{\psi} + \dot{\phi} \cos \theta). \qquad (7\text{-}31)$$

The nonconservative generalized forces, Q_θ, Q_ϕ, and Q_ψ are derived from the torque and damping force. Since the generalized coordinates are angles, the generalized forces have the units of torque (in-lb). They are obtained from the virtual work of the driving or shaft torque T_s, the disk load torque T_L, and the damping force F_d, as follows:

The total virtual work is

$$\delta W = \delta W_s + \delta W_L + \delta W_d \tag{7-32}$$

The virtual work of the shaft torque T_s is

$$\delta W_s = T_s[\delta\phi + \delta\psi]. \tag{7-33}$$

The virtual work of the disk load torque T_L is

$$\delta W_L = T_L[\cos\theta\delta\phi + \delta\psi]. \tag{7-34}$$

In general, T_L is proportional to some power n of the disk speed ω_z. Two cases are considered here: $n = 1$ (viscous), and $n = 2$ (aerodynamic).

For $n = 1$,

$$T_L = -C_L\omega_z = -C_L[\dot\phi\cos\theta + \dot\psi], \tag{7-35}$$

and for $n = 2$,

$$T_L = -\frac{\omega_z}{|\omega_z|}\overline{C}_L[\dot\phi\cos\phi + \dot\psi]^2, \tag{7-36}$$

where C_L and $\overline{C}_L$ are the disk load coefficients for viscous torque and aerodynamic torque, respectively.

The virtual work of the damping force F_d is

$$\delta W_d = F_{d_\theta}l\delta\theta + F_{d_\phi}l\sin\theta\delta\phi, \tag{7-37}$$

where F_{d_θ} and F_{d_ϕ} are the radial and tangential components of F_d, respectively.

If F_d is predominantly due to the drag of the disk in the working fluid, it will be proportional to some power n for each of the two cases above (viscous and aerodynamic drag, respectively).

For $n = 1$,

$$F_d = -C_d\sqrt{(l\sin\theta\dot\phi)^2 + (l\dot\theta)^2},$$

$$F_{d_\theta} = -C_dl\dot\theta, \tag{7-38}$$

$$F_{d_\phi} = -C_d(l\sin\theta)\dot\phi,$$

where C_d is the viscous damping coefficient.

For $n = 2$,

$$F_d = -\overline{C}_d\left[(l \sin \theta \dot{\phi})^2 + (l\dot{\theta})^2\right],$$

$$F_{d_\theta} = -\overline{C}_d l\dot{\theta} \sqrt{(l \sin \theta \dot{\phi})^2 + (l\dot{\theta})^2}, \qquad (7\text{-}39)$$

$$F_{d_\phi} = -\overline{C}_d l \sin \theta \dot{\phi} \sqrt{(l \sin \theta \dot{\phi})^2 + (l\dot{\theta})^2},$$

where $\overline{C}_d$ is the aerodynamic damping coefficient.

In all cases, the direction of F_d is assumed to be tangent to the path of the disk center, C, opposing the motion.

For the viscous case, substitution of (7-38) into (7-37), (7-35) into (7-34), and (7-33), (7-34), and (7-37) into (7-32) yields the total virtual work, which can be factored into the form

$$\delta W = Q_\theta \delta\theta + Q_\phi \delta\phi + Q_\psi \delta\psi. \qquad (7\text{-}40)$$

Substitution of equations (7-23), (7-24), and (7-25) into the expressions for Q_θ, Q_ϕ, and Q_ψ gives

$$Q_\theta = -C_d \frac{l^2}{I_{x'}} p_\theta, \qquad (7\text{-}41)$$

$$Q_\phi = T_s - \frac{C_d l^2}{I_{x'}} p_\phi + \left[\frac{C_d l^2}{I_{x'}} - \frac{C_L}{I_z}\right] p_\psi \cos \theta, \qquad (7\text{-}42)$$

$$Q_\psi = T_s - \frac{C_L}{I_z} p_\psi. \qquad (7\text{-}43)$$

For the aerodynamic case, a similar procedure using (7-36), and (7-39) gives

$$Q_\theta = -\overline{C}_d^3 \frac{p_\theta}{I_{x'}^2 \sin \theta} \sqrt{\left[p_\phi - p_\psi \cos \theta\right]^2 + p_\theta^2 \sin^2 \theta}, \qquad (7\text{-}44)$$

$$Q_\phi = T_s - \frac{p_\psi}{|p_\psi|} \frac{\overline{C}_L \cos \theta}{I_x^2} p_\psi^2 - \overline{C}_d l^3 \frac{p_\phi - p_\psi \cos \theta}{I_{x'}^2 \sin \theta}$$
$$\cdot \sqrt{\left[p_\phi - p_\psi \cos \theta\right]^2 + p_\theta^2 \sin^2 \theta}, \qquad (7\text{-}45)$$

$$Q_\psi = T_s - \frac{p_\psi}{|p_\psi|} \frac{C_L}{I_z^2} p_\psi^2. \qquad (7\text{-}46)$$

The differential equations of motion for the rotor of Fig. 7.23 are now given by (7-23)–(7-28), with Q_θ, Q_ϕ, and Q_ψ substituted from (7-41), (7-42), and (7-43) for the case with viscous load torque and damping, or from (7-44), (7-45), and

(7-46) for the case with aerodynamic load torque and damping. Note that the equations are nonlinear.

For both cases, an exact solution to the equations of motion is found which described nonsynchronous whirling at constant amplitude, as follows:

Viscous Case

The ratio of whirling speed to shaft speed is defined as

$$f = \frac{\dot{\phi}}{\omega_s}. \qquad (7\text{-}47)$$

Then

$$p_\theta = 0,$$
$$p_\phi = I_{x'} f \omega_s \sin^2 \overset{*}{\theta} + I_z \left[1 - f(1 - \cos \overset{*}{\theta}) \right] \omega_s \cos \overset{*}{\theta}, \qquad (7\text{-}48)$$
$$p_\psi = I_z \left[1 - f(1 - \cos \overset{*}{\theta}) \right] \omega_s,$$
$$\dot{p}_\theta = 0, \qquad \dot{p}_\phi = 0, \qquad \dot{p}_\psi = 0,$$

is a solution, where $\overset{*}{\theta}$ is the value of θ that satisfies

$$\frac{R_c(1 - \cos \theta)}{R_c(1 - \cos \theta)^2 + \sin^2 \theta}$$
$$= \frac{I_2 \pm \sqrt{I_2^2 + (4/\omega_s^2)(I_1 \cos \theta + I_2)(\Omega_g^2 + \Omega_K^2 (\theta/\sin \theta))}}{2(I_1 \cos \theta + I_2)}, \qquad (7\text{-}49)$$

and the whirling speed ratio f is then given by the right-hand side of (7-49). That is,

$$f = \frac{I_2 + \sqrt{I_2^2 + (4/\omega_s^2)(I_1 \cos \overset{*}{\theta} + I_2)(\Omega_g^2 + \Omega_K^2(\overset{*}{\theta}/\sin \overset{*}{\theta}))}}{2(I_1 \cos \overset{*}{\theta} + I_2)}. \qquad (7\text{-}50)$$

where

$$I_1 = \frac{I_{x'} - I_z}{I_{x'}}, \qquad I_2 = \frac{I_z}{I_{x'}}, \qquad (7\text{-}51)$$

$$\Omega_g^2 = \frac{Mgl}{I_{x'}} \qquad \Omega_K^2 = \frac{K_\theta}{I_{x'}},$$

and

$$R_c = \frac{C_L}{C_d l^2}. \qquad (7\text{-}52)$$

Aerodynamic Case

The equations of motion for this case are also satisfied by (7-48), with the whirling speed ratio f given by (7-50), where $\overset{*}{\theta}$ is now the value of θ that satisfies

$$\frac{\pm \sqrt{(1 - \cos \theta) \sin^3 \theta \overline{R}_c} - (1 - \cos \theta)^2 \overline{R}_c}{\sin^3 \theta - (1 - \cos \theta)^3 \overline{R}_c}$$

$$= \frac{I_2 \pm \sqrt{I_2^2 + (4/\omega_s^2)(I_1 \cos \theta + I_2)(\Omega_g^2 + \Omega_K^2(\theta/\sin \theta))}}{2(I_1 \cos \theta + I_2)} \qquad (7\text{-}53)$$

where

$$\overline{R}_c = \frac{\overline{C}_L}{\overline{C}_d l^3}, \qquad (7\text{-}54)$$

and I_1, I_2, Ω_g, Ω_K are defined the same as for the viscous case.

For both cases, the method of numerical solution is as follows:

1. Assume a value of $\overset{*}{\theta}$ and use it to calculate f from (7-50).
2. Calculate R_c or $\overline{R}_c$ from (7-49) or (7-53), respectively.
3. Calculate the shaft torque and/or damping from R_c and $\overline{R}_c$. Step 2 is simplified for both cases by substituting f for the right-hand side of (7-49) or (7-53), then solving (7-49) explicitly for R_c, or (7-53) for $\overline{R}_c$. The resulting expressions for R_c and $\overline{R}_c$ are

$$R_c = \frac{f \sin^2 \overset{*}{\theta}}{(1 - \cos \overset{*}{\theta})[1 - f(1 - \cos \overset{*}{\theta})]}, \qquad (7\text{-}55)$$

$$\overline{R}_c = \frac{f^2 \sin^3 \overset{*}{\theta}}{(1 - \cos \overset{*}{\theta})[1 - f(1 - \cos \overset{*}{\theta})]^2}. \qquad (7\text{-}56)$$

Equations (7-49)–(7-56), programmed on a digital computer, produced the results shown in Figs. 7.24–7.29. In general, small-angle assumptions were not made, even though the angles are small, since the precise magnitude of θ is of prime interest, and since some of the computations are strong functions[5] of θ.

---

[5]Wherever this is not the case, it is so indicated on the curves.

Figure 7.24 shows how the whirling speed ratio f varies with shaft speed. For any given rotor geometry, f becomes almost constant at high speeds, where the gyroscopic forces are strong. This causes the actual subsynchronous whirling frequency to become higher than the critical speed at high shaft speeds. This effect is independent of the type of loading or damping.

Figure 7.25 shows how the critical ratio of load torque to damping (R_c and $\overline{R}_c$) varies with whirling speed ratio f for both types of loading. The value of R_c or $\overline{R}_c$ read from the curves should be interpreted as the ratio of load torque to damping required to produce torquewhirl at a given frequency.

Figure 7.26 shows how the whirling amplitude θ increases with the ratio of load torque to damping. For a given rotor at a given speed, the appropriate value of f can be taken from Fig. 7.24. Note that Figs. 7.25 and 7.26 are independent of the mass or stiffness properties of the rotor (except insofar as they determine the value of f).

To illustrate some specific results that could be observed in a compressor or pump, a prototype machine is defined. The chosen design and performance parameters are believed to be representative of some modern high speed machines (although the model restricts the prototype to a single disk, or stage). These parameters are as follows:

Maximum horsepower (hp) = 2500 at 8000 rpm (1865 kW at 133.3 Hz)
Maximum speed $[\omega_s(60/2\pi)]$ = 8000 rpm (133.3 Hz)
Critical speed ($\sqrt{\Omega_K^2 + \Omega_g^2}$) = about 3200 rpm (53.3 Hz)
Disk weight (Mg) = 25 lb (11.35 kg)
Disk radius (R) = 7.5 in
Effective shaft stiffness (K_θ) = 10^5 in-lb/rad (11, 298 N-M/rad)

The damping and shaft length are treated as design variables, since they each could be used to suppress torquewhirl without compromising machine performance.

Figures 7.27, 7.28, and 7.29 show representative results for this particular machine.

Figure 7.27 gives the load horsepower required to produce various amplitudes of torquewhirl as a function of shaft speed. The corresponding whirling frequencies are also given. For example, at 8000 rpm (133.3 Hz), a load of about 1100 (821 kW) horsepower is required to produce a whirling amplitude of 10 mils (0.25 mm) at a frequency of 0.71 times shaft speed. This is for l = 5.5 in (14 cm) and 5 percent equivalent damping.

Since damping is usually computed on a linear basis, it is presented this way in Figs. 7.27 and 7.28. This makes the nonlinear damping coefficient $\overline{C}_d$ amplitude dependent, so the results are normalized to an amplitude of 47 mils (1.2 mm). For q percent damping, the coefficient is then given by

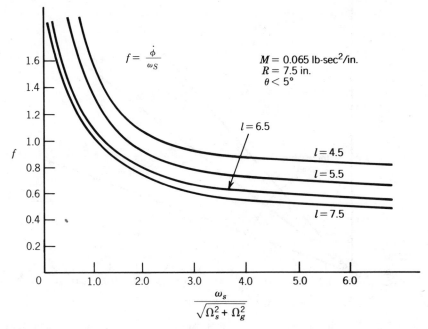

Figure 7.24. Whirling speed ratio versus dimensionless speed for both viscous and aerodynamic loadings. From Ref. [11].

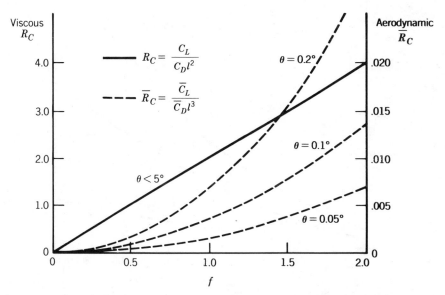

Figure 7.25. Dimensionless load torque required to produce torquewhirl versus whirling speed ratio: viscous and aerodynamic cases. From Ref. [11].

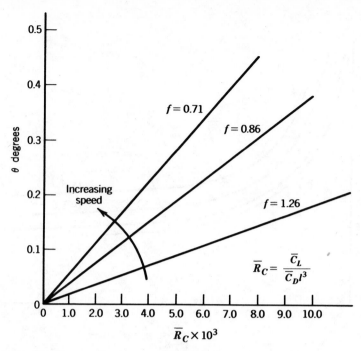

Figure 7.26. Whirling amplitude versus dimensionless load torque: aerodynamic case. From Ref. [11].

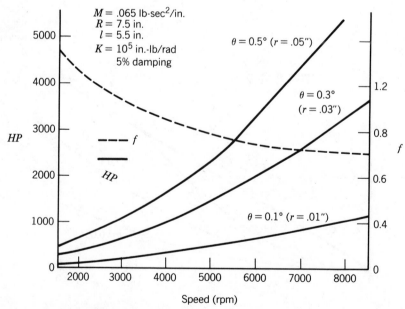

Figure 7.27. Horsepower required to produce torquewhirl and whirling speed ratio versus speed of prototype machine. From Ref. [11].

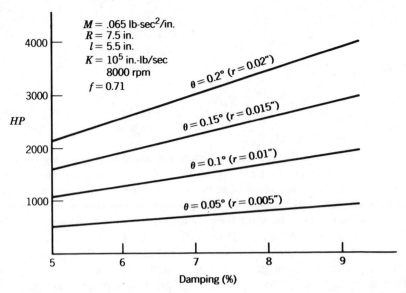

Figure 7.28. Horsepower required to produce torquewhirl versus percent of critical damping in prototype machine. From Ref. [11].

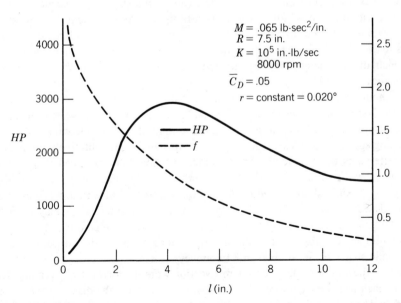

Figure 7.29. Horsepower required to produce torquewhirl and whirling speed ratio versus shaft length in prototype machine. From Ref. [11].

$$\overline{C}_d = \frac{C_d}{(0.047)\, f\omega_s}, \tag{7-57}$$

where

$$C_d = \frac{q}{100}\left[2M\sqrt{\Omega_s^2 + \Omega_g^2}\right]. \tag{7-58}$$

Figure 7.28 shows that the load horsepower required to produce a given amplitude of torquewhirl can be increased by increasing the damping. However, the amount of added damping required to produce significant reductions may become impractically large. For example, extrapolation of the curves shows that the damping would need to be approximately doubled to reduce the amplitude from 20 mils (0.51 mm) to 10 mils (0.25 mm), for a load of 2000 hp (1492 kW).

The most effective way of avoiding torquewhirl may be through optimization of the shaft length, l, as suggested by Fig. 7.29. The optimum shaft length is that which maximizes the required horsepower. A trade-off must be made with synchronous response to imbalance, however, since the curve shows that optimum l means operation at or near the critical speed ($f = 1$).

It is interesting to speculate as to how the torquewhirl characteristics demonstrated by this simple model will be manifested in machines of greater complexity. (The speculations can be confirmed using the transfer matrices given in a later section to simulate a multidisk rotor–bearing model). For example, a disk mounted between two bearings on a flexible shaft (not at midspan) can execute a coning motion with a potential for torquewhirl. Multiple disks on a flexible shaft are an extension of this case, in which torquewhirl can occur at a multiplicity of frequencies, each with a different required load torque and horsepower. The coning angle of each disk is determined by the mode shape associated with the particular frequency.

Although the results are not presented graphically, it was found that increasing the shaft stiffness in the model of Fig. 7.23 tends to increase the load torque or horsepower required to produce a given amplitude of torquewhirl, when the whirling speed ratio is less than unity. Since this analysis qualitatively predicts the characteristics of more complex machines, it should be expected that nonsynchronous whirl produced by high-load torque can be effectively suppressed by stiffening flexible shafts to reduce the coning angles and move the whirling speed ratio closer to 1.0 (see Fig. 7.29). Where this is not practical, it may be found helpful to selectively modify the shaft stiffness at specific locations, to reduce the coning angles of those disks with the largest load torque and/or misalignment in the mode shape at the troublesome frequency.

All of the solutions described in this section were for forward whirl in negative-work machines (i.e., compressors, pumps, etc.) in which the disk is driven by the shaft to do work on the fluid. No solutions were found for backward whirl of such machines.

For positive-work machines (i.e., turbines), backward whirl was found to be a solution with driving torque on the disk, but the equations used here for the torque–speed relationship probably do not represent a realistic model for turbines, so the solution is not presented.

In 1979, the torquewhirl theory was applied to the NASA space shuttle main engine turbopumps [43] in an attempt to analyze the nonsynchronous whirl instabilities that had been encountered in developmental testing. Under certain simplifying assumptions about the mode shape,[6] it was shown that torquewhirl could be a significant destabilizing force in the oxygen pumps if less than 5 percent of critical damping was present in the machine. The major source of damping in this machine is from the seals.

TORQUEWHIRL COEFFICIENTS FOR A LINEARIZED ANALYSIS

The basic idea underlying the torquewhirl theory just described is that the load torque vector on an impeller disk will tend to remain aligned with the disk (due to axisymmetry), and thus will have a component which acts on the whirl coordinate whenever the disk becomes misaligned by the mode shape of the flexible rotor. It can be shown that the generalized forces from torquewhirl produce cross-coupled stiffness coefficients, as follows.

For the rigid shaft, flexible joint, torquewhirl model, the disk rotation and translation coordinates are not independent (i.e., they are related by a kinematic constraint). The constraint equations are $R' = l\theta$ in polar coordinates or $X^2 + Y^2 = (l \sin \theta)^2$ in inertial $(X-Y)$ coordinates (see Fig. 7.23).

Since the equations and generalized forces are written in terms of Euler angles θ, ϕ, and ψ, a coordinate transformation is required to derive the stiffness and damping coefficients in $X-Y$ coordinates.

First, the generalized torque Q_ϕ of equations (7-42) and (7-45) be expressed as a tangential force

$$F_\phi - Q_\phi/R', \qquad (7\text{-}59)$$

and the generalized moment Q_θ can be expressed as a radial force

$$F_R = Q_\theta/l \cos \theta. \qquad (7\text{-}60)$$

It is Q_ϕ (or F_ϕ) which contains the destabilizing torquewhirl forces. The transformation to X and Y is

$$\begin{aligned} F_X &= F_R \cos \phi - F_\phi \sin \phi, \\ F_Y &= F_R \sin \phi + F_\phi \cos \phi, \end{aligned} \qquad (7\text{-}61)$$

[6]The author has since found that these assumptions are invalid, and torquewhirl is a smaller effect in this machine than predicted by Ref. [43].

where

$$\sin \phi = Y/(X^2 + Y^2)^{1/2},$$
$$\cos \phi = X/(X^2 + Y^2)^{1/2}. \tag{7-62}$$

For the aerodynamic case the generalized torque Q_ϕ is

$$Q_\phi = T_s - \overline{C}_L(\dot\psi + \dot\phi \cos\theta)^2 \cos\theta - C_d(l^3 \sin^3\theta)\dot\phi^2, \tag{7-63}$$

where

T_s = shaft torque
$\overline{C}_L$ = disk load coefficient
$\overline{C}_d$ = nonlinear damping coefficient

The shaft speed ω_s is

$$\omega_s = \dot\psi + \dot\phi \tag{7-64}$$

The shaft torque equals the disk load torque so that

$$T_s = \overline{C}_L(\dot\psi + \dot\phi \cos\theta)^2 \tag{7-65}$$

Therefore, the generalized torque Q_ϕ can be expressed as

$$Q_\phi = T_s[1 - \cos\theta] - \overline{C}_d(l^3 \sin^3\theta)\dot\phi^2. \tag{7-66}$$

The destabilizing part of Q_ϕ is the first term from the expression just above. The equivalent tangential force F_ϕ is

$$F_\phi = Q_\phi/R' = T_s(1 - \cos\theta)/R'. \tag{7-67}$$

The required relationships between R', θ, and α, β, and X, Y are

$$R' = (X^2 + Y^2)^{1/2},$$
$$X = \alpha l,$$
$$Y = -\beta l, \tag{7-68}$$
$$\theta = (\alpha^2 + \beta^2)^{1/2}.$$

Therefore, in terms of X and Y, we can write

$$F_\phi = \frac{T_s\{1 - \cos[(X^2 + Y^2)^{1/2}/l]\}}{(X^2 + Y^2)^{1/2}}. \tag{7-69}$$

Keeping only the first two terms of the cosine series yields

$$F_\phi = (T_s/2l^2)\,(X^2 + Y^2)^{1/2},$$

or   (7-70)

$$F_X = -F_\phi \sin \phi = -(T_s/2l^2)\,Y,$$
$$F_Y = F_\phi \cos \phi = (T_s/2l^2)\,X.$$

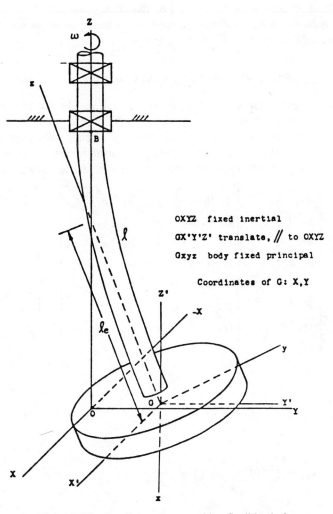

OXYZ fixed inertial
OX'Y'Z' translate, // to OXYZ
Oxyz body fixed principal

Coordinates of G: X,Y

Figure 7.30. Cantilevered rotor with a flexible shaft.

By inspection, it can be seen that the cross-coupled stiffnesses are

$$K_{XY} = T_s/2l^2,$$

$$K_{YX} = -T_s/2l^2. \tag{7-71}$$

It is interesting to note that these expressions for K_{XY} and K_{YX} have the same torque dependence as Alford's coefficients for the effect of tip clearance asymmetry in axial flow turbomachinery [7].

The author has found that the cross-coupled stiffness coefficients given by (7-71) can also be used for the model of Fig. 7.30, where the flexible joint is replaced by a elastic shaft. In this case l is taken as the axial distance l_e from the disk plane to the virtual pivot point as determined by the mode shape.

For a multidisk flexible rotor, the torquewhirl effect must be put into the transfer matrices, as shown in the following section.

TORQUE-DEPENDENT TRANSFER MATRICES

For a flexible rotor, the effect of tangential torque can be put into the transfer matrices for the shaft elements. Yim [44] has derived the torque-dependent transfer matrix. The free body diagrams of the station elements with tangential torque included are shown in Fig. 7.31. Omitting the bearing forces and shear effects for brevity, the transfer equation across the nth shaft element are as shown in (7-72):

$$
\begin{Bmatrix} X \\ Y \\ \alpha \\ \beta \\ V_X \\ V_Y \\ M_Y \\ M_x \end{Bmatrix}_{n+1}
=
\begin{bmatrix}
1 & 0 & l & 0 & e_{15} & e_{16} & e_{17} & e_{18} \\
0 & 1 & 0 & l & e_{25} & e_{26} & e_{27} & e_{28} \\
0 & 0 & 1 & 0 & e_{35} & e_{36} & e_{37} & e_{38} \\
0 & 0 & 0 & 1 & e_{45} & e_{46} & e_{47} & e_{48} \\
0 & 0 & 0 & 0 & 1 & 0 & 0 & 0 \\
0 & 0 & 0 & 0 & 0 & 1 & 0 & 0 \\
0 & 0 & 0 & 0 & e_{75} & e_{76} & e_{77} & e_{78} \\
0 & 0 & 0 & 0 & e_{85} & e_{86} & e_{87} & e_{88}
\end{bmatrix}_n
\begin{Bmatrix} X' \\ Y' \\ \alpha' \\ \beta' \\ V'_X \\ V'_Y \\ M'_Y \\ M'_X \end{Bmatrix}_n
\tag{7-72}
$$

where

$$e_{15} = e_{26} = \xi(l - \xi T \sin \gamma), \qquad e_{17} = e_{28} = e_{35} = e_{46}$$

$$= \xi(1 - \cos \gamma),$$

$$e_{16} = -e_{25} = \xi^2 T(1 - \cos \gamma) - \frac{l^2}{2T}, \qquad e_{18} = -e_{27} = e_{36}$$

$$= -e_{45} = \xi \sin \gamma - \frac{l}{T},$$

$$e_{37} = e_{48} = \frac{1}{T} \sin \gamma, \qquad\qquad\qquad e_{75} = e_{86} = \xi T \sin \gamma,$$

$$e_{38} = -e_{47} = \frac{1}{T} (\cos \gamma - 1), \qquad\qquad e_{76} = -e_{85} = \xi T(\cos \gamma - 1),$$

$$e_{77} = e_{88} = \cos \gamma \qquad\qquad\qquad e_{78} = -e_{87} = -\sin \gamma,$$

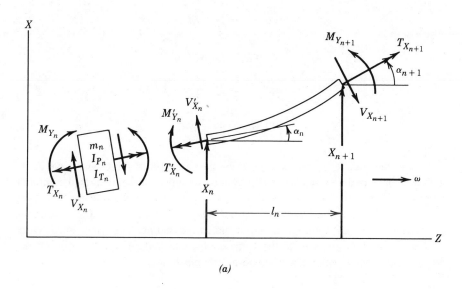

(a)

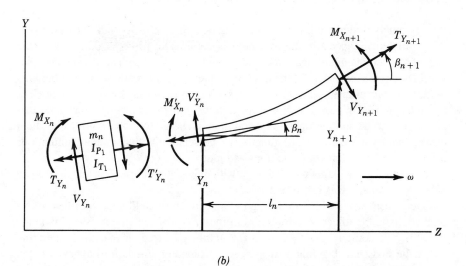

(b)

Figure 7.31. The nth rotor station with tangential torque included.

in which

$$\xi = \frac{EI}{T^2} \quad \text{and} \quad \gamma = \frac{Tl}{EI}.$$

For comparison with the transfer matrix $[T_{s_n}]$ of (7-14), L'Hospital's rule for the limits as $T \to 0$ give

$$\lim_{T \to 0} e_{15} = l_n^3/6EI_n,$$

$$\lim_{T \to 0} e_{17} = l_n^2/2EI_n,$$

$$\lim_{T \to 0} e_{75} = l_n,$$

$$\lim_{T \to 0} e_{77} = 1.$$

In deriving transfer matrices to describe the torquewhirl forces, we have come full circle back to the point of providing improved input for the linearized computer programs for eigenvalue analysis described earlier. For predicting thresholds of stability and the relative effects of various design modifications, these programs are currently the best tools available to the rotordynamics engineer. Much work remains to be done to provide the force coefficients required to describe all the destabilizing forces which act in modern turbomachinery.

THEORETICAL PREDICTIONS VERSUS FIELD OBSERVATIONS

It is extremely difficult at the present time to accurately predict rotordynamic instability problems, especially the threshold of stability, a priori. This is due to the lack of an accurate quantification of the destabilizing coefficients, of which torquewhirl is only one of many. In the case of torquewhirl, the assumption that the load torque remains aligned with the disk axis is probably not precisely accurate. It is also the author's opinion that destabilizing forces exist in centrifugal machines that have not yet been identified. This opinion is based on observations of field measurements and their lack of quantitative correlation with computer simulations using the currently known excitations as input.

A summary of the known excitations, with diagnostic information and the known effective "cures," was given in Table 7-1.

One area in which there is very good agreement of field data with theoretical predictions is the effect of rotor shaft stiffness. All of the cases of rotordynamic instability in compressors known to the author occurred in machines operating above the first shaft-bending critical speed. Computer simulations always predict that stiffening the shaft will raise the threshold speed of instability, and this has been verified by actual retrofit modifications to several different machines. It is interesting to note that a high ratio of shaft stiffness to bearing support stiffness also has a favorable effect on synchronous response problems (i.e., imbalance response) [20].

EXPERIMENTAL OBSERVATIONS FROM A SIMPLE TEST RIG

A simple test rig was set up in the Turbomachinery Laboratory at Texas A&M University to observe the effect of a working fluid, swirling in a housing, on the dynamics of an impeller with radial vanes. Figure 7.32 shows the impeller, supported vertically from a very flexible quill shaft. The shaft stiffness was kept low to produce a low critical speed and to allow the fluid dynamic effects on the impeller to predominate. The shaft is supported from ball bearings, so that there is no possibility of oil whip from fluid-film bearings as a destabilizing influence.

The impeller has been run both in the atmosphere, with no housing, and submerged in a working fluid (kerosene) contained in a cylindrical housing, open at the top. Variable speed is obtained with a DC gearmotor drive unit. The speed is measured with a proximity probe pulse tachometer and electronic digital counter. Figure 7.33 shows a time trace and frequency spectrum of the pendular vibrations

Figure 7.32. Vaned impeller in test rig.

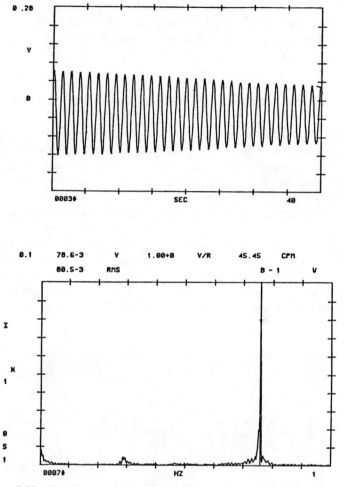

Figure 7.33. Time trace and frequency spectrum of test rig impeller vibrations.

(nonrotating) of the impeller suspended in the atmosphere. The frequency spectrum shows the natural frequency in air to be 45 cpm, and the time trace shows the logarithmic decrement to be 0.0198, which corresponds to 0.315 percent of critical damping, a very lightly damped system. Gyroscopic effects from rotation cause the actual critical speed to be slightly higher than 45 cpm. The damping comes from atmospheric drag and from internal hysteresis in the shaft assembly. The latter is known to be destabilizing on forward whirl, as described previously, if the internal damping is large enough relative to the external damping.

Rotating tests showed the impeller to be stable in the atmosphere up to a speed of 112 rpm, or 2.5 times the critical speed. Above 112 rpm, the impeller executes backward whirl with a small triangular orbit that neither grows larger nor dies out (log dec = 0). At no speed below 500 rpm is the impeller unstable in forward whirl in the atmosphere, thus showing internal hysteresis to be a relatively insignificant destabilizing influence in this test rotor.

Rotating tests with the impeller submerged in kerosene produced a critical speed of 40 rpm. The reduction in critical speed is caused by the virtual mass of the working fluid [45]. At supercritical speeds, the impeller is stable from 45 to 150 rpm (up to 3.75 times the critical speed). Above 150 rpm, the impeller becomes unstable in forward subsynchronous whirl, with an orbit that grows with time. At speeds between 314 and 360 rpm, the impeller executes either forward or backward nonsynchronous whirl, depending on whether a perturbation is supplied. In this speed range, the orbits reach a constant amplitude and do not grow further. Above 500 rpm, the impeller becomes unstable in backward whirl, with an orbit that grows with time.

Subsequent tests in the atmosphere, with a housing installed around the impeller, showed that the mere presence of the housing is sufficient to generate the instability with air as the working fluid. The threshold speeds observed with air were significantly higher than those with kerosene.

Thus, it is seen that the working fluid has a profound influence on the dynamics of a radial-vaned impeller, but it would be incorrect to imply much more than this from such a simple apparatus. It must be recognized that a centrifugal compressor develops much higher pressures around the impeller and has a diffuser.

However, the tests described point inescapably to the probability that the working fluid in a centrifugal compressor exerts destabilizing forces on its impellers. These forces must be considered, in addition to journal bearing effects and internal hysteresis, if accurate predictions of instability thresholds are to be made. An indirect effect from the fluid is torquewhirl, already discussed.

It is interesting to attempt a classification of the types of fluid forces that can act on impellers. For example, consider a shrouded impeller with a small axial clearance to the nonrotating diaphragm. This amounts to a very large "thrust bearing," which can generate unbalanced viscous shears on the impeller disk, or even circumferential pressure variations from the Reynolds' effect.

Similarly, a small radial clearance space around the radius of a thick rotating shroud could act as a large "journal bearing" and produce cross-coupled coefficients acting on the impeller.

Interestingly, it is neither the "thrust bearing" nor the "journal bearing" effect that produced the nonsynchronous whirling observed in the simple test rig. The clearances were too large to produce significant film pressures[7] (1.27 cm, or 0.5 in). The phenomena in the test rig (and in some of the unstable centrifugal machines) must be produced by more complex interactions between the fluid and the impeller.

The radial and tangential components of the hydrodynamic force on a centrifugal impeller have been measured and analyzed by researchers [46, 47] at the California Institute of Technology. Results of this work to date have not revealed

---

[7]Although the unbalanced viscous shear forces probably are significant.

destabilizing forces of the magnitude required to explain the instabilities observed in the field.

Childs [48, 49] has analyzed the hydrodynamic force generated by pressures in the clearance behind impeller shrouds. His initial results show that the destabilizing force peaks in a narrow speed range and is difficult to characterize by a cross-coupled stiffness due to the nonlinear dependence on speed.

REFERENCES

1. Gunter, E. J., Jr., *Dynamic Stability of Rotor–Bearing Systems*, NASA SP-113, pp. 14–16 (1966).
2. Newkirk, B. L., and Taylor, H. D., "Shaft Whipping Due to Oil Action in Journal Bearings," *General Electric Review*, **28**, 559–568 (1925).
3. Childs, D. W., "The Space Shuttle Main Engine High-Pressure Fuel Turbopump Rotordynamic Instability Problem," ASME Paper No. 77-GT-49, presented at the Gas Turbine Conference, Philadelphia, March 27–31, 1977.
4. Pinkus, O., and Sternlicht, B., *Theory of Hydrodynamic Lubrication*, McGraw-Hill, New York, 1961.
5. Kirk, R. G., and Miller, W. H., "The Influence of High Pressure Oil Seals on Turbo-Rotor Stability," *ASLE Transactions*, **22**(1), 14–24 (1979).
6. Ehrich, F. F., "Shaft Whirl Induced by Rotor Internal Damping," *Journal of Applied Mechanics*, pp. 279–282 (June 1964).
7. Alford, J. S., "Protecting Turbomachinery from Self-Excited Rotor Whirl," *Journal of Engineering for Power*, pp. 333–344 (October 1965).
8. Ehrich, F. F., "The Influence of Trapped Fluids on High Speed Rotor Vibration," *Journal of Engineering for Industry*, pp. 806–812 (November 1967).
9. Ehrich, F. F., "The Dynamic Stability of Rotor/Stator Radial Rubs in Rotating Machinery," *Journal of Engineering for Industry*, pp. 1025–1028 (November 1969).
10. Benckert, H., and Wachter, J., "Flow-Induced Spring Coefficients of Labyrinth Seals for Application in Rotordynamics," *Proceedings of the Workshop on Rotordynamic Instability Problems in High-Performance Turbomachinery, Texas A&M University, May 12–14, 1980*, NASA CP-2133, pp. 189–212.
11. Vance, J. M., "Torquewhirl: A Theory to Explain Nonsynchronous Whirling Failures of Rotors with High Load Torque," *Journal of Engineering for Power*, pp. 235–240 (April 1978).
12. Vance, J. M., "Rotordynamic Instability in Centrifugal Compressors: Are all the Excitations Understood?" *Journal of Engineering for Power*, pp. 288–293 (April 1981).
13. Ehrich, F. F., "An Aeroelastic Whirl Phenomenon in Turbomachinery Rotors," ASME Paper No. 73-DET-97, presented at the Design Engineering Technical Conference, Cincinatti, September 9–12, 1973.
14. Ehrich, F. F., "Identification and Avoidance of Instabilities and Self-Excited Vibrations in Rotating Machinery," ASME Paper No. 72-DE-21, presented at the Design Engineering Conference and Show, Chicago, May 8–11, 1972.
15. Wachel, J. C., "Nonsynchronous Instability of Centrifugal Compressors," ASME

Paper No. 75-Pet-22, presented at the Petroleum Mechanical Engineering Conference, Tulsa, OK, September 21–25, 1975.

16. Fowlie, D. W., and Miles, D. D., "Vibration Problems with High Pressure Centrifugal Compressors," ASME Paper No. 75-Pet-28, presented at the Petroleum Mechanical Engineering Conference, Tulsa, OK, September 21–25, 1975.

17. Dimentburg, F. M., *Flexural Vibrations of Rotating Shafts*, Butterworths, London, 1961, pp. 42–50 (USSR translation).

18. Macchia, D., "Accelerating of an Unbalanced Rotor Through the Critical Speed," ASME Paper No. 63-WA-9, presented at the Winter Annual Meeting, Philadelphia, November 17–22, 1963.

19. Childs, D. W., Dressman, J. B., and Childs, S. B., "Testing of Turbulent Seals for Rotordynamic Coefficients," *Proceedings of the Workshop on Rotordynamic Instability Problems in High-Performance Turbomachinery, Texas A&M University, May 12–14, 1980*, NASA CP-2133, pp. 121–138.

20. Barrett, L. E., Gunter, E. J., and Allaire, P. E., "Optimum Bearing and Support Damping for Unbalance Response and Stability of Rotating Machinery," *Journal of Engineering for Power*, pp. 89–94, (January 1978).

21. Tison, J. D., *Dynamic Stability Analysis of Overhung Rotors with High Load Torque*, Thesis for Master of Engineering, University of Florida (1977), pp. 4–10.

22. Pipes, L. A., *Applied Mathematics for Engineers and Physicists*, 2nd ed., McGraw-Hill, New York, 1958, pp. 239–242.

23. Taylor, H. D., "Critical Speed Behavior of Unsymmetrical Shafts," *Journal of Applied Mechanics*, pp. A-71–A79 (June 1940).

24. Childs, D. W., "Fractional Frequency Rotor Motion Due to Nonsymmetric Clearance Effects," ASME Paper No. 81-GT-145, presented at the Gas Turbine Conference, Houston, March 9–12, 1981.

25. Foote, W. R., Poritsky, H., and Slade, J. J., Jr., "Critical Speeds of a Rotor with Unequal Shaft Flexibility," *Journal of Applied Mechanics*, pp. A-77–A-84 (June 1943).

26. Yamamoto, T., and Ōta, H., "On the Unstable Vibrations of a Shaft Carrying an Unsymmetrical Rotor," *Journal of Applied Mechanics*, pp. 515–522 (September 1964).

27. Eshleman, R. L., and Eubanks, R. A., "Effects of Axial Torque on Rotor Response: An Experimental Investigation," ASME Paper No. 70-WA/DE-14, presented at the Winter Annual Meeting, New York, November 29–December 3, 1970.

28. Lund, J. W., "Stability and Damped Critical Speeds of a Flexible Rotor in Fluid-Film Bearings," *Journal of Engineering for Industry*, **96**(2), 509–517 (1974).

29. Vance, J.M., and Laudadio, F.J., "Experimental Measurement of Alford's Force in Axial Flow Turbomachinery", *Journal of Engineering for Gas Turbines and Power*, **106** (3), 585–590 (July 1984).

30. Kirk, R. G., and Gunter, E. J., "Transient Response of Rotor–Bearing Systems," *Journal of Engineering for Industry*, Ser. B, **96**(2), 682–693 (1974).

31. Murphy, B. T., and Vance, J. M., "An Improved Method for Calculating Critical Speeds and Rotordynamic Stability of Turbomachinery," *Journal of Engineering for Power*, **105**(3), 591–595 (July 1983).

32. Forsyth, G. E., Malcolm, M. A., and Moler, C. B., *Computer Methods for Mathematical Computations*, Prentice-Hall, Englewood Cliffs, NJ, 1977, pp. 17–19.

33. Barrett, L. E., Gunter, E. J., and Allaire, P. E., "The Stability of Rotor–Bearing Systems Using Linear Transfer Functions," Report No. UVA/464761/ME76/133, School of Engineering and Applied Science, University of Virginia, December 1976.

34. Wachel, J. C., "Rotordynamic Instability Field Problems," *Proceedings of the 2nd Workshop on Rotordynamic Instability Problems in High-Performance Turbomachinery, Texas A&M University, May 10–12, 1982*, NASA CP-2250, pp. 1–19.

35. Lund, J. W., and Saibel, E., "Oil Whip Whirl Orbits of a Rotor in Sleeve Bearings," *Journal of Engineering for Industry*, pp. 813–823 (November 1967).

36. Tondl, A., *Some Problems of Rotor Dynamics*, Publishing House of the Czechoslovak Academy of Sciences, Prague, 1965.

37. Tison [21], pp. 28–42.

38. Vance, J. M., and Yim, K. B., "Experimental Verification of Torquewhirl—The Destabilizing Influence of Tangential Torque," *Rotating Machinery Dynamics—Volume One*, Proceedings of the 1987 ASME Conference on Mechanical Vibration and Noise, Sept. 27–30, Boston, Mass., pp. 11–17.

39. Bousso, D., "A Stability Criterion for Rotating Shafts," *Israel Journal of Technology*, **10**(6), 409–423 (1972).

40. Goldstein, H., *Classical Mechanics*, Addison-Wesley, Reading, MA, 1965, pp. 107–109.

41. Vance, J. M., and Sitchin, A., "Derivation of First-Order Difference Equations for Dynamical Systems by Direct Application of Hamilton's Principle," *Journal of Applied Mechanics*, pp. 276–278 (June 1970).

42. Vance, J. M., and Sitchin, A., "Numerical Solution of Dynamical Systems by Direct Application of Hamilton's Principle," *International Journal for Numerical Methods in Engineering*, **4**, 207–216.

43. Nelson, C. C., "A Torquewhirl Analysis of the Space Shuttle Main Engine High Pressure Turbopumps," ASME Paper No. 79-DET-76, presented at the 7th Vibrations Conference, St. Louis, September 9–12, 1979.

44. Yim, K. B., *Load-Induced Rotordynamic Instability in Turbomachinery*, Ph.D. Dissertation in Mechanical Engineering, Texas A&M University (December 1984).

45. Walston, W. H., Ames, W. F., and Clark, L. G., "Dynamic Stability of Rotating Shafts in Viscous Fluids," *Journal of Applied Mechanics*, pp. 291–299 (June 1964).

46. Brennen, C. E., Acosta, A. J., and Caughey, T. K., "A Test Program to Measure Fluid Mechanical Whirl-Excitation Forces in Centrifugal Pumps," *Proceedings of the Workshop on Rotordynamic Instability Problems in High-Performance Turbomachinery, Texas A&M University, May 12–14, 1980*, NASA CP-2133, pp. 229–235.

47. Adkins, D. R., and Brennen, C. E., "Origins of Hydrodynamic Forces on Centrifugal Pump Impellers," *Proceedings of the Workshop on Rotordynamic Instability Problems in High-Performance Turbomachinery, Texas A&M University, June 2–4, 1986*, NASA CP-2443, pp. 467–491.

48. Childs, D. W., "Force and Moment Rotordynamic Coefficients for Pump-Impeller Shroud Surfaces," *Proceedings of the Workshop on Rotordynamic Instability Problems*

in High-Performance Turbomachinery, Texas A&M University, June 2–4, 1986, NASA CP-2443, pp. 503–529.

49. Childs, D. W., ''Fluid-Structure Interaction Forces at Pump-Impeller-Shroud Surfaces for Rotordynamic Calculations,'' *Rotating Machinery Dynamics—Volume Two*, Proceedings of the 1987 ASME Conference on Mechanical Vibration and Noise, Sept. 27–30, Boston, Mass., pp. 581–594.

Chapter VIII

Measurements

This chapter will describe some of the instrumentation and measurement techniques which make it possible to verify and improve the predictions of rotordynamic analysis and which now make turbomachinery troubleshooting more of a science than an art. Significant portions of the text, and a number of the figures, are drawn from Ref. [1].[1]

MEASUREMENT OBJECTIVES

Rotordynamic measurements will usually be made with one or two of the following objectives in mind. They will often determine the type, quantity, and required quality of the data to be obtained.

1. *Developmental Testing.* In the development of new machines, measurements on test rigs and prototypes will determine if the design requirements have been met and if the machine can be expected to be reliable. For example, critical speeds in the operating speed range may need to be modified or damped. The instrumentation used here is often a "one-shot" project, without a requirement for long life.

2. *Commissioning.* New machines in process or petrochemical plants are commissioned when delivered by the manufacturer to the user. A test program is designed to satisfy the user that the machine meets all specifications. For rotor-

[1]Permission to use this material by the Hewlett-Packard Co. is gratefully acknowledged.

dynamics, the important specifications are API 617 and 670 [2] or equivalent, as well as specifications added by the user at the time of purchase.

3. *Diagnostics.* Rotordynamic measurements can be used to identify the cause of machine failures or malfunctions. Unless the machine has permanent instrumentation installed, the primary requirements for this application are portability and ease of installation. Effective troubleshooting often requires intelligent comparisons of measured data with computer simulations.

4. *Maintenance Programs.* It has been demonstrated that a well-designed program of machinery maintenance, using computer storage and analysis of rotordynamic data taken continually or periodically, can effect large savings in plant operation costs. The instrumentation requirements are unique to this application: long life, day-to-day repeatability, and computerized data acquisition, storage, and analysis. Computer simulations may also be used to aid in fault diagnosis, but this has not been done to a great extent to date. Most of the maintenance programs rely heavily on recorded operating histories, stored over a period of time and used to compare with current measurements. These programs work best with relatively simple turbomachinery, such as small compressors driven by electric motors. More development is needed for successful application to large complex machinery such as that of an electric utility power plant.

INSTRUMENTATION

Transducers convert the measured quantity into a voltage, which can be displayed on an oscilloscope, recorded, plotted, or analyzed by a computer. The voltage output of most transducers is an "analog signal," i.e., it is a time-varying continuous voltage analogous to the quantity being measured. Only a scale factor ("calibration constant") is required to determine the magnitude of the measured quantity at any time.

Although, strain, force, pressure, or temperature occasionally need to be measured for rotordynamic evaluation, the two most commonly measured variables are vibration and rotor speed. Commercially available transducers can be used to measure vibratory displacement, velocity, or acceleration (Fig. 8.1). The three signals usually look quite different, and the best one to use depends on the application. The important factors influencing the decision are the expected range of frequencies and the mechanical mobility (inverse of impedance) of the rotor or structure where the measurement is to be made.

Most vibration measurements in turbomachinery are made at the bearings, and the mechanical mobility determines the relative amplitudes of motion between the shaft, the bearing, and the bearing housing or foundation. Generally, measurements should be made where amplitudes are largest, and if relative motion occurs it should be measured, along with the absolute motion of at least one of the participants.

The velocity and acceleration parameters of the vibration are offset in phase

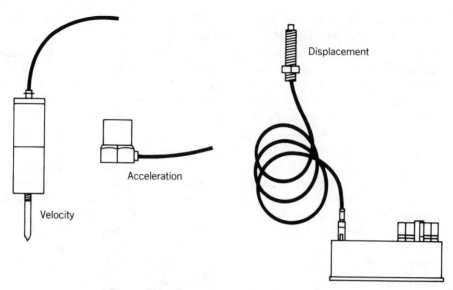

Figure 8.1. Three types of vibration transducer.

relative to displacement—an important consideration when one is using phase for analysis. Phase relationships are shown in Fig. 8.2.

Velocity, for example, is offset from displacement by 90° (one complete cycle is 360°). At point *B*, when the displacement is maximum, the velocity is zero. At point C, when displacement is zero, velocity is maximum. Following the same reasoning, acceleration can be shown to be offset 90° from velocity and thus 180° from displacement.

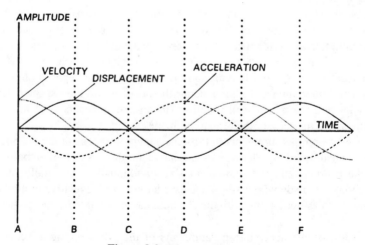

Figure 8.2. Phase relationships.

The amplitude of the vibration parameters also varies with frequency—an important consideration in transducer selection. Velocity increases in direct proportion to frequency, while acceleration increases with the square of frequency. This variation with frequency, and the phase relationships shown in Fig. 8.2, are illustrated in the equations below. In these equations, which apply only to vibration of a single frequency $\omega = 2\pi f$, A is the vibration amplitude and f is the frequency in hertz:

$$\text{displacement} = A \sin (2\pi ft), \tag{8-1}$$

$$\text{velocity} = 2\pi f A \cos (2\pi ft), \tag{8-2}$$

$$\text{acceleration} = -(2\pi f)^2 A \sin (2\pi ft). \tag{8-3}$$

The three vibration parameters are thus closely related and, in fact, can be derived from each other by a signal analyzer. However, the variation in vibration amplitude with frequency, as well as transducer limitations, often mean that only one of the parameters will supply the information necessary for analysis.

The impact of variation in amplitude with frequency is illustrated in Fig. 8.3. In this example, potentially dangerous vibration levels are present in a low speed fan and a high speed gearbox. The two items to note are (1) displacement and acceleration levels differ widely, and (2) velocity is relatively constant.

From the first, we can conclude that frequency considerations are important in selecting a vibration transducer. Acceleration is not a good choice for very low frequency measurements, while displacement does not work well for high frequencies. Note that these are limitations of the vibration parameter, not the transducer itself. Frequency range limitations of specific types of transducers are also an important consideration, and specification sheets should be consulted.

The fact that velocity is a good indicator of damage, independent of machine speed, implies that it is a good parameter for general monitoring work. That is, a

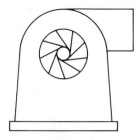

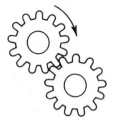

Case 1: 600 rpm fan

Displacement: 10 mils p-p
Velocity: 0.3 in/sec
Acceleration: 0.1 g

Case 2: 15kHz gear mesh

Displacement: 1.2 mils p-p
Velocity: 0.12 in/sec
Acceleration: 30 g's

Figure 8.3. Frequency considerations.

vibration limit can be set independent of frequency. (Velocity tends to remain constant with damage level because it is proportional to the energy content of the vibration.) However, the upper frequency limitation of velocity transducers can be a problem for gear and high speed blade analysis, and velocity transducers cannot be attached to a rotating shaft.

Displacement Transducers

Noncontacting displacement transducers (also known as proximity probes) are used to measure relative shaft motion directly. A high frequency oscillator is used to set up eddy currents in the shaft without actually touching it. As the shaft moves relative to the sensor, the eddy current energy changes, modulating the oscillator voltage. This signal is demodulated, providing an output signal proportional to displacement. This is illustrated in Fig. 8.4, and a typical application with two probes is shown in Fig. 8.5.

Key characteristics of displacement transducers are as follows:

1. Displacement transducers measure relative motion between the shaft and the mount, which is usually the machine housing. Thus, vibration of a stiff shaft/ bearing combination that moves together is difficult to measure with a displacement transducer alone.

2. Signal conditioning is included in the electronics. Typical outputs are on the order of 200 mV/mil or 8 mV/μm [1 mil is 0.001 in; 1 μm (micrometer) is 0.001 mm].

3. Shaft surface scratches, out-of-roundness, and variation in electrical properties all produce a signal error. Surface treatment and run-out subtraction can be used to solve these problems.

4. Installation is sometimes difficult, often requiring that a hole be drilled in the machine.

5. The output voltage contains a DC offset of approximately -8 V, requiring the use of AC coupling or a bucking voltage for sensitive measurements; AC coupling is a feature of most signal analyzers and simply means that an input capacitor is used to block DC. The practical disadvantage of AC coupling is reduced instrument response below 1 Hz (60 rpm).

Noncontacting displacement probes are used on virtually all industrial turbomachinery because their flexible bearings and heavy housings result in small response at external locations. They are required by American Petroleum Institute (API) specifications in petrochemical plants. Some gas turbines, especially those in aircraft engines, use relatively stiff rolling element bearings and can thus use housing-mounted transducers (velocity and acceleration) effectively, but measurements of shaft motion are still useful.

Displacement transducers are also commonly used as tachometer signals and as a phase reference, by detecting the passage of a keyway ("keyphasor").

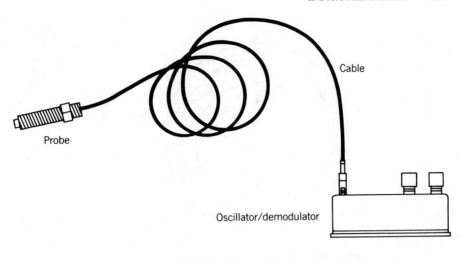

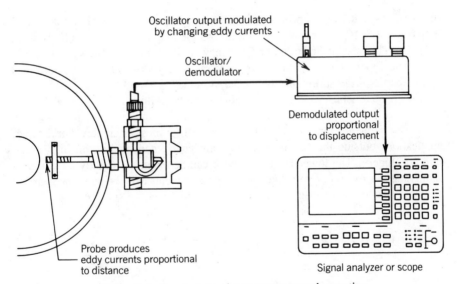

Figure 8.4. Proximity probe components and operation.

Velocity Transducers

Velocity transducers were the first vibration transducer, and virtually all early work in vibration severity was done using velocity criteria. Velocity transducer construction is shown in Fig. 8.6. The vibrating coil moving through the field of the magnet produces a relatively large output voltage that does not require signal conditioning. The size of the voltage is directly proportional to the velocity of the vibration. As

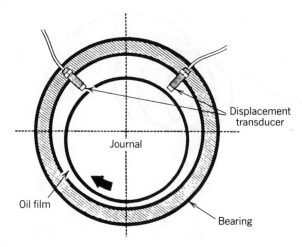

Figure 8.5. Typical application of proximity probes.

shown in Fig. 8.7, the spring–mass–damper system is designed for a natural frequency of 8–10 Hz or lower, which allows the magnet to stay essentially fixed in space. This establishes a lower frequency limit of approximately 10 Hz (600 rpm). The upper frequency limit within the range of 1000–2000 Hz is determined by the inertia of the spring–mass–damper system.

Key characteristics of velocity transducers are as follows:

1. The frequency range of 10–1000 Hz is ideal for most machinery work. Major applications outside this frequency range are gears, and blading on high speed turbomachinery. The lower frequency limit can be extended slightly by compensating for the roll-off. (Note that phase measurements will be in error near the resonant frequency because of the 180° phase shift.)

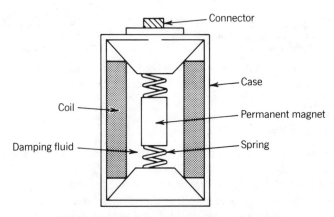

Figure 8.6. Velocity transducer construction.

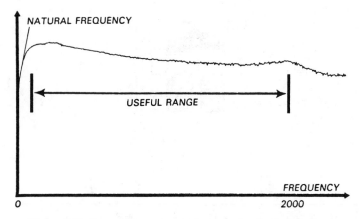

Figure 8.7. Frequency response of a typical velocity transducer.

2. Installation of velocity transducers is relatively noncritical, and extension probes and magnetic mounts work well. In addition, no signal conditioning is required.

3. Because they are electromechanical devices with moving parts, velocity transducers can change calibration over time and wear out. High temperature operation can also cause changes in calibration.

Velocity transducers were once the standard for vibration monitoring work on machines other than turbomachinery, and they are especially good for hand-held measurements. Their popularity has declined somewhat because accelerometers are typically more rugged and offer wider frequency response.

Accelerometers

Accelerometers are a popular transducer for general vibration analysis because they are accurate and rugged, and available for a wide range of applications. Construction of a simple accelerometer is shown in Fig. 8.8. The vibrating mass applies a force on the piezoelectric crystal that produces a charge proportional to the force (and thus to acceleration). A charge amplifier converts the charge to voltage.

The frequency response of a typical accelerometer is shown in Fig. 8.9. Note that the natural frequency is above the operating range (it is below the operating range in velocity transducers). Operation should be limited to about 30 percent of the natural frequency.

Accelerometer sensitivity is largely dependent on the size of the mass, with a larger mass producing more output. High output is especially important for increasing the usability of accelerometers at low frequencies. However, it should be noted that natural frequency decreases as mass increases. Thus increased sensitivity tends to move the operating range down in frequency.

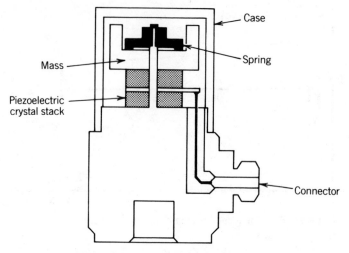

Figure 8.8. Accelerometer construction.

Accelerometer output is a low level, high impedance signal that requires signal conditioning. The traditional method has been to use a charge amplifier, as mentioned above and shown in Fig. 8.10. However, accelerometers are now available with built-in signal conditioning electronics that require only simple current-source supply. These accelerometers, sometimes referred to as ICP (for integrated circuit piezoelectric), can be directly connected to a compatible signal analyzer. Another advantage of the ICP accelerometer is that expensive low-noise cable required to connect traditional accelerometers to the charge amplifier is not required. This is especially important when long cables are involved.

The following are key characteristics of accelerometers:

1. Accelerometers offer the broadest frequency coverage of the three transducer

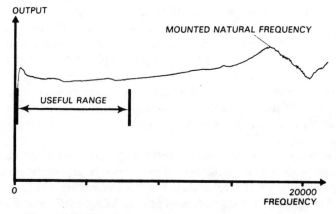

Figure 8.9. Frequency response of a typical accelerometer.

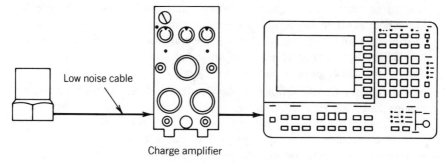

Figure 8.10. Charge amplifier installation.

types. Their weakness is at low frequency, where low levels of acceleration result in small output voltages. Their large output at high frequencies also tends to obscure lower frequency signals when overall level is measured (this is not a problem with the wide dynamic range of modern high-quality signal analyzers).

2. Accelerometers require signal conditioning; however, as noted above, the development of the ICP accelerometer has eliminated this disadvantage.

3. The low frequency response of piezoelectric accelerometers is limited to approximately 1 or 2 Hz. Piezoresistive and cantilever-beam accelerometers are available that respond down to DC, although it is worth remembering that acceleration level is low at very low frequencies.

4. Accelerometers are very sensitive to mounting, and should never be hand-held. They should be securely attached with a threaded stud, high strength magnet, or industrial adhesive. The mounting surface should be flat and smooth—preferably machined.

The following steps and considerations will be helpful in choosing the best transducer for a particular application:

Step 1: Determine the Variable of Interest. If you are interested in monitoring a critical clearance or relative displacement, the only choice is a displacement transducer. Although acceleration and velocity can be converted to displacement, it will be an absolute measurement rather than the relative measurement given by a displacement transducer. Also, if the relative vibration of a rotor or shaft is to be measured, a proximity probe must be used. If the variable is a quantity other than a clearance or relative displacement, and if the measurement is not on a rotating shaft, go on to the next step.

Step 2: Mechanical Impedance Considerations. If the rotor vibration is not well transmitted to the machine housing (which will be the case for most well-designed industrial turbomachinery), you must use a displacement transducer to measure the shaft directly. If the shaft is not accessible (as an internal shaft in a gearbox) or if the bearing is very stiff, you should use a casing mounted velocity or acceleration transducer. In borderline cases, it may be appropriate to use both absolute and relative motion transducers.

If Steps 1 and 2 indicate a displacement transducer, it is the one that will provide the best results. If a housing-mounted acceleration or velocity transducer is indicated, go on to Step 3.

Step 3: Frequency Considerations. If the frequency of the expected vibration is greater than 1000 Hz, you must use an accelerometer. If the vibration will be in the range of 10–1000 Hz, either velocity or acceleration transducers can be used. The vibration nomograph of Fig. 8.11 can be used to determine whether an accelerometer will produce sufficient output.

In general, use a velocity transducer if (a) overall level is being measured to detect defects; (b) the transducer will be hand-held; and (c) the machinery being analyzed is low speed (i.e., less than 1200 rpm).

Use an accelerometer if (a) high frequency (above 1000 Hz) blading or gear defects are being analyzed; (b) structural response is being measured; (c) long transducer life (more than 2 years) is required; and (d) high temperatures are encountered (although high temperature velocity transducers are available). Note that the operating temperature range of ICP accelerometers is usually limited to 250°F (120°C).

After the type of transducer has been determined, consult transducer manufacturers for specific model recommendations.

After the transducer has been selected, it must be properly installed for best results. Table 8-1 shows some suggested transducer locations for several types of machines. In general, the number of transducers used on a machine combination is determined by how critical the machine is to the process and how expensive it is to repair or replace. Table 8-1 is intended to show typical applications and should be used only as an introductory guide.

Proper mounting of the transducer to the machine is also critical, especially with displacement and acceleration transducers, and the manufacturer's recom-

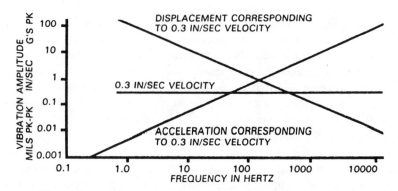

Figure 8.11. Vibration nomograph for amplitude versus frequency.

TABLE 8-1. Suggested Transducer Locations

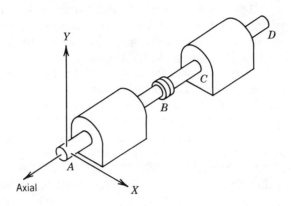

Machine Description	Transducer	Location
Steam turbine/large pump or compressor with fluid film bearings	Displacement	Radial horizontal and vertical at A, B, C, D. Redundant axial at A and D.
Gas turbine or medium size pump	Displacement	Radial horizontal and vertical at A and B.
	Velocity	Radial horizontal or vertical at A and B.
Motor/fan, both with fluid-film bearings.	Displacement or Velocity	One radial at each bearing. One axial displacement to detect thrust wear.
Motor/pump or compressor with rolling element bearings.	Velocity or Acceleration	One radial at each bearing. One axial, usually on motor, to detect thrust wear.
Gearbox with rolling element bearings.	Acceleration	Transducers mounted as close to each bearing as possible.
Gearbox shafts with fluid film bearings.	Displacement	Radial horizontal and vertical at each bearing. Axial to detect thrust wear.

mendations should be followed closely. *One particular caution:* the transducer should never be mounted to a sheet metal cover, since local resonances may easily be in the operating speed range.

For measurement of rotor critical speeds and mode shapes, computer predictions can show the best axial locations to mount proximity probes. Obviously, nodal points should be avoided. Usually, practical considerations dictate the location of proximity probes in bearing housings or where the shaft is exposed. Very hot locations must be avoided.

SIGNAL ANALYSIS AND RECORDING

The signal obtained from a machinery vibration transducer is a complex combination of responses to multiple internal (and sometimes external) forces. The key to effective analysis is to reduce this complex signal to individual components, each of which can then be correlated with its source.

Two perspectives are available for analyzing vibration signals: (1) the time domain view of vibration amplitude versus time and (2) the frequency domain view of vibration amplitude versus frequency. While the time domain provides insight into the physical nature of the vibration, the frequency domain is ideally suited to identifying its components. The advantage of modern signal analyzers for machinery analysis is their ability to work in both domains.

A time domain presentation of the vibration signal shows how amplitude varies with time. The time domain display of Fig. 8.12 shows how vibration due to an unbalanced rotor varies with time, using a displacement transducer and assuming the speed is well below the critical speed. The amplitude of the signal is proportional to the amount of imbalance [see equation (1-4)], and the cycle repeats once per revolution. This signal is easy to analyze because it is an idealized example with a single source of vibration; real world vibration signals are much more complex.

When more than one vibration component is present, analysis in the time domain becomes more difficult. This situation is illustrated in Fig. 8.13, where two sine wave frequencies are present. The result of this combination is a time domain display in which the individual components are difficult to derive.

The time domain is a perspective that feels natural and provides physical insight into the vibration. It is especially useful in analyzing impulsive signals from bearing and gear defects, and truncated signals from looseness. The time domain is also

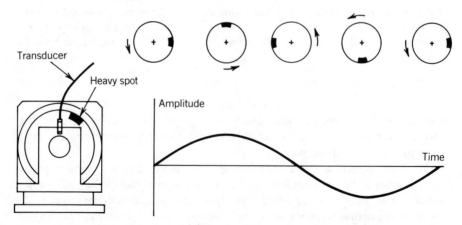

Figure 8.12. A single-frequency time domain signal caused by imbalance.

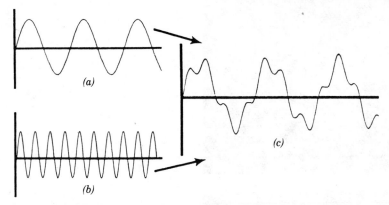

Figure 8.13. A complex time domain signal produced by two sine waves added together.

useful for analyzing vibration phase relationships. For time domain displays, the oscilloscope is the traditional instrument. Modern digital scopes allow storage of the waveforms.

The oscilloscope also facilitates a display of the orbit being executed by the rotor or bearing journal. This requires two transducers at the same axial location, spaced at 90° circumferentially. One signal is connected to the vertical amplifier of the scope, the other to the horizontal amplifier (the time sweep must be turned off). Both amplifiers should be set to the same gain. The orbits are also called Lissajous patterns, and as one gains experience they can become quite useful as diagnostic guides. For example, Fig. 8.14(a) shows an orbit produced by synchronous whirl (response to imbalance), and Fig. 8.14(b) shows an orbit with both synchronous and subsynchronous frequencies in the signal (the rotor is executing a whirling motion at both frequencies simultaneously). With additional frequencies added, the Lissajous patterns can become complex and difficult to analyze.

In the time domain, individual components of complex signals are difficult to determine, whether viewed as orbits or with a time sweep. A perspective that is better suited to analyzing these components is the frequency domain.

Figure 8.15(a) is a three dimensional graph of the signal shown in Fig. 8.13. Two of the axes are time and amplitude as seen in the time domain. The third axis is frequency, which allows the components of the waveform to be separated. When the graph is viewed along the frequency axis, the complex signal of Fig. 8.13 is seen. It is the summation of the two sine waves, which are no longer recognizable.

However, if we view the graph along the time axis as in Fig. 8.15(c), the frequency components area readily apparent. In this view of amplitude versus frequency, each frequency component appears as a vertical line. Its height represents its amplitude, and its position represents its frequency. This frequency domain representation of the signal is called the spectrum of the signal.

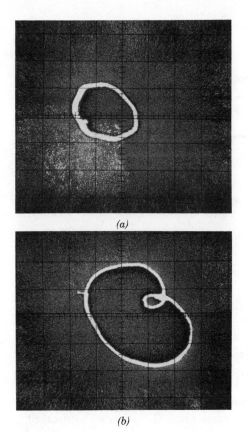

(a)

(b)

Figure 8.14. Oscilloscope photographs of (a) a synchronous rotor orbit (one frequency); (b) an orbit with both a synchronous and a subsynchronous frequency present.

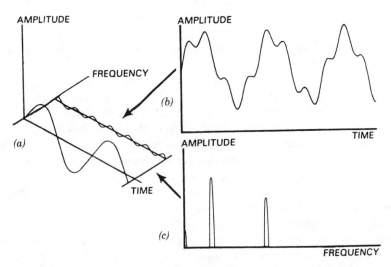

Figure 8.15. The relationship between time and frequency domain.

The power of the frequency domain lies in the fact that any real world signal can be generated by adding up sine waves of various frequencies. Thus, just as in the example we used to illustrate the frequency domain began as a summation of sine waves, we could perform a similar reduction back to sine wave components for any machinery vibration signal. It is important to understand that the frequency spectrum of a vibration signal completely defines the vibration—no information is lost by converting to the frequency domain (provided phase information is included).

Figure 8.16 shows the time and frequency domain of four signals that are common in machinery vibration.

a. The frequency spectrum of a pure sine wave is a single spectral line. For a sine wave of period T seconds, this line occurs at $1/T$ Hz.

b. A square wave, which is much like the truncated signal produced by mounting or bearing cap looseness, is made up of an infinite number of odd harmonics. Harmonics are components that occur at integer multiples of a fundamental frequency. In rotating machinery analysis, the fundamental frequency is shaft speed (although whirl instability is usually subsynchronous). Because square wave type signals from machinery are not ideal, their spectra often contain both odd and even harmonics.

c. Rolling-element bearing and gear defects usually produce impulsive signals that are typified by harmonics in the frequency domain. These harmonics are spaced at the repetition rate of the impulsive.

d. Finally, some signals are modulated by residual imbalance. The frequency spectrum of a modulated signal consists of the signal being modulated (the carrier), surrounded by sidebands spaced at the modulating frequency. (Modulation is described in the final section of this chapter on torsional vibration measurements.)

Spectral maps (also known as cascade plots or "waterfalls") usually consist of a series of vibration spectra measured at different speeds. A variety of other parameters, including time, load, and temperature, are also good third dimensions for maps. The most common method for mapping the variation in vibration with revolutions per minute is to measure successive spectra while the machine is coasting down or running up in speed. The map in Fig. 8.17 was made on a signal analyzer with this capability built in.

In addition to showing how vibration changes with speed, spectral maps quickly indicate which components are related to rotational speed. These components will move across the map as speed changes, while fixed frequency components move straight up the map. This feature is especially useful in recognizing rotordynamic instabilities that occur at fixed (or only slightly changing) frequencies.

The complete frequency domain representation of a signal consists of an amplitude spectrum and a phase spectrum. While the amplitude spectrum indicates signal level as a function of frequency, the phase spectrum shows the phase relation between spectral components. In machinery vibration analysis, phase is required

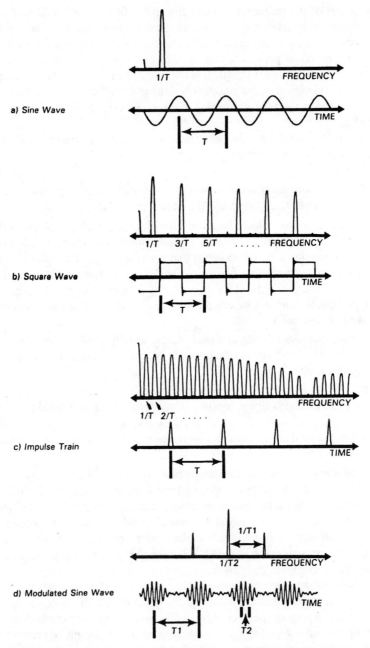

Figure 8.16. Examples of frequency spectra common in turbomachinery drivetrains.

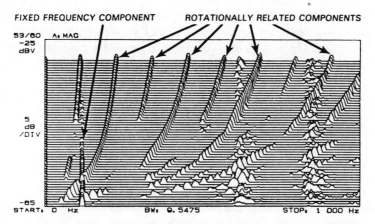

Figure 8.17. A spectral map, or waterfall display of frequency components changing with time.

for most balancing techniques (see Chapter V). It is also useful in differentiating between faults that produce similar amplitude spectra. Not all analyzers produce both amplitude and phase spectra, however.

When a signal analyzer with phase capability is not available, a synchronous tracking filter (to be described below) will give phase data and is useful for balancing.

A phase shift between two signals A and B is illustrated in Fig. 8.18. The signals are separated by one-quarter of a circle, or 90°. The phase of the trace A leads that of trace B because its peak occurs first.

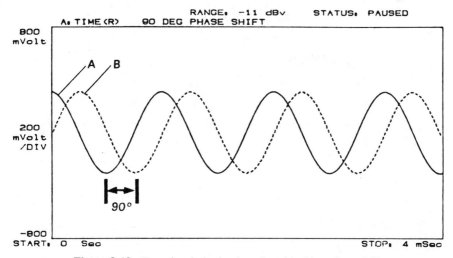

Figure 8.18. Two signals in the time domain with a phase shift.

In the frequency domain, each amplitude component has a corresponding phase with respect to a ''trigger'' signal or with respect to another vibration signal of the same frequency. Figure 8.19 is a spectrum display from a signal analyzer with phase capability indicating a 90° phase relationship between the frequency component and the trigger signal (amplitude is shown as a dashed line). The phase is −90° because the peak of the signal occurs after the trigger. In this case signal A is used as the trigger. In balancing, the voltage pulse from a keyphasor is used as a trigger.

Frequency Domain Analyzers

Instruments that display the frequency spectrum are generally referred to as spectrum analyzers, real time analyzers (RTA), or fast Fourier transform analyzers (FFT). However, only digital signal analyzers based on a microcomputer are capable of true RTA–FFT operation. Early analyzers, and some present-day inexpensive analyzers, use parallel or swept filters.

A simple block diagram of a parallel filter analyzer is shown in Fig. 8.20. These analyzers have several built-in filters that are usually spaced at $\frac{1}{3}$- or 1-octave intervals. This spacing results in resolution that is proportional to frequency. For a $\frac{1}{3}$-octave analyzer, resolution varies from around 20 Hz at low frequencies to several thousand hertz (kHz) at high frequency. A variation of the parallel filter analyzer that is sometimes used in machinery work has several filters that can be individually selected.

Parallel filter analyzers are usually relatively low in cost and battery operated, but their resolution is not nearly good enough for most machinery analysis. This is especially true for gear and bearing analysis, where frequency resolution of a

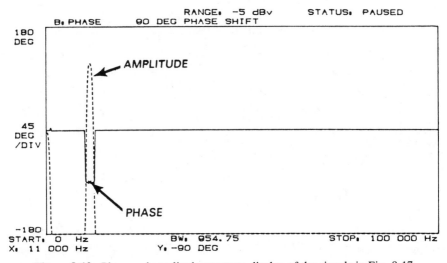

Figure 8.19. Phase and amplitude spectrum display of the signals in Fig. 8.17.

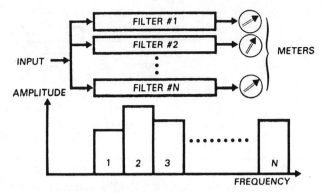

Figure 8.20. Frequency analysis with parallel filters.

few hertz is often required, and for sideband analysis of torsional vibration. Parallel filter analyzers are used mostly for acoustic measurements.

Swept filter analyzers use a tunable filter, much like a radio receiver. The block diagram for this type of analyzer is shown in Fig. 8.21. The frequency resolution of these instruments is on the order of 5 Hz—better than parallel filter analyzers, but not good enough for all machinery vibration analysis applications. Their operation is much slower than the parallel filter analyzers. This is because transient responses in the filter must be allowed to settle at each new frequency. This slow operation increases the time required to complete a vibration survey. Also, because of the time required to sweep through the frequency spectrum, swept-filter analyzers can miss short duration events.

Digital signal analyzers (Fig. 8.22) use digital techniques to effectively synthesize a large number of parallel filters. The large number of filters (typically 400) provides excellent resolution, and the fact that they are parallel means that measurements can be made quickly. Some digital signal analyzers also provide both time and phase spectrum displays and can be connected directly to computers

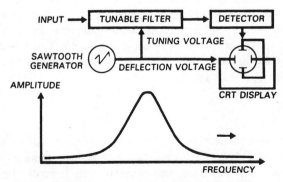

Figure 8.21. Frequency analysis with a tunable filter.

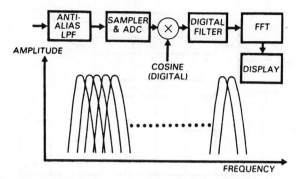

Figure 8.22. Frequency analysis with a digital signal analyzer.

for automated measurement. The time domain display capability allows the analyzer to double as an oscilloscope. Waterfall plots, such as Fig. 8.17, can be made only with a digital signal analyzer that incorporates the spectral mapping capability. Most dual-channel digital signal analyzers also have the capability to compute and display frequency-dependent transfer functions such as mechanical impedance or mobility. Digital instruments that incorporate all of these features are the best choice for analyzing turbomachinery measurements. They are expensive, but turbomachinery failures are usually much more expensive.

Synchronous Tracking Filters

For balancing, for measuring only synchronous vibration, and for plotting synchronous response to imbalance versus shaft speed, a synchronous tracking filter ("digital vector filter," or "trim balance analyzer") is useful. These instruments pass the vibration signal through a narrow band pass filter, which is automatically kept tuned to the frequency of shaft speed. Their operation is illustrated by Fig. 8.21 (although the filter may be a digital device, i.e., a microcomputer), where the tuning voltage is controlled by a tachometer signal (i.e., the band pass frequency is continuously determined by shaft speed) and the CRT display may be replaced by digital readouts. Typical inputs to the instrument are a once-per-revolution tachometer signal from a keyphasor, and one or two channels of vibration signal from proximity probes. Typical (digital) outputs are vibration amplitude (band pass filtered or unfiltered), phase lag of the vibration signal behind the keyphasor pulse, and shaft speed.

Recorders

Measurements on turbomachinery can be very expensive, and even more expensive to repeat. The "downtime" available to make measurements on operational machinery under controlled conditions is usually very limited. The probability of getting a number of channels of data analyzed on an RTA, with all the buttons and knobs for calibration and range set correctly, is not very high on the first try. For these reasons, an FM analog tape recorder is valuable for data storage and retrieval. The FM mode allows accurate recording of frequencies below 50 Hz. After recording data from a machine, it can be played back through the signal analyzer any number of times until the desired results are extracted properly.

Analog/Digital Data Acquisition

Digital signal analyzers employ analog/digital (A/D) data acquisition, but most do not provide for permanent storage of the data. Instruments are available that convert 16 or more channels of analog signal to digital form in almost real time. ''Almost real time'' means that the channels are multiplexed at rates up to 1 MHz. The digital data is stored on magnetic disks. This type of A/D capability is also available as plug-in cards for some types of personal computers.

The author has found this capability to be especially useful for measurements on laboratory test rigs where extensive computer analysis of the data can be performed as it is acquired.

SOME MEASUREMENT TECHNIQUES

For parameter identification to normalize computer models, for balancing, diagnostics, and troubleshooting, the author has found the following procedures to be useful. Many of them were developed in the experimental project at Shell Westhollow Laboratories described in Chapter IV [3]. The reader is encouraged to improvise so as to meet his/her own particular requirements.

Techniques of torsional vibration measurement are described in a separate section at the end of this chapter.

Static Stiffness Measurements

Measurement of the structural stiffness of bearing supports and, in the case of rolling-element bearings, the bearings themselves is generally a straightforward procedure that yields useful numerical values for computer modeling of rotor–bearing systems. The basic idea is to apply a known static load to the shaft or housing structure and measure the resulting deflection at the same point and in the same direction.

Proximity probes can be used to measure the deflection, but they should be mounted on a surface that does not deflect under the applied load, or else the measurement is a relative (not absolute) deflection. Dial indicators can also be used, with the same caveat.

Vertical loads are most easily applied directly with known weights. Horizontal loads can be applied by a cable taken around a pulley 90° and down to a hanging (known) weight, or the cable can be stretched out horizontally and attached to a structural member of the building. With the cable fairly taut, weights are hung at midspan. A load cell is incorporated in the cable near the point of attachment. The geometry can be adjusted until the cable is horizontal at the point of attachment (to the shaft or bearing housing).

Bump Tests

The natural frequencies of rotors, both "free–free" and mounted in rolling-element bearings, can be measured by striking the rotor with a soft hammer and putting the measured vibration signal into a FFT signal analyzer. RMS averaging of the signals from many rapidly repeated bumps helps to build up the spikes on the screen of the signal analyzer to easily recognizable amplitudes at the natural frequencies. Several hard and soft hammers should be tried until the best free vibration signal is obtained.

The rotor supports will affect the natural frequencies. Fluid-film bearings will have a radically different support stiffness when the rotor is at rest than when it is running at speed. The stiffness of rolling-element bearings will also change with speed, but to a much smaller extent.

The measured free–free natural frequencies of several rotors are reported in Chapter IV and compared with computer-predicted values. The free–free frequencies are an excellent way to test the accuracy of a computer model for a rotor, since the often uncertain bearing support values are not involved. To measure the free–free frequency, the rotor is hung from the ceiling by two long (at least one-third the rotor length) cables or ropes. The impulses are applied in a horizontal plane, normal to the rotor. The long vertical support rope provides negligible horizontal stiffness ($K = W/2l$, where W is the rotor weight and l is the cable length) at the point of attachment to the rotor. In the measurements reported in Chapter IV, both impact (using a rubber or plastic hammer) and random "white noise" excitation (using a shaker) were employed.

Random shaker and impact excitations were found to produce comparable results, but the impact method required more time and work on the part of the operator since a large number of averages were required. The only disadvantage of the random excitation was that a white noise generator was needed and a suitable shaker had to be attached to the rotor. For some applications, this disadvantage would be a major one.

Experimental data gathering and frequency analysis were greatly facilitated by the use of a Hewlett-Packard 5420A digital signal analyzer, with a 54470B digital filter and a 54410A analog/digital filter.

With the single exception of the 3-in uniform shaft, which was measured with proximity probes, the vibration was measured using B&K Type 4344 accelerometers amplified by B&K Type 2635 charge amplifiers. The accelerometers were very small, weighing only 2.7 g, and had a negligible effect on the natural frequencies of the rotors. The accelerometers were attached to the shaft with machine screws through holes in adjustable stainless steel hose clamps.

Calculated and measured mode shapes were plotted automatically on an HP-9872 plotter, controlled by an HP-9845 computer.

For natural frequency measurements, an accelerometer was mounted on one end of the rotor. (The rotor end is never a node in free–free vibration.) The vibration signal from the accelerometer charge amplifier was input to one channel of the spectrum analyzer, which was preset to the desired number of averages, frequency range of interest, type of excitation, etc. It was found expedient to first produce a spectrum over a broad frequency range, to identify the first four or five natural frequencies, and then to utilize the "zoom expansion" feature of the analyzer to measure each of the identified frequencies with high resolution (small bandwidth). Early in the project, it was found that failure to use a small bandwidth would result in errors up to four percent.

Mode shapes were measured by determining the transfer function between two accelerometers mounted on the rotor. One accelerometer was kept at one end of the rotor as a normalizing "input," and the other accelerometer was moved along the rotor as a "rover" to define the mode shape.

To maximize confidence in the accuracy and precision of the mode shape measurements, the two accelerometers were calibrated together simultaneously on the same clamp and the coherence function was measured during each transfer magnitude and phase measurement.

In almost all of the mode shape measurements, the transfer function phase angle was found to vary less than 3° from 0° or 180°. This indicated very low internal damping within the rotor. (Damping will produce a phase shift.)

The data on rotor dimensions, required to generate accurate computer models, were obtained by taking micrometer measurements of the rotor outer diameter at various axial locations along the shaft (typical spacing between measurements was 2 in) and at all disks and other changes in diameter.

The rotors were weighed on a scale, which together with the dimension measurements confirmed the weight density of 0.283 lb/in^3 used in the computer models.

Bump tests can also be used to measure system damping. If a rotor–bearing system does not have fluid-film bearings or squeeze film dampers, the system damping can be measured with a bump test while the rotor is at rest (zero speed). Figure 8.23 shows the free vibration signal from a laboratory rotor–bearing system captured on an oscilloscope with "memory." (Most modern scopes with memory use digital signal storage for this capability.) Figure 8.24 illustrates the calculation of a modal viscous damping ratio ξ_m from the logarithmic decrement δ of the decaying waveform.

Note that Fig. 8.23 is a pure single-frequency waveform. Care must be taken to excite only the mode of interest. This can usually be done by experimenting with various impact locations on the rotor or structure. Computer-predicted mode shapes will show the approximate location of antinodes where the impact should be applied. If multiple-frequency waveforms cannot be avoided, sophisticated signal analysis codes are required to extract the logarithmic decrement of each frequency.

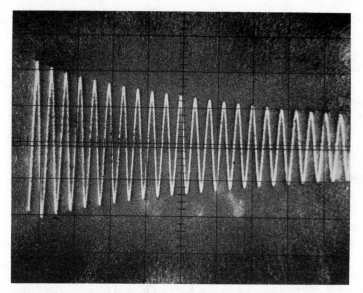

Figure 8.23. Free vibration signal on an oscilloscope.

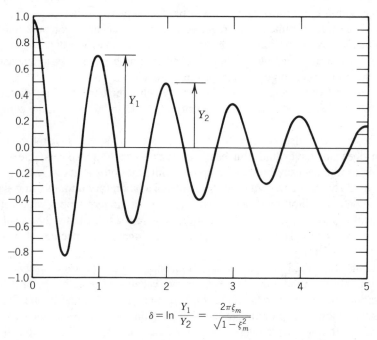

$$\delta = \ln \frac{Y_1}{Y_2} = \frac{2\pi \xi_m}{\sqrt{1 - \xi_m^2}}$$

Figure 8.24. Calculation of the modal damping ratio from the measured logarithmic decrement.

Coast-Downs

A coast-down measurement is made by bringing the rotor up to a predetermined speed, turning off power, and taking measurements continually as the rotor decelerates to rest. The deceleration rate is generally repeatable, as opposed to a run-up, and an advantage if the driver is electric is that there are no stray electric fields or currents to disturb the measurements.

Coast-downs are most commonly used to generate plots of synchronous response to rotor imbalance versus speed. In this case the vibration signal and the tachometer signal are fed into a synchronous tracking filter and the outputs are used to drive an X-Y plotter, where X is the rotor speed in revolutions per minute and Y is synchronous vibration amplitude.

In a variation of this procedure a spectrum analyzer (FFT–RTA) is used in place of the synchronous tracking filter. The analyzer is set to the "peak hold" function, which stores the maximum amplitude measured at each frequency window during the measurement period. The resulting amplitude versus frequency plot will be the synchronous response to imbalance only if the nonsynchronous vibration amplitude never exceeds the synchronous amplitude at every frequency window during the measurement period.

Figure 8.25 shows a coast-down measurement made on the laboratory rotor (at Shell Westhollow Research Center) described in Chapter IV, using a Hewlett-Packard 3582A spectrum analyzer on peak hold. In this case the nonsynchronous amplitudes are relatively large, so the plot is not of synchronous response.

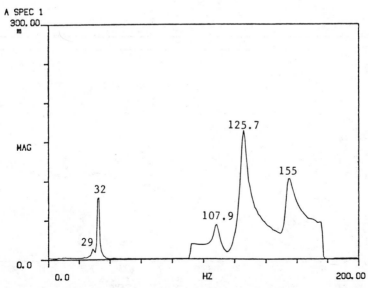

Figure 8.25. "Peak hold" spectrum from a coast-down through the second, third, and fourth critical speeds of the laboratory rotor–bearing system at Shell Westhollow Research Center (from Ref. [3]).

However, the nonsynchronous vibration was excited at the eigenvalues of the rotor-bearing system (which is usually the case), so the plot can be used to identify the natural frequencies.

Figure 8.26 shows a coast-down of the labyrinth seal test rotor mentioned in Chapter VI, measured with a Bently-Nevada DVF2 synchronous tracking filter. This is a true plot of synchronous response to imbalance. This type of plot is the most reliable way to identify critical speeds. The first critical speed in Fig. 8.26 is 3400 rpm, and the second is above the range of the plot. In turbomachines with enough damping to suppress the peaks (e.g., most utility power-plant turbines), the synchronous response plot will reveal this fact.

If the peaks are well defined and fairly well separated, the synchronous response plot can be used to estimate the critical damping ratio ξ. If the rotor approximates one of the simple models described in Chapter I with known imbalance u, equation (1-4) or (1-12) can be used together with the peak response r_s at the critical speed $\omega_{cr} \approx \omega_n$ to solve for the damping coefficient c or the critical damping ratio ξ. In

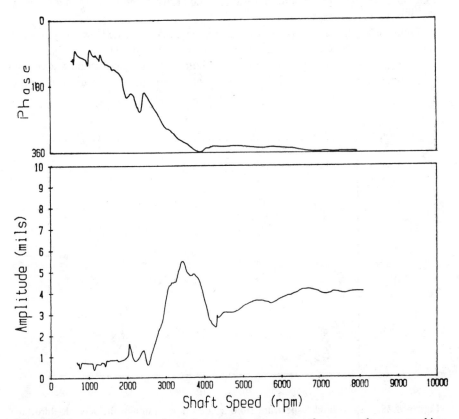

Figure 8.26. Coast-down plot for the labyrinth seal test rotor, from a synchronous tracking filter.

most practical cases, however, the "bandwidth method" must be used. Figure 8.27 illustrates application of the bandwidth method to calculate the modal damping ratio ξ_m.

Constant Speed Measurements

Many industrial turbomachines run all of the time or most of the time at only one speed. In such cases it may be required to make all measurements at a constant speed. The amount of information that can be obtained from a frequency spectrum taken at constant speed is sometimes surprising, especially if computer predictions of the eigenvalues are available for guidance. Most turbomachines have misalignment, bearing imperfections, two-phase working fluids, and external perturbations from other machines, all of which produce multiple-frequency excitation even at constant speed.

Figure 8.28 shows a frequency spectrum measured on the single-stage steam turbine described in Chapter IV, driving a process compressor in the region of surge at 9500 rpm. This is the measurement used to identify the eigenvalues for comparison with the computed predictions, as shown in Table 4-15. The synchronous vibration at 159 Hz is not necessarily a critical speed, although its large amplitude suggests the possibility. Its first harmonic is at 318 Hz. Electrical noise at 60 Hz and all its harmonics would mask critical speeds at 60, 120, 180, and 360 Hz. The most likely critical speeds are then 140, 205, and 219 Hz with a possibility near 158 Hz (the frequency window width of the analyzer is 1 Hz for the 0–400 Hz range). This partially corresponds with past experience with this turbine in which start-ups show the machine going through one critical speed around 8000 rpm (133 Hz) and approaching another at its top speed of around 11,000 rpm (183 Hz). Figure 8.29 is another spectrum taken from the same machine at

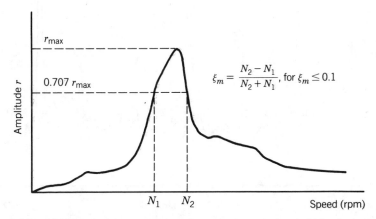

Figure 8.27. Calculation of the modal damping ratio from the bandwidth of a coast-down plot (synchronous response to imbalance).

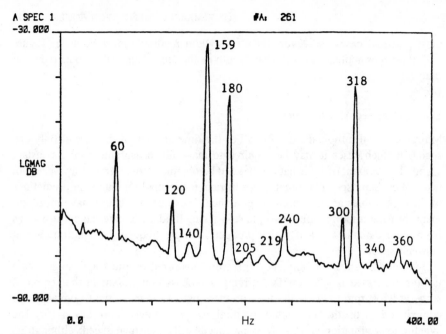

Figure 8.28. Vibration spectrum of a single-stage gas turbine running at 9500 rpm (from Ref. [3]).

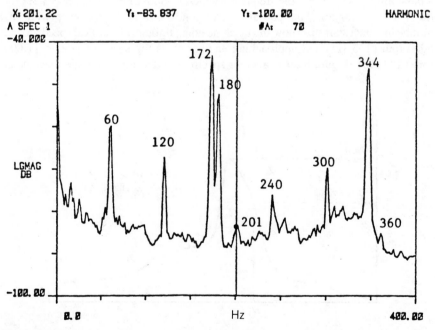

Figure 8.29. Steam turbine vibration spectrum with the 201-Hz component more pronounced (from Ref. [3]).

a later time, showing the spike just above 200 Hz more clearly and the spike near 220 Hz much less pronounced. For comparisons with the computer predictions of Chapter IV, the critical speeds of the single-stage steam turbine were taken as 140 and 205 Hz, with other possibilities at 158–159 and 180 Hz. Looking at Fig. 8.28, one might observe that none of these frequencies would be identified positively as eigenvalues without the computer predictions for comparison and guidance.

The most common application of constant-speed measurements is for balancing. Several methods for in-place balancing of rotors are described in Chapter V. Those methods requiring phase measurements are greatly facilitated by a synchronous tracking filter ("trim balance analyzer") such as the Bently-Nevada DVF2 or the Spectral Dynamics (Scientific-Atlanta) SD-119C. The operation of these instruments was described earlier.

DIAGNOSIS OF SOME COMMON CONDITIONS IN TURBOMACHINERY

Most machine defects or conditions produce a unique set of vibration components that can be used for identification. This section describes these vibration patterns (or "signatures") for some common machinery defects. Where appropriate, frequency calculation formulas and details of spectrum generation are also included.

Imbalance

Rotor imbalance exists to some degree in all machines and is characterized by sinusoidal vibration at a frequency of once per revolution. In the absence of high resolution analysis equipment, imbalance is usually first to get the blame for excessive once-per-revolution vibration, but this type of vibration can also be caused by several other faults.

Phase can play a key role in detecting and analyzing imbalance, and it is important to remember the phase shifts associated with the "critical speed inversion" (see Chapter I). A state of imbalance occurs when the center of mass of a rotating system does not coincide with the center of rotation. It can be caused by a number of things, including incorrect assembly, material buildup, and rotor sag. As described in Chapter V, the imbalance can be in a single plane (static imbalance) or multiple planes (couple imbalance). The combination is referred to as dynamic imbalance. The result is a vector that rotates with the shaft, producing the classic once-per-revolution vibration characteristics.

The key characteristics of the vibration caused by imbalance are (1) it is sinusoidal at a frequency of once-per-revolution ($1x$), (2) it is a rotating vector, and (3) amplitude changes with speed. These characteristics are useful in differentiating imbalance from faults that produce similar vibration.

The vibration caused by pure imbalance is a once-per-revolution sine wave, sometimes accompanied by low level harmonics. The faults commonly mistaken for imbalance usually produce high level harmonics or occur at a higher frequency. Usually, if the signal has large-amplitude harmonics above once-per-revolution, the fault is not imbalance. However, high level harmonics can occur with large imbalance forces, when horizontal and vertical support stiffnesses differ by a large amount, or when large clearances exist in rolling-element bearings.

The following faults are often mistaken for imbalance because they result in increased levels of vibration at running speed. However, each has other distinctive characteristics that can be used for identification.

Misalignment. The key characteristics that identify misalignment are a large second harmonic component and a high level of axial vibration. Bent shafts and improperly seated bearings are special cases of misalignment and produce similar vibration. The relative phase of axial vibration due to misalignment, measured at the ends of the shaft, is typically 180°. These signals are usually in phase when the shaft is out of balance.

Load Variation. Uneven loading and retained fluid in pumps result in imbalance. Higher torque loading can also cause an increase in running speed vibration level. The key to correct interpretation of these increases is a good understanding of the machine's operating characteristics. When measuring baseline vibration, it is important to check variation with key operating parameters such as load, pressure, and temperature. If the variation is significant, it can be roughly characterized by taking baseline spectra under a variety of operating conditions.

Mechanical Looseness. The vibration spectrum that results from looseness almost always includes higher harmonics. If the looseness is directional, it can be identified by relatively large amplitude changes with transducer location or orientation.

Critical Speed or Resonance. A critical speed (natural frequency) or local resonance at running speed will produce a high $1x$ vibration level. Resonance is usually identifiable because the vibration level will be significantly reduced at frequencies higher or lower than resonance. Resonances may be the result of installation faults or may be a design defect (see Chapter IV).

Excessive Clearance in Fluid-Film Bearings. The increase in running speed vibration level is usually accompanied by higher frequency harmonics. However, an increased clearance in a fluid-film bearing can also produce a subsynchronous whirl instability called oil whip (see below and Chapters VI and VII).

Backward Whirl

Backward whirl is not a defect in itself, but it does occur and is usually indicative of very light damping and operation between two critical speeds. Very severe rotor rubs can produce violent backward whirl that can be quite destructive.

Backward whirl can be identified by noting the phase relationships of signals from two proximity probes mounted on perpendicular axes (as in Fig. 8.5, where the right transducer signal will lead the left transducer when the whirl is backward), or by noting the transfer function of the two signals displayed on a spectrum analyzer.

The lab rotor at Shell Westhollow Research Center described in Chapter IV provides a good example of a backward whirl mode. Figure 8.25 shows the frequency spectrum obtained during coast-down through the second, third, and forth critical speeds. The two spikes at the first critical speed are there in the form of subsynchronous whirling. The short spike at 29 Hz is the backward component of the first mode. The spike at 32 Hz is the forward component and would normally be called the first critical speed. It would be difficult, at best, to interpret this from an orbit displayed on an oscilloscope screen. However, the same two displacement probes normally used to display an orbit were fed into a two-channel FFT analyzer. The transfer function (Fig. 8.30) of the P2 signal divided by the P1 signal then indicates the whirl direction. At a frequency of 32 Hz probe 2 leads probe 1 by 90°, indicating forward whirl. At 29 Hz probe 2 lags by 90°, indicating backward whirl at this frequency.

The phase angle at the other three critical speeds are seen to be near either 0° or +180°. This indicates modes that are very close to being planar (i.e., the "orbit" is a straight line). The actual whirl directions of these modes are not particularly important because they are so very nearly planar. The forward rotating imbalance force can excite them regardless of their respective whirl directions. The orbital directions of whirl modes can be different at different axial locations along the rotor, and straight-line orbits occur at the transitions from forward to backward whirl.

Rolling-Element Bearing Conditions

Rolling-element (antifriction) bearings are the most common cause of failure in small turbomachines. Overall vibration level changes are difficult to detect in the early stages of deterioration. However, the unique frequency characteristics of rolling-element bearing defects make vibration analysis an effective tool for both detection and analysis.

The specific frequencies that result from bearing defects depend on the defect, the bearing geometry, and the speed of rotation.

The required bearing information to predict the characteristic frequencies is shown in Fig. 8.31. This information is usually available from the bearing manufacturer. A computer code in BASIC that computes the expected frequencies, given bearing parameters and rotational speed, is listed in Table 8-2. *One caution:* parameters for the same model number bearing can change with manufacturer.

The major problem in detecting the early stages of failure in rolling-element bearings is that the resulting vibration is low level and often masked by higher level vibration from other sources. If monitoring is performed with a simple vibration meter (or in the time domain), these low levels will not be detected and unpredicted failures will occur. A good solution is regular monitoring of critical

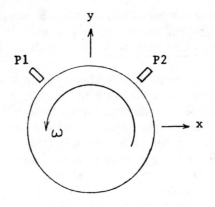

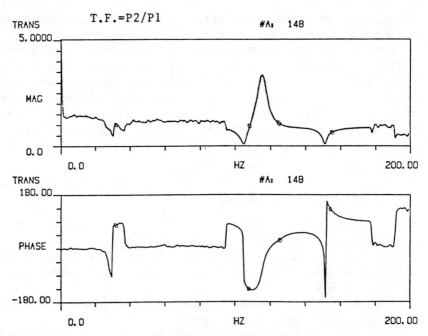

Figure 8.30. Transfer function of signal P2 to signal P1, measured on a spectrum analyzer (from Ref. [3]).

machinery with a high quality spectrum analyzer, since the high resolution and dynamic range can show components as small as $1/1000$ the amplitudes of higher level vibration.

Formulas for calculation of the frequencies resulting from bearing defects are given in Table 8-3 (refer to Fig. 8.31). The formulas assume a single defect, rolling contact, and a rotating shaft with fixed outer race. The results can be

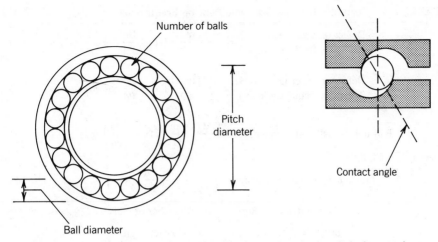

Figure 8.31. Required bearing information to predict characteristic frequencies.

expressed in orders of rotation by leaving out the (rpm/60) term. The BASIC program in Table 8-2 will compute the bearing frequencies automatically. Again, remember that bearing parameters can change with manufacturer.

If bearing dimensions are not available, inner and outer race defect frequencies can be approximated as 60 and 40 percent of the number of balls multiplied by

TABLE 8-2.   BASIC Computer Program for Characteristic Bearing Frequencies

```
10 ! Bearing frequency calculation program for HP 85
20 !
30 DIM D$[32]
40 DEG @ CLEAR
50 DISP "Enter bearing description:"
60 INPUT D$
70 DISP "Enter ball diameter, pitch dia- meter:"
80 INPUT B,P
90 DISP "Enter contact angle, # of balls:"
100 INPUT A,N
110 DISP "Enter RPM (0 for ORDERS)"
120 INPUT F
130 IF F=0 THEN L$=" Orders" ELSE L$=" Hertz"
140 IF F=0 THEN F=60
150 F=F/60 ! Convert RPM to Hz
160 PRINT D$
170 PRINT USING "14A,2DZ.3D,7A" ; "Ball diameter:",B," inches"
180 PRINT USING "15A,2DZ.3D,7A" ; "Pitch diameter:",P," inches"
190 PRINT USING "14A,2DZ.D,8A" ; "Contact angle:",A," degrees"
200 PRINT USING "16A,DD" ; "Number of balls:",N
210 IF F>1 THEN PRINT USING "6A,5DZ.DD,4A" ; "Speed:",F*60," rpm"
220 PRINT "   --------------"
230 IMAGE 17A,4DZ.2D,7A
240 PRINT USING 230 ; "Ball pass--outer:",F/2*N*(1-B/P*COS(A)),L$
250 PRINT USING 230 ; "Ball pass--inner:",F/2*N*(1+B/P*COS(A)),L$
260 PRINT USING 230 ; "Ball spin:",F/2*(P/(2*B))*(1-(B/P*COS(A))^2),L$
270 PRINT USING 230 ; "Fund. train:",F/2*(1-B/P*COS(A)),L$
280 PRINT USING "4/"
290 END
```

TABLE 8-3.   Formulas for Characteristic Bearing Frequencies

$$\text{Defect on outer race (ball pass frequency outer)} = \frac{(n)}{2}\frac{(\text{rpm})}{60}\left(1 - \frac{\text{Bd}}{\text{Pd}}\cos\phi\right) \tag{1}$$

$$\text{Defect on inner race (ball pass frequency inner)} = \frac{(n)}{2}\frac{(\text{rpm})}{60}\left(1 + \frac{\text{Bd}}{\text{Pd}}\cos\phi\right) \tag{2}$$

$$\text{Ball defect (ball spin frequency)} = \frac{(\text{Pd})}{2\text{Bd}}\frac{(\text{rpm})}{60}\left[1 - \left(\frac{\text{Bd}}{\text{Pd}}\right)^2\cos^2\phi\right] \tag{3}$$

$$\text{Fundamental train frequency} = \frac{1}{2}\frac{\text{rpm}}{60}\left(1 - \frac{\text{Bd}}{\text{Pd}}\cos\phi\right) \tag{4}$$

Pd = Pitch diameter n = Number of balls

Bd = Ball diameter ϕ = Contact angle

running speed, respectively. This approximation is possible because the ratio of ball diameter to pitch diameter is relatively constant for rolling-element bearings.

The following two observations will help to shed light on the bearing formulas in Table 8-3: (1) Since the balls contact both the shaft speed inner race and the fixed outer race, the rate of rotation relative to the shaft center is the average, or one-half the shaft speed. This is the reason for the factor of $1/2$ in formulas 1 through 4. (2) The term in parenthesis is an adjustment for the diameter of the component in question. For example, a ball passes over defects on the inner race more often than those on the outer race because the linear distance (which is proportional to diameter) is shorter.

The fundamental train frequency, which occurs at a frequency lower than running speed, is usually caused by a severely worn cage.

Rolling-element bearing frequencies are transmitted well to the machine case (because the bearings are stiff) and are best measured with accelerometers or velocity probes. For bearings that provide axial support, axial measurements often provide the best sensitivity to detect vibration (because machines are usually more flexible in this direction).

While the computation of characteristic bearing frequencies is straightforward, several factors can modify the vibration spectrum that results from bearing defects, as follows:

1. Bearing frequencies are usually modulated by residual imbalance, which will produce sidebands at running frequency. Other vibration can also modulate (or be modulated by) bearing frequencies, and bearing spectra often contain components of these frequencies.

2. As bearing wear continues and defects appear around the entire surface of the race, the vibration will become much more like random noise, and discrete spectral peaks will be reduced or disappear completely. This will also be the case

with roughness caused by abrasive wear or lack of lubrication. Another variation that occurs in advanced stages is concentration of the defect energy in higher harmonics of the bearing characteristic frequency.

3. Some of the characteristic frequencies will appear in the vibration spectrum of a good bearing. This is usually due to production tolerances and does not imply incipient failure. Comparison with a baseline spectrum will help to avoid misinterpretation.

4. To modify the formulas for a stationary shaft and rotating outer race, we can change the signs in formulas 1 and 2 of Table 8-3.

5. Contact angle can change with axial load, causing small deviations from calculated frequencies.

6. Small defects on stationary races that are out of the load zone will often only produce noticeable vibration when loaded by imbalance forces (i.e., once-per-revolution).

The example spectrum of Fig. 8.32 is the result of a defect in the outer race. Note the sidebands at running speed, which are characteristic of most bearing spectra.

The spectrum in Fig. 8.33 is also the result of a defect in the outer race. In this example, the characteristic band pass frequency has disappeared but its harmonics remain. The component around 200 Hz is gear mesh vibration.

Conditions to Look for with Oil-Film Bearings

Oil-film bearings generally provide high damping, so machines employing them generally show broadly smeared low amplitude critical speed peaks and gradual

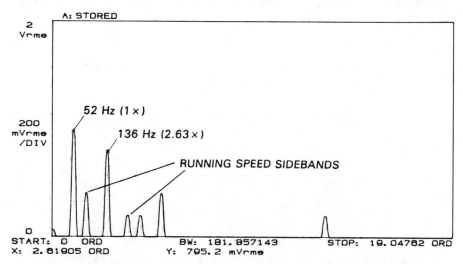

Figure 8.32. Vibration spectrum indicating a defect in the outer race of a ball bearing.

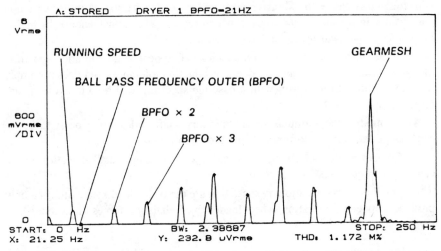

Figure 8.33. Vibration spectrum showing outer race defect from BPFO harmonics only.

phase changes on coast-down plots. The most serious condition that can be produced by oil-film bearings is the subsynchronous whirl instability called oil whip, as described in Chapters VI and VII.

When oil whip is imminent, a subsynchronous frequency (sometimes called "oil whirl") can appear even before the threshold speed of instability is reached. This frequency is usually 0.43–0.49 of running speed. Once the threshold speed is reached (often about twice the lowest eigenvalue or critical speed), the whirling frequency "locks in" to the eigenvalue even as shaft speed continues to increase (assuming the rotor–bearing system has not wrecked). The cascade plot of Fig. 8.34 illustrates this phenomenon.

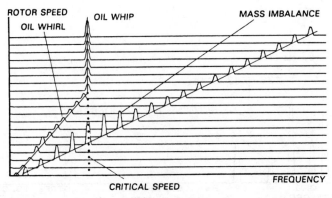

Figure 8.34. Cascade plot from a rotor run-up showing oil whirl and oil whip.

Misalignment Conditions

Vibration due to misalignment is usually characterized by a 2x running speed component and high axial levels. Phase, both end to end on the machine and across the coupling, is a useful tool for differentiating misalignment from imbalance.

Misalignment takes three basic forms: (1) preload from a bent shaft or improperly seated bearing, (2) radial offset of the shaft centerlines of machines in the same train, and (3) angular offset of the shaft centerlines in the same train. Flexible couplings increase the ability of the train to tolerate misalignments; however, they are not a cure for serious alignment problems.

The axial component of the force due to misalignment is shown in Fig. 8.35. Machines are often more flexible in the axial direction, with the result that high levels of axial vibration usually accompany misalignment. These high axial levels are a key indication of misalignment.

Figure 8.36 illustrates a typical frequency spectrum caused by misalignment, with the high amplitude component at twice running speed.

As shown in Fig. 8.35, the axial vibration at each end of the machine (or across the coupling), is 180° out of phase. This relationship can be used to differentiate misalignment from imbalance, which produces in-phase axial vibration. This test cannot be used in the radial direction, since imbalance phase varies with the type of imbalance. Relative phase can be measured with some single-channel spectrum analyzers, using a keyphasor reference, or directly with a dual-channel spectrum analyzer that has transfer-function capability.

If the misalignment frequency is dominant, the relative phase of the two end signals can also be observed by displaying them simultaneously on a dual-trace oscilloscope.

Several notes of caution relative to phase measurements are appropriate at this point:

1. Machine dynamics will affect phase readings, so that the axial phase relationship may be 150° or 200° rather than precisely 180°.

2. Transducer orientation is important. Transducers mounted axially to the

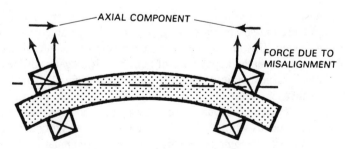

Figure 8.35. Axial forces due to misalignment.

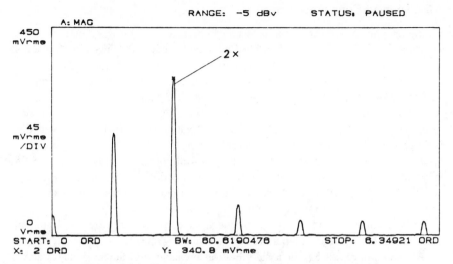

Figure 8.36. Frequency spectrum caused by misalignment.

outside of the machine will most often be oriented in opposite directions. If this is the case, a 180° phase relationship will be measured as 0°.

3. The phase relationships described hold only at speeds below the first critical speed. Using phase for diagnostics at higher speeds requires knowledge of the system rotordynamics, as determined by an accurate computer simulation.

4. Great care must be exercised when one is measuring relative phase with a single-channel spectrum analyzer (assuming it has the capability). Two measurements are required, each referenced to the shaft with a keyphasor. These measurements must be made at the same speed unless trigger delay or external sample control is used. In general, several measurements should be made at each point to ensure that phase readings are repeatable.

Rotor/Stator Rubs

Rubs occur most often at the rotating seals where clearances are small. They are usually intermittent, and so are classified as one of the "parametric excitations" described in Chapter VII. The vibration characteristic of an intermittent rub at subcritical rotor speeds is the appearance of harmonics (integer multiples of rotor speed) of relatively large amplitude in the frequency spectrum [4]. At supercritical rotor speeds a subsynchronous frequency is produced that tracks the rotor speed at some exact integer fractional frequency (usually $1/2$ but sometimes $1/3$ or $1/5$). When the subsynchronous whirling frequency approaches a lightly damped eigenvalue the amplitude can become large very quickly, thus simulating the behavior of a whirl instability.

Mechanical Looseness Conditions

Mechanical looseness usually involves mounts or bearing caps, and almost always results in a large number of harmonics in the vibration spectrum. Components at integer fractions of running speed may also occur and may be falsely identified as a subsynchronous whirl instability. Looseness tends to produce vibration that is directional, a characteristic that is useful in differentiating looseness from rotational defects such as imbalance. A technique that works well for detecting and analyzing looseness is to make vibration measurements at several points on the machine (velocity transducers work well for this). Measured vibration level will be highest in the direction and vicinity of the looseness.

The harmonics that characterize looseness are a result of impulses and truncation (limiting boundaries) in the machine response. Consider the bearing shell in Fig. 8.37. When it is tight, the response to imbalance at the transducer is sinusoidally varying. When the mounting bolt is loose, there will be truncation when the looseness is taken up. Although these waveforms are idealized, the mechanism

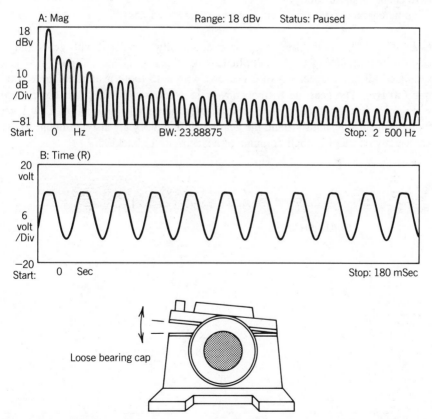

Figure 8.37. Frequency spectrum and truncated waveform caused by a loose bearing cap.

for producing harmonics should be clear. In terms of the rotor–bearing mathematical model, it is a nonlinearity in the support stiffness.

Gear Conditions

Gear problems are characterized by vibration spectra that are typically easy to recognize but difficult to interpret. The difficulty is due to two factors: (1) it is often difficult to mount the transducer close to the problem, and (2) the number of vibration sources in a multigear drive result in a complex assortment of gear mesh, modulation, and running frequencies. Because of the complex array of components that must be identified, the high resolution provided by a high quality spectrum analyzer is a virtual necessity. It is helpful to detect problems early through regular monitoring, since the advanced stages of gear defects are often difficult to analyze. Baseline vibration spectra are helpful in analysis because high level components are common even in new gearboxes. Baseline spectra taken when the gearbox is in good condition make it easier to identify new components, or components that have changed significantly in level.

Some characteristic gear frequencies are discussed next.

Gear Mesh. This is the frequency most commonly associated with gears and is equal to the number of teeth multiplied by rotational frequency. Figure 8.38 is a simulated vibration spectrum of a gearbox with a 15-tooth gear running at 3000 rpm (50 Hz). The gear mesh frequency is 15 × 50 = 750 Hz. This component will appear in the vibration spectrum whether the gear is bad or not. Low level running speed sidebands around the gear mesh frequency are also common. These are usually caused by small amounts of eccentricity or backlash.

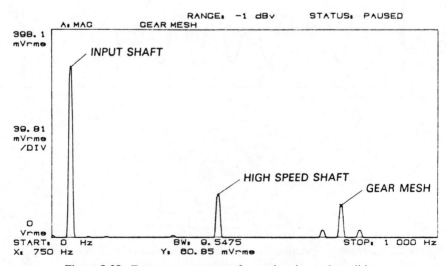

Figure 8.38. Frequency spectrum of a gearbox in good condition.

The amplitude of the gear mesh component can change significantly with operating conditions, implying that gear mesh level is not a reliable indicator of condition. On the other hand, high level sidebands or large amounts of energy under the gear mesh or at the gear natural frequency components (Fig. 8.39) are a good indication that a problem exists.

Natural Frequencies. The impulse that results from large gear defects usually excites the natural frequencies of one or more gears in a set. Often this is the key indicator of a fault, since the amplitude of the gear mesh frequency does not always change. In the simulated vibration spectrum of Figure 8.39, the gear mesh frequency is 1272 Hz. The broadband response around 600 Hz is centered on a gear natural frequency, with sidebands spaced at the running speed of the bad gear. The high resolution zoomed spectrum of Fig. 8.39(b) shows this detail.

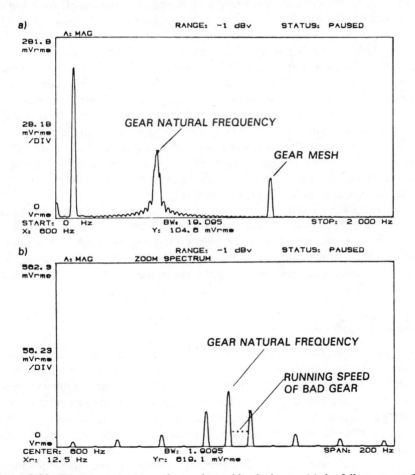

Figure 8.39. Frequency spectrum of a gearbox with a bad gear: (a) the full spectrum; (b) zoom expansion.

Sidebands. Frequencies generated in a gearbox can be modulated by backlash, eccentricity, loading, bottoming, and pulses produced by defects. The sidebands produced are often valuable in determining which gear is bad. In the spectrum of Fig. 8.39(b), for example, the sidebands around the natural frequency indicate that the bad gear has a running speed of 12.5 Hz. In the case of eccentricity, the gear mesh frequency will usually have sidebands spaced at running speed.

Conditions Produced by Blades and Vanes

Problems with blades and vanes are usually characterized by high fundamental vibration or a large number of harmonics near the blade or vane passing frequency. Some components of passing frequency (number of blades or vanes × speed) are always present, and levels can vary markedly with load. This is especially true for high speed machinery and makes recording of operating parameters for historical data critical. It is very helpful in the analysis stage to have baseline spectra for several operating levels.

If a blade or vane is missing, the result will typically be imbalance, resulting in high $1x$ vibration. For more subtle problems such as cracked blades, changes in the vibration are both difficult to detect and difficult to quantify. Detection is a problem, especially in high speed turbomachinery, because blade vibration is difficult to measure directly. Strain gauges can be used, but the signal must be either telemetered or transferred through slip rings. Doppler detection techniques show promise but have not been sufficiently developed for practical use. Indirect detection produces a spectrum that is the result of complex interactions that may be difficult to explain. This, combined with the large variation of levels with load, makes spectra difficult to analyze quantitatively.

One characteristic that often appears in missing- or cracked-blade spectra is a large number of harmonics around the blade passing frequency. Figure 8.40 shows how a space in the vibration signal greatly increases the number of harmonics without changing the fundamental frequency. The spectrum and time trace at the top are from the machine with a missing blade; the bottom records are from a normal machine.

Resonance Conditions

In addition to the critical speeds of turbomachinery rotor–bearing systems described in Chapter IV, problems with local resonances can occur when the machine housing or attached structures are excited by running speed (or harmonics of running speed). These problems are usually easy to identify because levels drop appreciably when running speed is raised or lowered. Spectral maps are especially useful for detecting resonance vibration because the strong dependence on rotational speed is readily apparent.

Phase is also a useful tool for differentiating resonances from rotationally related components. Say, for example, that a high level of vibration at 16 times running

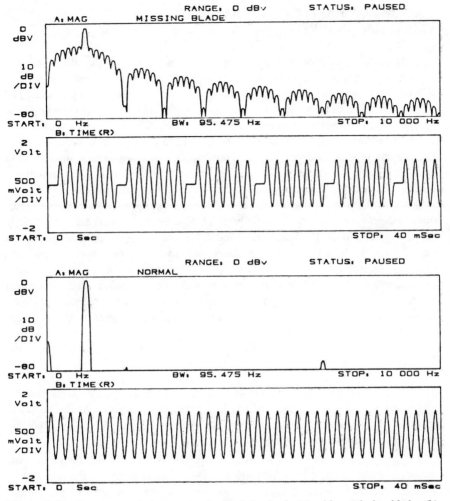

Figure 8.40. Spectrum and time trace of a bladed wheel: (a) with a missing blade; (b) normal.

speed is encountered. If the vibration is rotationally related (e.g., a blade passing frequency), the phase relative to a keyphasor signal or residual imbalance will be constant. This is a useful technique when it is not practical to vary the speed of the machine.

Piping is one of the most common sources of resonance problems. When running speed coincides with a natural frequency, the resulting vibration will be excessive and strain on both the pipe and the machine can lead to early failure. The most logical approach is to change the natural frequency of the pipe. It can be raised by making the pipe shorter or stiffer (e.g., by adding a support), or it can be lowered by making the pipe longer. The same rules apply to any attached structure.

MACHINERY MAINTENANCE BASED ON VIBRATION ANALYSIS

The objective of maintenance is to keep machines running, especially those that are critical to plant production or safety of personnel. Unexpected catastrophic failures cause loss of production, large repair bills, or loss of human life. The classic maintenance strategy for avoiding such failures is to periodically disassemble critical machines for inspection and rebuilding. This process results in costly downtime, often used to inspect machines that are in perfect working order. Because the process is expensive, it is only applied to a few critical machines. In addition, faulty reassembly or damage in transit from the repair shop sometimes results in a machine in worse condition than before the maintenance.

A more effective approach is to schedule repairs on the basis of machine condition, as determined by vibration analysis. This "predictive" maintenance strategy can be applied to all the major machines in a plant or fleet and has proven its effectiveness in hundreds of maintenance organizations. In a typical program, overall vibration level is measured regularly with a vibration meter and compared to established severity limits or past readings. The vibration level of critical machinery is often monitored on a continuous basis and compared against preset limits. If an excessive level is detected, a signal analyzer is used to determine the severity and nature of the problem.

Signal analysis to determine the frequency components is a critical part of the process for several reasons:

1. Overall vibration level can change with load and operating speed, thereby presenting a misleading picture of machine condition. Analysis of the vibration spectrum indicates whether or not a serious problem exists. This is an important step in avoiding unnecessary repairs.

2. Taking a machine out of service for repairs can rarely be done without some impact on production, so it is important to know just how severe the problem is. Analysis can help the diagnostician decide, for example, whether the machine can be run until the next scheduled plant shutdown. Thus signal analysis is valuable in maximizing the effectiveness of a maintenance program.

3. Repair time is minimized because the nature of the problem is known. Technicians will not spend valuable time looking for the fault, and the necessary replacement parts can be ordered prior to disassembly.

The advantages of signal analysis make it appropriate in some cases to monitor the levels of individual frequency components (rather than overall vibration level) to detect faults. Monitoring individual components also gives earlier warning of failure. This is an especially important consideration in highly loaded machines using rolling-element bearings, whose condition can deteriorate rapidly.

The benefits of a predictive maintenance program based on vibration analysis are described further by Jackson [5] and Mitchell [6], who also provide details on establishing a program of predictive maintenance.

TORSIONAL VIBRATION MEASUREMENTS

Torsional vibration currently is not a common measurement made for turbomachinery diagnostics or for a general maintenance program. One reason is that torsional vibration is a difficult measurement to make. However, torsional vibration is probably the most common cause of gear tooth breakage, and it has also been identified as a cause of blade fatigue in steam turbines. In turbomachinery trains driven by electric motors and in turbogenerator sets, torsional vibration is often produced by variations in electromagnetic torque (see Chapter III).

The author has conducted a study of torsional vibration measurement techniques with R. S. French and P. M. Barrios. The remainder of this chapter is adapted from that study, first reported in Ref. [7]; the final two subsections are based on data of Professor Barrios.

In contrast to the large number of complete instrumentation systems that can be purchased off the shelf to measure lateral (translational) vibrations in rotating machinery, only a few systems are available commercially for torsional vibration measurement [8].

Most of the commercially available systems for torsional vibration measurement either require significant modifications to the rotating machine (which may be unacceptable for reasons of safety, reliability, or expense) or have significant limitations of application or performance. Consequently, engineers requiring measurements of torsional vibration often have no recourse but to design their own systems, using commercially available subcomponents wherever possible and tailoring the performance specifications to the particular application. Examples can be found in Refs. [9] and [10].

The oldest and probably the most widely known method of measuring torsional vibration utilizes strain gages bonded to the surface of the rotating shaft, oriented along the directions of principal strain. The method is accurate when properly installed and calibrated, but has some severe practical disadvantages. First, the signal must be transferred from the rotating shaft to a nonrotating frame, either by the use of slip rings or by radiotelemetry. In the field, dirty brushes or radio interference can produce low signal/noise ratios. Improper applications of the strain gages can produce signals related to shaft bending rather than torsion. The gage bonding agent may be deteriorated by environmental conditions, such as temperature or process chemicals. If all of these problems are to be avoided, the requisite installation time for an accurate and reliable system may become prohibitive, especially in a troubleshooting situation.

Alternatively, prefabricated and precalibrated shaft sections are commercially available with preinstalled strain gages and slip ring assemblies, or with radiotelemetry antennae installed. These devices are meant to replace some existing section of shaft, or shaft coupling, in the machine. However, retrofitting such devices into a machine often raises questions of a possible loss of machine reliability or safety. In many cases, space is not available for the instrumented shaft section with its couplings, slip rings, or antennae.

Another type of measurement system which was commercially available in past years was known as a "torsiograph." This instrument produced a voltage proportional to the oscillatory angular velocity of the shaft. It had the disadvantage of requiring an exposed end of the shaft for its attachment, and it did not respond accurately at frequencies below 10 Hz [8]. This latter limitation is critical for heavy machinery and also applies to some of the instrumentation systems commercially available today.

It is instructive to note that the two measurement systems just described (strain gage and angular velocity measurement) will produce radically different measurements when applied at the same shaft location. The strain gages measure the twist in the shaft and consequently will produce a maximum signal in regions where the variation of angular velocity is minimum.[2] Therefore, the selection of transducer types and locations should be guided by knowledge of the torsional mode shapes, or else a sufficient number of measurement locations should be used to determine the mode shape experimentally.

In machinery drivetrains, sometimes the first evidence of torsional vibration can be damage or breakage of gear teeth. In fact, gear damage is probably the most common first incentive for measuring torsional vibration in rotating machinery. If gears are present, they offer an ideal source of "carrier signal" for most of the measurement systems to be described here. Figure 8.41 shows the type of carrier signal that can be produced, for example, by a magnetic transducer excited by the passing gear teeth. Frequency or amplitude modulation of the signal can yield the torsional vibration characteristics, provided the frequencies of interest are much lower than the carrier wave frequency, which is typically the case.

A distinction must be made between steady-state and transient measurements. Transient measurements (when shaft speed and/or torsional vibration are changing with time) are much more difficult to make and to analyze. An application where transient measurements are required is the start-up of drivetrains using large synchronous electric motors as the driver [9–11]. These motors can produce torque pulsations of large amplitude and variable frequency during start-up that excite torsional vibration superposed onto the acceleration schedule, as illustrated in Fig. 8.42, and described mathematically in Chapter III.

The ideal torsional vibration measurement system would yield accurate measurements for both steady-state and transient conditions, would be applicable to any drivetrain with or without gears, and would be quickly and easily installed in the field even if only a short length of exposed rotating shaft or coupling spacer was accessible. The prototype of such a system, called TIMS for "time interval measurement system," is described below after several more conventional systems have been described. In all, five different types of instrumentation systems (including the strain gages, used as a benchmark), were assembled by French and used to measure torsional vibration in two experimental test rigs designed and constructed in the Turbomachinery Laboratories at Texas A&M University [7, 14].

---

[2]That is, at the node of the torsional vibration mode shape.

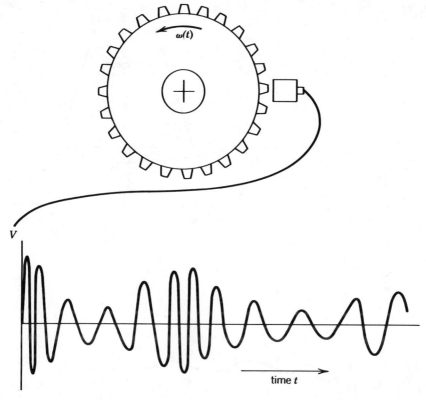

Figure 8.41. Carrier signal for torsional vibration measurement.

Strain Gages

The electrical resistance strain gage operates on the basic principle that certain metals exhibit a change in electrical resistance proportional to a change in mechanical strain. Within the thermal and mechanical limits of the particular strain gage, this relationship is linear according to Hooke's law. For precision measurement of the small resistance changes associated with the elastic strain of steel, one or more gages are bonded to the strained element and incorporated into a Wheatstone bridge. With an input of about 10 V, the output voltage is linearly proportional to the strains. When carefully installed and calibrated, this type of system has proven to be accurate, reliable, and repeatable over a long history of use [12].

For torsional vibration measurements, a full Wheatstone bridge is bonded to the surface of the shaft for measurement of the cyclic strain. The gages are oriented along the directions of principle strains (zero shear), which are at 45° to the shaft axis as shown in Fig. 8.43. This particular arrangement eliminates the effects of bending strains and temperature changes at the gage location.

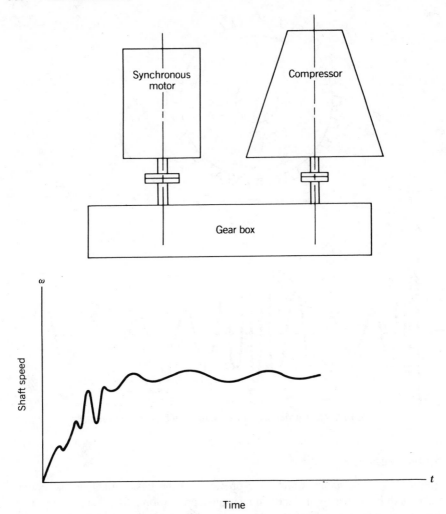

Figure 8.42. Torsional vibration excited by synchronous motor start-ups.

In the study by French, the gages used were Micro-Measurements MA-06-125TD-120, assembled in a 45° rosette. The power supply and signal conditioner were connected to the bridge through a commercially available slip ring assembly, and a voltage amplifier was used to boost the signal to a more convenient level.

Carrier Signal Transducers

If torsional vibration is to be measured in a drivetrain containing gears, either a magnetic transducer or a proximity probe can be installed close to the teeth of a

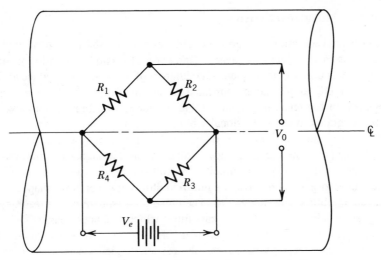

Figure 8.43. Orientation of strain gages on shaft.

selected gear to produce a carrier signal (Fig. 8.41). The predominant frequency in the carrier signal is the gear tooth passing frequency.

Since a proximity probe produces output voltage proportional to the instantaneous probe–tooth gap, the amplitude of its carrier signal will not be modulated by torsional vibration; only the tooth passing frequency will be modulated. Conversely, the output signal from a magnetic transducer is both amplitude and frequency modulated, since the transducer produces a voltage proportional to the instantaneous velocity of the gear tooth in its close field. Instrumentation systems based on amplitude modulation of the carrier signal therefore require a magnetic transducer.

Advantages of the magnetic pickup are that it requires no external power source, its output signal is generally clean with low noise levels, and it produces a strong signal which often requires no amplification.

In French's study, two B&K MM0002 transducers were used, one for each gear so as to determine the relative amplitudes, for comparison with the strain gage measurements of twist in the shaft.

For the AM system described below, the magnetic transducers were velocity-calibrated over the speed range 1000–2500 rpm to 0.48 MV (peak)/(degree/sec). In general, this calibration constant will be affected by the transducer gap to the gear tooth, so this setting must not be changed during a series of tests.

An FM carrier signal can also be optically transduced from alternating light and dark lines around the shaft, using light reflected to a photocell with fiberoptics. A special reflective tape, which was crafted to meet this requirement, is described later. It was used with the time interval measurement system (TIMS) and with one of the FM systems.

Frequency-Modulated Systems

Since torsional vibration is simply a cyclic variation of shaft speed, it produces a variation of carrier signal frequency. One class of torsional vibration instrumentation systems utilizes the frequency modulation of this signal to produce an analog signal with DC amplitude proportional to instantaneous shaft speed. The AC component of the resulting signal is therefore torsional vibration velocity, which can be integrated to obtain torsional vibration displacement.

The central and most critical component of this type of system is the electronic "box" that performs the frequency demodulation and signal conditioning. The general principle of a typical circuit is as follows: Each pulse, or cycle, of the carrier signal triggers a monostable multivibrator that generates a shaped pulse of precision amplitude and time width. The pulse train (square wave) resulting from gear rotation is integrated by a low pass filter, which produces a steady DC voltage output when the gear speed is constant. If there is a torsional vibration at a frequency within the filter pass band, the carrier frequency varies about its mean value and the DC output of the low pass filter varies accordingly.

This type of electronic process is called frequency-to-voltage conversion, with the output voltage in this case proportional to the instantaneous rotational velocity of the shaft.

There are, or course, differences from one FM-type instrument to another in the filtering and calibration techniques, but the general design concept is similar for all.

In French's study, two commercially available FM-type instruments were used. Both worked well, using a carrier wave produced by the magnetic transducer.

One of these instruments is marketed specifically for torsional vibration applications. It is manufactured by Econocruise, Ltd., in England and is designated Model TV-1. It has two channels, with a relative phase meter, and has internal filtering. All calibration is preset at the factory for a 120-tooth gear, which can be verified or changed by procedures documented in the technical manual.

The other FM instrument used was a general-purpose frequency-to-voltage converter, Model FC-62 manufactured by the Validyne Engineering Corporation. The FC-62 is calibrated with a range adjustment screw, using a variable frequency signal generator. For example, with the 120-tooth gear used in these tests, the gear tooth passing frequency was 4000 Hz at a shaft speed of 2000 rpm. Using a 4000-Hz sine wave from the signal generator, the DC output of the FC-62 can be set to any desired value up to 10 V. The higher the DC set level, the greater the sensitivity to speed variations. This setting determines the calibration constant in millivolts per degree per second, calculated by dividing the DC output by the speed in degrees per second. If torsional vibration measurements are to be made at constant shaft speed using the FC-62, the DC level corresponding to the mean carrier frequency (shaft speed) can be suppressed to zero with a potentiometer on the front panel, leaving only the variations representing torsional vibration. In the tests by French, the mean shaft speed varied so an external high pass filter (AC coupling) was used instead that passes only the variations in voltage above a preset (low) frequency. This allows high sensitivity settings to be used without exceeding the range of readout instruments.

An Amplitude-Modulated System

It has already been pointed out that a magnetic transducer installed close to passing gear teeth produces a carrier signal with amplitude (as well as frequency) modulated by the instantaneous gear velocity. An electronic process known as envelope detection can be used to produce a voltage analogous to the torsional vibration velocity.

Figure 8.44 illustrates the basic idea of envelope detection. In a typical circuit, the carrier signal is full wave rectified with a bipolar precision diode detector and a differential amplifier. The rectified waveform then passes through a low pass filter where its level is averaged. The result is a DC level proportional to amplitude analogous to the DC proportional to frequency described in the previous subsection. The output is finally routed through another low pass filter to reduce the running speed component of the signal induced by the inevitable minute deviations of gear teeth dimensions or magnetic properties.

With a moderate capability for wiring and packaging electronic circuits, it is entirely practical to custom-build an envelope detection system. Schematics are

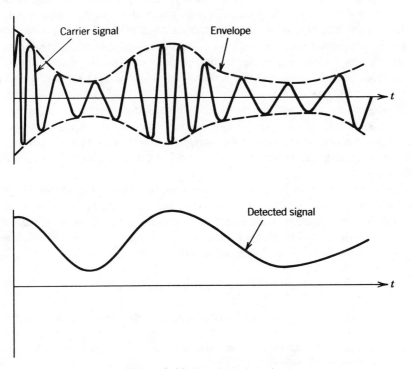

Figure 8.44. Envelope detection.

readily available, and the cost of components is low compared to the purchase of a commercial instrument.

A commercial instrument was obtained and used for this study since it had additional capabilities for other applications as well. The instrument used was the Model 223A vibration envelope detector (VED), manufactured by the Shaker Research Corporation.

The amplitude of the input waveform to the VED, and consequently its output waveform, is a function of the voltage output of the magnetic transducer. The VED itself was also found to have a slight gain, which was ascertained by delivering constant amplitude sine wave inputs from a signal generator and measuring the DC output. The overall calibration constant for the system used in this study, including the transducer and filters, was found to be 0.63 MV/(deg-sec), thus allowing the torsional vibration velocity to be calculated from the output voltage signal.

Frequency Analysis and the Sideband System

A frequency spectrum analyzer was used with each of the instrumentation systems tested except for the TIMS, which has a digital output. Figure 8.45 shows an example of a frequency-transformed signal, copied from the display of the Nicolet 660 dual-channel analyzer used in this study. The spectrum of Figure 8.45(b) showing the RMS amplitude of each frequency component in the signal is usually easier to interpret and more useful than the time trace of Fig. 8.45(a), as described earlier in this chapter. In measurements made on a typical drivetrain, the dominant frequency components are expected[3] to be torsional excitation frequencies or natural frequencies; such a component can become destructively large when it is both of these.

If one accepts the spectrum analyzer (with zoom capability) as a necessary component of any of the analog-based instrumentation systems, then it can be stated that the sideband system is the simplest and least expensive. It consists of nothing more than a transducer to produce an FM carrier signal and the spectrum analyzer.

The sideband system has been used for years to analyze the unwanted speed fluctuations (''wow'') of tape recorders [13] and other precision constant-speed devices. It was suggested to the author for measurement of torsional vibration in rotating machinery by Mr. Henry Bickel. The principle of operation is as follows:

The simplest form of an FM carrier signal is

$$V(t) = V_0 \cos(2\pi f_c t + \beta \sin 2\pi f_m t), \qquad (8\text{-}4)$$

where f_c is the basic carrier signal frequency, f_m is the frequency of modulation

---

[3]The identification, or at least confirmation, of these frequencies is usually accomplished by computer modeling as described in Chapter III.

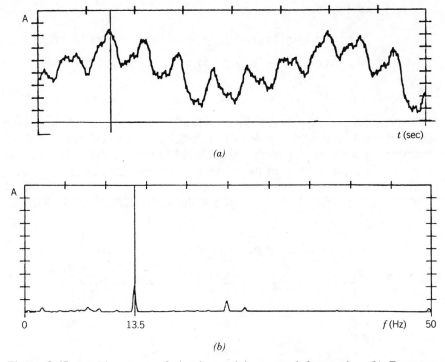

Figure 8.45. (a) Time trace of signal containing several frequencies. (b) Frequency spectrum of above signal.

(torsional vibration frequency), and β is proportional to the torsional vibration amplitude. If the carrier signal is generated from a gear with N_T teeth running at a mean shaft speed of N rpm, and with an angular displacement of torsional vibration given by

$$\theta(t) = \theta_0 \cos \omega t, \qquad (8\text{-}5)$$

then it must be that

$$f_c = N_T N / 60, \qquad (8\text{-}6)$$

$$f_m = \omega / 2\pi, \qquad (8\text{-}7)$$

and

$$\beta = N_T \theta_0. \qquad (8\text{-}8)$$

Equation (8-4) can be expanded into mathematical series known as Bessel functions to give

$$V(t) = V_0 \Big\{ J_0(N_T\theta_0) \cos 2\pi f_c t$$

$$+ J_1(N_T\theta_0) \cos 2\pi(f_c + f_m) t - J_1(N_T\theta_0) \cos 2\pi(f_c - f_m) t$$

$$+ J_2(N_T\theta_0) \cos 2\pi(f_c + 2f_m) t + J_2(N_T\theta_0) \cos 2\pi(f_c - 2f_m) t$$

$$+ \cdots \Big\}. \tag{8-9}$$

The quantity $\beta = N_T\theta_0$ is called the modulation index of the $J_i(N_T\theta_0)$, which are the Bessel functions of ith order, $i = 1, 2, 3, \ldots$. Numerical values of Bessel functions are tabulated in a number of published handbooks and therefore do not have to be calculated. Some of the values useful to this application are given in Table 8-4.

With this knowledge, one can now use the spectrum analyzer to determine

TABLE 8-4. Bessel Functions of Zero and First Order

β	$J_0(\beta)$	$J_1(\beta)$
0.0	1.0000	0.0000
0.1	0.9975	0.0499
0.2	0.9900	0.0995
0.3	0.9776	0.1483
0.4	0.9604	0.1960
0.5	0.9385	0.2423
0.6	0.9120	0.2867
0.7	0.8812	0.3290
0.8	0.8463	0.3688
0.9	0.8075	0.4059
1.0	0.7652	0.4401
1.1	0.7196	0.4709
1.2	0.6711	0.4983
1.3	0.6201	0.5220
1.4	0.5669	0.5419
1.5	0.5118	0.5579
1.6	0.4554	0.5699
1.7	0.3980	0.5778
1.8	0.3400	0.5815
1.9	0.2818	0.5812
2.0	0.2239	0.5767
2.1	0.1666	0.5683
2.2	0.1104	0.5560
2.3	0.0555	0.5399
2.4	0.0025	0.5202

frequency and magnitude of torsional vibration very accurately without the aid of any accessory equipment. The spectrum analyzer must have the capability to expand about any center frequency with a narrow observation window. This is commonly referred to as zoom. About a 12 : 1 improvement in resolution is ample for most circumstances. By connecting the transducer signal directly to the FFT analyzer and setting the center frequency of the expansion window equal to the carrier frequency, a spike will appear in the center of the screen, flanked by a symmetrical array of spikes of decreasing magnitude. Figure 8.46 is an example of such a spectrum. The bordering components of the spectrum are called sidebands. They exist as a result of the Fourier series expansion of the frequency modulated waveform. An examination of equation (8-9) will help clarify this phenomenon, as there are components at $(f_c + f_m)$, $(f_c - f_m)$, $(f_c + 2f_m)$, etc. The frequency of torsional vibration can be readily determined as the difference in frequency from the carrier to either of the first-order sidebands, both being equidistant from the carrier frequency.

Returning to the matter of Bessel functions, the magnitudes of all these spikes, including the center, or carrier frequency spike, are related in that they are all Bessel functions of the same modulation index, $N_T\theta_0$. To determine the amplitude θ_0 of torsional vibration, one needs only to obtain the ratio of the amplitude of the first-order sideband to the center or zero-order spike for a Bessel function of the first kind. This ratio of sideband amplitudes is the quotient of $J_1(\beta)/J_0(\beta)$, where β is equal to $N_T\theta_0$. By searching a table of Bessel functions, as shown in Table 8-4, for the combination of $J_0(\beta)$ and $J_1(\beta)$ that are in this ratio, the modulation index β is determined. Since β is equal to $N_T\theta_0$, and the number of teeth N_T on the gear is known, the peak amplitude of torsional vibration in radians is found.

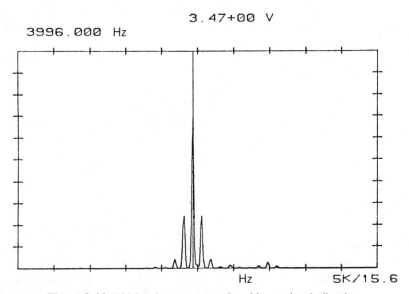

Figure 8.46. Sideband spectrum produced by torsional vibration.

Amplitude modulation of the FM signal that occurs when a magnetic transducer is used introduces a small error into the above development. In this case, V_0 in equation (8-4) is no longer a constant but is a function of time $V_0(t)$, linearly proportional to the angular velocity of the gear. The effect on the amplitude of the sidebands is quite small. Neglecting the AM effect results in a modulation index (β) error that is on the order of 0.1 percent. If desired, the error can be eliminated completely by averaging the amplitudes of the two first-order sidebands. The complete development of the Bessel function form of the magnetic transducer output, including the amplitude modulation mentioned above, is shown in Appendix B of Ref. [14].

Test Results

Table 8-5 is a summary of results from steady-state tests of the various systems for six different excitation amplitudes, with an excitation frequency of 8 Hz and a rotational speed of 2000 rpm.

Tests were also conducted with other excitation frequencies, but in all cases the predominant test rig response was at 8 Hz. (The excitation torque was not purely harmonic.) Best results were therefore obtained at the resonance where the responding mode shape was well defined and the strain gage signal could be accurately converted into an amplitude at the outboard gear.

Measurements were made and recorded under both steady-state and transient conditions. To make a typical steady-state measurement, the rig was brought up to a constant speed, a constant-amplitude excitation voltage was applied to the field of the motor, and the vibration allowed to stabilize. The strain gage signal was then read into one channel of the spectrum analyzer, and the signal from one of the test instruments was read into the other channel. Figure 8.47 shows an example, with the spectrum from the FC-62 FM system shown above the spectrum from the strain gages. The rotational speed is 2000 rpm, and the excitation frequency is 8 Hz (resonant). The spikes on the spectral plots at 8 and 33.3 Hz are torsional vibration and lateral run-out (or shaft whirling), respectively. Note that the 33.3-Hz component is synchronous with shaft speed and is produced by both the FM system and the strain gages. Since this is typical for measurements made on any rotating machine, it is fortunate that large-amplitude torsional excitations are usually not synchronous. Whenever they are, a method to separate the torsional vibration from the lateral vibration at the same frequency must be devised. One way to eliminate the lateral signal when using gear-excited transducers is to install two transducers on opposite sides (180° apart) of the same gear and sum the two signals electronically.

Transient tests were conducted for a start-up condition in which the test rig was accelerated from 0 to 2500 rpm in 15–20 sec. Three different torsional excitation formats were used. In the first, a constant amplitude excitation at 8 Hz was applied during acceleration. In the second, a constant amplitude excitation at 13.5 Hz was applied. In the third and last phase of the transient testing, the frequency of excitation was swept downward from 100 Hz at the start to 5 Hz as the speed reached

TABLE 8-5. Steady-State Tests Results, 2000 rpm, 8-Hz Excitation

Test No.	TV-1		FC-62		VED		Sidebands	
	True Vibration Amplitude Deg. Peak	(Fraction of True) Error Deg.	True Vibration Amplitude Deg. Peak	(Fraction of True) Error Deg.	True Vibration Amplitude Deg. Peak	(Fraction of True) Error Deg.	True Vibration Amplitude Deg. Peak	(Fraction of True) Error Deg.
1	0.029	(0.97) 0.001	0.035	(1.03) 0.001	0.038	(1.16) 0.006	0.034	(1.00) 0
2	0.084	(1.01) 0.001	0.079	(1.00) 0	0.074	(1.03) 0.002	0.078	(1.00) 0
3	0.129	(1.04) 0.005	0.131	(1.00) 0	0.134	(1.18) 0.025	0.133	(1.00) 0
4	0.162	(1.04) 0.006	0.164	(1.00) 0	0.171	(1.11) 0.019	0.169	(1.00) 0
5	0.324	(0.96) 0.013	0.319	(1.01) 0.002	0.328	(1.17) 0.057	0.315	(1.00) 0.001
6	0.676	(0.96) 0.024	0.683	(1.00) 0.001	0.688	(1.14) 0.094	0.674	(1.02) 0.014

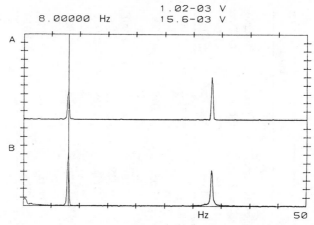

1.02-03 V
8.00000 Hz   15.6-03 V

Figure 8.47. Frequency spectrum of the FC-62 (FM system) signal (above) and of the corresponding strain gage signal (below).

2500 rpm. This is an approximate simulation of a synchronous motor start-up that has a twice-slip frequency of torsional excitation as it accelerates to speed.

In all three phases of the transient testing, the predominant torsional vibration response of the test rig was at 8 Hz regardless of the excitation frequency. The 8 Hz mode was so lightly damped that it could be excited by acceleration alone, without any imposed vibratory excitation.

Table 8-6 shows the test results for the three series of transient tests. Since spectrum averaging is not an appropriate procedure for transient measurements, the ''peak hold'' function of the spectrum analyzer was employed to obtain the data presented in Table 8-6. This function, which is available on most spectrum analyzers, stores the maximum amplitude measured at each frequency during the

TABLE 8-6.   Transient Test Results, 0–2500 rpm Start-Up

	TV-1		FC-62		VED	
	True Vibration Amplitude	(Fraction of True)	True Vibration Amplitude	(Fraction of True)	True Vibration Amplitude	(Fraction of True)
Excit. Freq. Hz	Deg. Peak	Error Deg.	Deg. Peak	Error Deg.	Deg. Peak	Error Deg.
8	0.235	(1.00)	0.245	(1.00)	0.241	(1.17)
		0.001		0.001		0.40
13.5	0.041	(1.05)	0.104	(0.94)	0.077	(3.70)
		0.002		0.006		0.208
Sweep	0.069	(0.82)	0.051	(0.90)		
		0.012		0.005	N/A[a]	

[a]Not available.

start-up (as described earlier in this chapter). Care must be taken to ensure that the sampling rate is sufficient to capture the true peak at each frequency during the sweep. This is governed by the sweep rate.

Evaluation of the Measurement Systems

In Table 8-5 it can be seen that three of the systems tested measure steady-state amplitudes of torsional vibration quite accurately, with a maximum error of 4 percent. The VED system (amplitude modulated) had larger errors, as high as 18 percent.

The two FM systems were comparable, with the FC-62 slightly more accurate. It should be pointed out, however, that the TV-1 produces an output signal that is an order of magnitude higher than the FC-62. This could eliminate the need for an amplifier and improve the signal/noise ratio.

The best accuracy in steady-state testing was achieved with the sidebands system. This should not be surprising, since it employs the least number of components.

The relatively poor results from the VED (AM) system are due primarily to the effect of lateral whirling, or "run-out" (electrical or mechanical asymmetry of the gear teeth with respect to the shaft centerline), which has a much greater effect on amplitude modulations of the carrier wave than it does on the frequency modulations. The resulting large signal at running speed and its harmonics tend to mask the torsional vibration signal and put it down into the realm of "noise." To put this limitation into perspective, it should be remembered that care was taken to minimize lateral motion in the test rig and this is usually not possible in the field. The other three systems clearly must be considered superior.

From the transient test results in Table 8-6, it can be seen that only the two FM systems produced acceptable results. The sidebands system is not applicable to transient measurements with the electronics available today, since the center frequency moves rapidly during a start-up and is generally independent of the torsional vibration frequencies.

Both of the FM systems measure torsional vibration amplitudes accurately (0–6 percent error) when the shaft speed is changing (40 rpm/sec) and the vibration frequency is constant. For the swept frequency tests, it can be seen that the FC-62 is more accurate. Each row in Table 8-6 is the result of a number of repetitions for each test, and this result was checked to ensure that it represented a factual trend. The earlier statement about the stronger output signal from the TV-1 holds in this case also.

A Special Tape for Optical Transducers[4]

When a gear or bladed wheel is not available for transducing a carrier signal, a reflective tape with alternating light and dark lines can be used to excite a fiberoptic

---

[4]The information and tests described in these last two subsections was developed by Professor P. M. Barrios of Zulia University, Maracaibo, Venezuela, while visiting the Turbomachinery Laboratories at Texas A&M University.

photocell transducer. The tape can be wrapped around any section of accessible shafting in a drivetrain where torsional vibration is to be measured.

A tape with equally spaced reflective and nonreflective lines was not found to be commercially available at the time of this study, so it was made. Utilizing a computer with graphics capability and a digital plotter, black lines were drawn on a white nonstretching fiberglass adhesive tape. The computer can be programmed to produce optimum line widths for each application, depending on the shaft speed and diameter. Since the black ink tends to come off, it was coated with a clear varnish. This degraded the performance of the fiberoptic tachometer due to its high reflectivity. Nevertheless, useful results were obtained.

Figure 8.48. Optical tape on disks of test rig.

For testing the various tape configurations, a smaller test rig was built that was easier and less time-consuming to operate. Figure 8.48 shows this test rig with the optical tape installed on both of the 4-in outer diameter inertial disks, which is a typical size for shafting in industrial machinery. The natural frequency is also typical of heavy machinery at 10.2 Hz. In this rig the torsional excitation is provided by gating a power transistor in the motor armature circuit with a signal generator.

One question which was resolved by tests on this apparatus is the possible effects of the tape overlap at the ends, as it is not practical to achieve perfect light/dark line spacing for every possible shaft circumference. Using the FC-62 system to demodulate the carrier signal, it was found that overlap (one unequally spaced line) produces a signal composed only of synchronous (shaft speed) frequency and its harmonics. This is the same effect that shaft run-out produces from a magnetic transducer. By varying the overlap, it was found that the amplitude of the unwanted synchronous signal can be predicted from the amount of overlap. This would allow subtraction of the overlap signal from the total synchronous component to give the torsional vibration in cases where the latter is synchronous.

Figure 8.49 shows the steady-state torsional response at 10.2 Hz measured with this system. Figure 8.50 is a time trace of torsional resonance while the test rig is accelerating at about 2000 rpm/min.

A Time Interval Measurement System

The development of the special reflective tape for use with a fiberoptic–photocell transducer, to produce the carrier signal, led to the design and testing of a promising

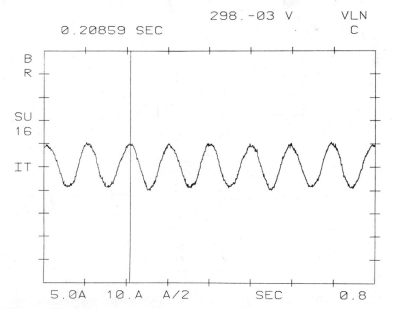

Figure 8.49. Steady-state response at 10.2 Hz from optical tape. From Ref. [7].

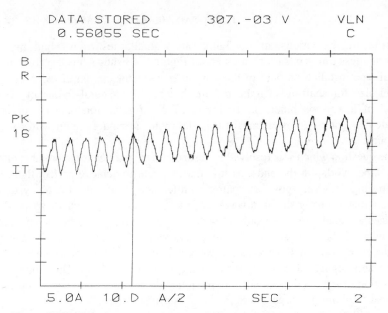

Figure 8.50. Transient resonant response from optical tape. From Ref. [7].

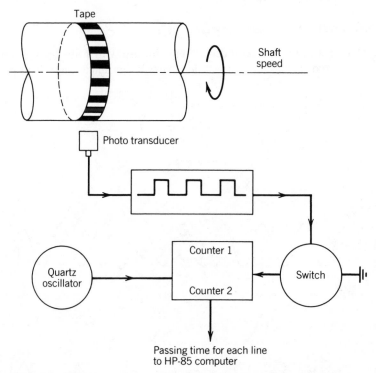

Figure 8.51. The TIMS concept. (a) Schematic of TIMS components. (b) Time trace of counting scheme. From Ref. [7].

prototype instrumentation system. This system, designated TIMS (time interval measurement system), is especially well suited to measurement of rapidly changing transients.

The concept originated from a digital circuit for a tachometer designed by Dr. Mark Darlow at Rensselaer Polytechnic Institute.

Figure 8.51 illustrates the system conceptually. Figure 8.52(a) is a schematic

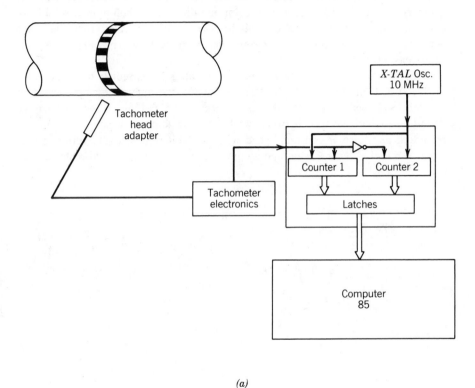

(a)

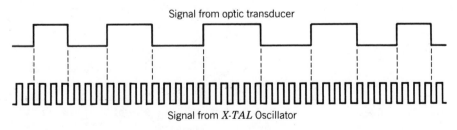

(b)

Figure 8.52. The TIMS concept. (a) Schematic of TIMS components. (b) Time trace of counting scheme. From Ref. [7].

of the components. The fiberoptic transducer produces a square wave that turns the counters on and off. While on, each counter counts the number of oscillations from a 10-MHz oscillator [see Fig. 8.52(b)]. The latches store these numbers for input to the computer interface. The digits stored are equivalent to the passing times for each line on the tape. They are inverted by a computer code to produce angular velocity versus time.

Although it is possible to implement this system using commercially available counters (see Ref. [15]), cost considerations dictated a shop-built assembly of chips, latches, oscillators, etc. on a circuit board. The particular circuit built was applicable only to the speeds and frequencies of the test rig. The objective was to show feasibility.

Even with the computer-spaced lines on the tape, the extremely high resolution of this system produced a digital "noise" due to the uneven line spacing. This problem was magnified by the reflective varnish, which reduced the contrast and definition of the interface between white and black lines. It was overcome by recording the noise at constant speed with no excitation and subtracting it later from each digital record, leaving only the torsional vibration as "corrected data."

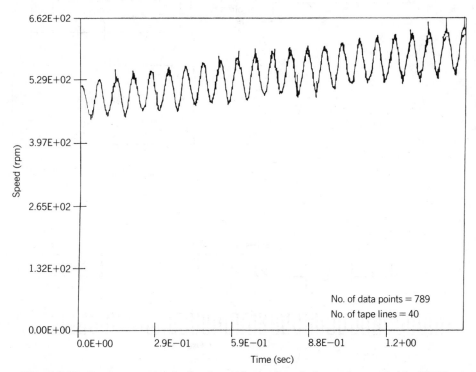

Figure 8.53. Resonant torsional vibration with transient shaft speed, recorded by TIMS. From Ref. [7].

Figure 8.53 is an example of transient torsional vibration recorded from the test rig by feeding the corrected data to a digital plotter. The resonant (10.2-Hz) torsional vibration is shown superposed onto the accelerating speed of the test rig.

REFERENCES

1. "Dynamic Signal Analyzer Applications," Application Note 243-1, Hewlett-Packard Co.

2. American Petroleum Institute Standard 617 (Section 2.8) and Standard 670.

3. Vance, J. M., Murphy, B. T., and Tripp, H. A., "Critical Speeds of Turbomachinery—Computer Predictions Versus Experimental Measurements," *Proceeedings of the 13th Turbomachinery Symposium, Texas A&M University*, pp. 105–130 (1984).

4. Beatty, R. F., "Differentiating Rotor Response Due to Radial Rubbing," *Journal of Vibration, Acoustics, Stress, and Reliability in Design*, **107,** 151–160 (April 1985).

5. Jackson, Charles, *The Practical Vibration Primer*, Gulf Publishing, Houston, 1979.

6. Mitchell, John S., *Machinery Analysis and Monitoring*, Penn Well Books, Tulsa, OK, 1981.

7. Vance, J. M., and French, R. S., "Measurement of Torsional Vibration in Rotating Machinery," *Journal of Mechanisms, Transmissions, and Automation in Design*, **108**(4), 565–577 (December 1986).

8. Verhoef, W. H., "Measuring Torsional Vibration," *Instrumentation Technology*, pp. 61–66 (November 1977).

9. Mruk, G. K., Halloran, J. D., and Kolodziej, R. M., "New Method Predicts Startup Torque," *Hydrocarbon Processing*, pp. 229–234 (May 1978).

10. Ramey, D. G., and Harold, P. F., "Measurements of Torsional Dynamic Characteristics of the San Juan No. 2 Turbine-Generator," *Journal of Engineering for Power*, pp. 378–384 (July 1977).

11. Sohre, J. S., "Transient Torsional Criticals of Synchronous Motor-Driven, High-Speed Compressor Units," ASME Paper 65-FE-22, presented at the Applied Mechanics and Fluids Engineering Conference, Washington, DC, June 7–9, 1965.

12. Dally, J. W., and Riley, W. F., *Experimental Stress Analysis*, McGraw-Hill, New York, 1978, pp. 141–199.

13. Savage, D., "Effects of Flutter on Tape Recorded Data," *Sound and Vibration*, pp. 18–24 (November 1969).

14. French, R. S., *An Experimental Study of Torsional Vibration Measurement*, M.S. Thesis, Texas A&M University (August 1981).

15. "Fundamentals of Time Interval Measurements," Application Note 200-3, Hewlett-Packard Co.

Index